THE
NATIONAL
AIR
AND SPACE
MUSEUM

C.D.B. BRYAN

THE NATIONAL AIR AND SPACE MUSEUM

SECOND EDITION

Art Directed and Designed by
DAVID LARKIN

Photographs by
MICHAEL FREEMAN, ROBERT GOLDEN, and DENNIS ROLFE

Second Edition Photographs by
JONATHAN WALLEN

Harry N. Abrams, Inc., Publishers, New York

Second Edition 1988

Project Editor: Edith M. Pavese
Assistant Editor for the First Edition: Margaret Donovan
Designer for the Second Edition: Gilda Hannah

Library of Congress Cataloging-in-Publication Data
Bryan, C.D.B. (Courtlandt Dixon Barnes)
  The National Air and Space Museum.

  Bibliography: p.
  Includes index.
  1. National Air and Space Museum.   I. Title
TL506.U6W373   1988   629.1′074′0153   88-3383
ISBN 0–8109–1380–1

A Times Mirror Company

Printed and bound in Japan

To the last crew of Space Shuttle *Challenger*

Francis R. Scobee
Michael J. Smith
Robert E. McNair
Judith A. Resnik
Gregory B. Jarvis
Sharon Christa McAuliffe
and my friend
Ellison S. Onizuka

Contents

ACKNOWLEDGMENTS TO THE FIRST EDITION

My participation in this volume grew out of an idle dinner conversation in Chicago with Ian Ballantine who, by casually mentioning old airplanes, inadvertently touched upon a childhood fascination with all things that flew that has continued in me up to the present. Our talk inevitably turned from there to one of my favorite places, the Smithsonian Institution, for I had been a steady visitor to the old Air Museum ever since it had opened in a Quonset off to one side of the "Castle." Moreover, my interest was clearly shared by more than twenty million people who had visited the new National Air and Space Museum in its first two years. When Ian mentioned that Harry N. Abrams, Inc., had an agreement with the Smithsonian Institution for publishing an authorized book on their new Museum, I immediately volunteered to write the text. Ever since that evening I have been assisted, educated, instructed, amused, and indulged by highly specialized men and women of the National Air and Space Museum staff whose knowledge and expertise are but imperfectly reflected by my text.

I could not have undertaken this work without the assistance and encouragement of Ian Ballantine and Harry N. Abrams, Inc.'s President Andrew Stewart and Digby Diehl, his editor-in-chief; nor would the work have progressed so pleasurably and smoothly without the extraordinarily able and professional Abrams' staff—in particular editor Edith Pavese and designer Nai Chang; but the book was made possible only by the complete and eager cooperation we all received from the staff of the National Air and Space Museum. In behalf of us all I would like to single out certain individuals whose unstinting giving of their time and expertise are owed recognition in print.

Michael Collins, Under Secretary of the Smithsonian Institution, and Melvin B. Zisfein, Deputy Director of NASM, lent their authority and approval to this project and made it possible for me to work from the Museum galleries' concept scripts. This pattern of management support continues under the new Director, Dr. Noel W. Hinners. Executive Officer Walter J. Boyne personally guided me through the Silver Hill Museum complex and the *Enola Gay*'s bomb bay, and subsequently guided the text through its rockier channels. Public Information Officer Lynne Murphy, her successor Rita Bobowski, and Staff Assistant Louise Hull provided invaluable contacts and administrative assistance.

It is said that a museum is only as good as its collection; but a collection is only as good as its curators. Paul Garber, NASM's Historian Emeritus, was responsible for many of the Museum's earliest acquisitions. Charles H. Gibbs-Smith, NASM's first Lindbergh Professor of Aerospace History, was extremely helpful with information on the Wrights. In the Aeronautics section, however, our gratitude is owed Assistant Director of Aeronautics Donald S. Lopez; Curator of Aircraft Louis S. Casey; Curator of Propulsion Robert B. Meyer; and Curators

Robert C. Mikesh and Edmund T. Wooldridge. A very special thanks to Assistant Curator Claudia M. Oakes whose uncluttered writing and thinking, encouragement, knowledge, and humor made things far easier than they would otherwise have been. Thanks, too, to Museum Technician Jay P. Spenser and Museum Specialist Supervisor Elmont J. Thomas of the Aeronautics section.

Frederick C. Durant III, Assistant Director of Astronautics, was embarrassingly patient in answering my questions about spacecraft principles and generous in sharing his knowledge and appreciation of space pioneer Robert H. Goddard. If the space sections of this text appear at all learned they reflect the invaluable assistance and expertise of Walter H. Flint and Tom D. Crouch, Curators of Astronautics, and Gregory P. Kennedy and Walter J. Dillon, Assistant Curators.

Curator of Art James Dean was gracious enough to permit some of his art treasures to be sprinkled throughout this book despite the fact that the Art Gallery displays the collection more as a unified whole. And we thank, too, Bill Good and Mary Henderson of the Art Department for so generously giving their time.

I would like to acknowledge a special "writer's debt" to the Museum's Reference Librarian Dominick A. Pisano who tracked down some wonderful stories and leads, and Karl P. Suthard, a Technical Information Specialist in the Library. And for his help at the Silver Hill Museum, our gratitude to Edward Chalkley, that Museum complex's Assistant Chief.

Hernan Otano, Chief of the Audiovisual Unit, Richard Wakefield, A/V Supervisor, Electronic Technicians John Hartman and Daniel Philips, Exhibit Section's Chief of the Production Unit Frank Nelms, Exhibit Specialist Sylvandous Anderson, Painter Daniel Fletcher and Carpenter Milan Tomasevich all generously shared their materials, equipment, experience, and time.

Howard Wolko, Assistant Director of the Science and Technology section, and Curators Richard Hallion and Paul Hanle helped me feel almost comfortable in writing about what had previously been incomprehensible. I am grateful, too, to Robert W. Wolfe, a geologist with the Center for Earth and Planetary Studies, for his assistance and Von Del Chamberlain, Chief of the Presentations Division, and Patricia Woodside for their help on the Albert Einstein Spacearium section.

Joseph Davisson, NASM's Building Manager, Claude Russell, Assistant Building Manager, Supervisor Mary Whittaker, and Artifacts Crewman Larry Johnson helped make the inaccessible accessible. Captain Preston Herald III, Captain of the Museum's Protection Division, and his staff saw that we were able to freely wander about.

And finally, I would like to acknowledge my genuine debt to a great many writers who have so generously permitted me to quote from their articles and books.

C.D.B. Bryan
Guilford, Conn.

ACKNOWLEDGMENTS TO THE SECOND EDITION

First of all, our thanks must go to Museum Director Martin Harwit for generously allowing free access to staff and materials for the updating of this book. We also thank his secretary, Tiffany German, for her help.

One of the nicest aspects of revising the previous edition of this book was the opportunity of renewing old friendships. Chief among them would be Donald S. Lopez who, ten years ago, was the Museum's Assistant Director of Aeronautics and today is Deputy Director of the Museum. It was in this capacity that Don gave so much assistance in coordinating our work on this book by introducing us to curators, seeing that research materials were made available, and suggesting possible leads. He was as generous with his library as he was with his time and knowledge. This would be a good place to express my appreciation to Don Lopez's special assistant, Toni Thomas, who, despite her dismay that I was not Mr. Abrams, willingly gave of her time to gather pounds and pounds of research material for me.

Special thanks must go to Rita Cipalla, Chief, Public Affairs and Museum Services, who helped us every step of the way coordinating Museum staff with that of Abrams with efficiency, diplomacy, and charm. Thanks, too, to Holly Haynes, Public Affairs Specialist, who facilitated our request for archival photographs.

Among the gallery curators whose help I greatly appreciated, a special thanks has to go to my longtime friend Claudia M. Oakes, the distinguished and knowledgeable Curator of the Department of Aeronautics and of the Golden Age of Flight gallery and formerly curator of Early Flight. Claudia lent me books, time, her office, and her warm and witty companionship at lunch. Thanks, also, to the current curator of Early Flight, Peter Jakab, for his help in tracking down sources and stories.

Tim Wooldridge, Assistant Director for Museum Operations, guided me through the history of jet aviation, entertained me with his experiences as a Navy pilot, and showed me where to look at the Keith Ferris paintings for that artist's inside jokes. (Is the bombardier in the B-17 *really* sighting down a beer can?)

Priscilla Strain, curator of Looking at Earth, was extremely helpful and I am grateful for her assistance with the text as well as for locating additional photographs for this section. David DeVorkin, former curator of the Stars gallery was patient with my mundane questions. And curator R.E.G. Davies, the learned authority on air transportation, made the story of the Comet jet disasters come alive.

Edna Owens, Writer-Editor in the Exhibits Department, guardian and protector of all the exhibits' written materials, was extremely kind in permitting me to rummage through her files. Greater trust has no man or woman.

Larry Wilson, Technical Information Specialist, made the library use easy and was especially helpful in leading us through the intricacies of the Museum's videodisc system and to the black-and-white photos we might need. We thank Mary Henderson, curator of art, and her assistant, Susan Lawson-Bell, for their help in updating the art images in the revision.

Patricia Graboske and David Romanowski of the Publications Office coordinated the updating of the Chronology and Technical Appendix of this book.

We thank, too, Richard Horigan, Foreman of the Restoration Shop at the Garber Facility, Ronald Wagaman and Steven Fitch at the Samuel P. Langley Theater, and James Sharp, Chief of the Albert Einstein Planetarium, for their help in updating these sections.

And we give special thanks to Dave Heck of the Audio/Visual Unit and to Capt. Adolph Smith and the men of Company E For their help during the photography for this revision. Without their help we could not have completed this task in so timely a fashion.

Meanwhile, back in New York, among those at Harry N. Abrams, Inc., who saw this book through, I am grateful to Barbara Lyons, Director of Rights and Reproductions, for her help with last-minute questions and to Neil Ryder Hoos who had the job of tracking down the publishers of some long-out-of-date materials and getting permission to quote from them. It was a difficult job, but, I hope not thankless.

Gilda Hannah was the designer of the new chapters and the one who seamlessly incorporated all the new material. How can I not be appreciative of someone who makes what I write look good?

Jonathan Wallen our very talented photographer helped us make this Second Edition every bit as visually exciting as the first. Thanks, too, to his assistant, Allen Folsom.

To Paul Gottlieb, President and Publisher of Harry N. Abrams, Inc., I continue to be grateful for his friendship and guidance, and I again offer my thanks.

But there is no one at the Museum or at Abrams to whom I owe more than I do to my friend and editor, Edith M. Pavese. Her intelligence, taste, humor, and perceptiveness infuse every word and image of this text. This is the second time we have done this book together and, although it doesn't get any easier, it does get better. I feel especially lucky to have Edith to work with and I look forward to the future projects we will do.

The dream *is* alive.

C.D.B. Bryan
Guilford, Conn.
May, 1988

FORD
TRI-MOTOR

PITCAIRN
MAILWING

Foreword to the First Edition

At the end of World War II, General Hap Arnold worried that his war birds would all be converted to scrap metal. One of each type, he thought, should be saved for posterity, so he talked it over with Congressman (now Senator) Jennings Randolph of West Virginia. From this conversation emerged Public Law 722 (August 12, 1946) which created the National Air Museum, whose name was amended in 1966 to include Space. In the stiff congressional language of the 1946 Act, the fledgling museum was instructed to "memorialize the national development of aviation; collect, preserve, and display aeronautical equipment of historic interest and significance; serve as a repository for scientific equipment and data pertaining to the development of aviation; and provide educational material for the historical study of aviation." This book records, thirty-three years later, the results of this mandate.

As several chapters in this book make clear, the Smithsonian Institution's interest in aviation goes back to the nineteenth century, and the 1896 aerodrome of Professor Samuel Pierpont Langley, the third Secretary of the Smithsonian, hovers proudly over one of Robert Goddard's rockets in the Milestones of Flight gallery. Paul Garber, who joined the Smithsonian staff in 1920 and who is still at the Museum, has the instincts of a mother squirrel facing a tough winter. Over the intervening years, Paul built up the aeronautical collection to the point that today it is the finest in the world. In similar fashion, a cornucopia of space artifacts has been made available, so that when $40 million for the new building was finally approved in 1971 (thanks to people like Dillon Ripley, Barry Goldwater, and James Webb), the Museum staff had the luxury of selecting the very finest artifacts from a superb collection.

One fundamental question in planning the new building's interior was what slice of the pie to devote to air, and what to space. As a former astronaut and presumably with an emotional tilt toward space, I sometimes had the feeling that the aeronautics staff was watching me with a jaundiced eye. But in fact I've spent thousands of hours flying airplanes and only hundreds aboard spacecraft, and I feel just as comfortable in the gallery on Flight Testing as I do in that devoted to the Apollo program. I relied upon Melvin Zisfein, the Deputy Director of the Museum, to create a master plan which divided our air and space heritage into manageable packages, and assign each to a gallery. Generally Mel's plan followed thematic lines but in some cases the ever-accelerating pace of aeronautics and astronautics was highlighted. In the Milestones of Flight gallery, for instance, the original Wright flyer and the Apollo 11 Command Module, whose flights were separated by a mere 66 years, were placed next to each other. Nearby, and spanning the gap, Lindbergh's *Spirit of St. Louis* was hung on thin wires, and before we knew it, opening day was upon us.

July 1, 1976, was a proud day. "A perfect birthday present from the American people to themselves," President Ford called the Museum, as he and Chief Justice Warren Burger (the Chancellor of the Regents of the Smithsonian Institution) watched a red, white, and blue ribbon being cut by a signal radioed back over nearly 200 million miles from a Viking spacecraft approaching the planet Mars. Like an expectant father, the architect, Gyo Obata, watched as the first visitors streamed inside. Within two years more than twenty million had followed them, making it the busiest museum in the world.

For those who have not visited the Museum, or for those who want to savor that visit with a tangible reminder, this book not only documents its principal exhibits, but conveys the mood and spirit of the Museum as well. Through photographs and text, a vicarious visit unfolds, and one is transported across the English Channel in a balloon, over the sands of Kitty Hawk with the Wrights, across the Atlantic with Lindbergh, onto the Moon with Armstrong and, finally, far out into the future where only the mind can reach.

The National Air and Space Museum is, I think, a cheery and friendly place, and this beautiful volume captures that spirit of optimism. It also makes clear some of the processes by which our daily lives are touched by advances in aeronautics and astronautics. I hope that future events justify the enthusiasm expressed for humanity's love affair with air and space, and I trust that exciting new chapters will be added to future editions as we continue to push ever upward and outward.

Michael Collins
Under Secretary,
Smithsonian Institution

Foreword to the Second Edition

Ten years ago, when the first edition of this volume was published, the National Air and Space Museum's new building on the Mall was only two years old, although twenty million visitors had already streamed through its doors. The Museum's staff, working with director and ex-astronaut Michael Collins, had managed, with one bold thrust, to create the most popular museum in Washington, D.C., and perhaps the world. Over the intervening years a further hundred million visitors have toured the Museum; Paul Garber, who has been at the Smithsonian since 1920 and who built up most of our aeronautical collection, still comes to work every day; and the Wright Flyer, the Apollo 11 Command Module, and the *Spirit of St. Louis* still grace the Milestones of Flight gallery.

And yet, much has changed. Most of the changes may not be apparent to the visitor, but they take the form of requests and inquiries concerning significant omissions. The Museum now owns the first of NASA's Shuttle spacecraft, the *Enterprise*. Because it is too large to be brought to the Museum, it sits at the Washington Dulles International Airport, where it was brought piggyback on its mother craft, a specially equipped Boeing 747. Surrounded by other airplanes and spacecraft too large to be brought into the Museum, the *Enterprise* sits in a provisory hangar, where it awaits the construction of a Museum Annex that we hope to begin building in the next few years. Its construction is essential if we are to keep our visitors informed about air and space developments far into the twenty-first century.

The Air and Space Museum must be able to evolve in order to fulfill its tacit promise to the public. Evolution implies an ability to adjust to changes in the aerospace field—changes that uniformly tend toward larger size as commercial aircraft ferry growing numbers of passengers across oceans, and heavy boosters lift increasingly massive payloads into orbit. Size, by itself, is not a criterion of merit, but visitors come to the Museum expecting to see airplanes and spacecraft that have played pivotal roles in the history of our country.

The single, most frequently expressed wish in letters addressed to the Museum is the request to see the *Enola Gay*, the bomber that dropped the atomic bomb on Hiroshima and changed the nature of warfare for all time. The massive Boeing B-29 bomber currently is being restored in the Museum's Paul E. Garber Facility in Suitland, Maryland, and can be viewed there by visitors on guided tours. But the Suitland facility is only visited by 30,000 people a year, contrasted to the nine million that will visit the Museum on the Mall this year. We hope also to exhibit the *Enola Gay* at a Museum Annex as soon as it is restored.

The *Enola Gay* embodies a second, more subtle change the Museum needs to undergo in response to changing public attitudes. No longer is it sufficient to display sleek fighters as we do in our World War II gallery. Visitors are asking about how our way of life has been changed by aviation and space technology, both for the good and the bad. We reap the benefits of travel by air on vacations in distant countries; instant participation, via satellite television link, at Olympic games on the other side of the Earth; and improved three-day weather forecasts that warn of impending storms. On the debit side we live in a world of newly coined phrases—mutually assured destruction, nuclear winter, and megadeaths—brought into our vocabularies by the proliferation of sophisticated nuclear arms and the powerful aircraft and missiles designed to deliver them. We cannot and should not ignore these. The Museum can and should be fun for youngsters, but if our children and grandchildren are to inherit our world intact, they must also be taught to see clearly technology's place in human activities. The Air and Space Museum can serve to teach them the changes that aviation and space flight have brought about in the course of this century.

Although we live in an age that takes aviation and space flight for granted, most of us can think of members of our immediate families who were born before man ever learned to fly in heavier-than-air craft. But three generations are hardly enough time for our nation to assess the profound impact of aviation and space technologies on our lives. As we read through this Second Edition of *The National Air and Space Museum*, we may want to reflect on the moral dilemmas that these technologies have introduced.

Martin Harwit
Director,
National Air and
Space Museum

A Note on the Photography

If the exhibits as they appear in this book seem as real and lively to the readers as the experience of visiting the Museum, then the three photographers and myself have achieved our objective. The photographers never had previously the opportunity, or the challenge, of attempting to represent such a vast, diverse museum experience through the camera, from scratch. The effort proved an extraordinary experience in itself.

It was necessary, for instance, for one of the photographers to spend a week photographing the huge murals that dominate the main entrance to the Museum. To do this he had to perch in a cage 50 feet above the ground, holding his breath and taking pictures with a 45-second exposure, patiently guessing that he would miss the tremor caused by the new Metro under the Museum—a tremor which could be felt only on his high, extended platform.... Photographing the insides of the Apollo Command Module brought home the recognition that it had not been designed for photography. But the acute discomforts to the photographer of maneuvering in that restricted space to get his shots demonstrated, as nothing else could, the truly astonishing feat of the three men who lived in that Module for a full week.

During the actual shooting of the many exhibits we were mindful of the fact that in a true peoples' museum we had to conduct our work without hindering the enjoyment of the visitors. We hope we succeeded in this. It was interesting to note how many people were themselves enjoying the fun of photography. And we noticed, with amusement, that whenever we were attempting to photograph the public itself, people would observe our attempts and curtail their own examination of the exhibit, politely stepping aside, not realizing that their fascination with the riches the Museum has to offer was exactly what we were after. Nevertheless, our thanks to those adults and children who do appear in this book, for making that part of our task such a pleasure. Our thanks, too, to all the Museum staff who so ably assisted us. But the photographers especially would like to thank E.J. Thomas, who helped us to get into position for many particularly difficult shots, and who was an unfailing mainstay throughout the months of our work in the Museum.

David Larkin
London

Editor's Note: The National Air and Space Museum is an evolving museum whose exhibits change periodically to reflect advances in aeronautics and astronautics. The contents of this Second Edition, which were formulated in the Fall of 1987, reflect the Museum as it was organized at that time.

Introduction

John H. Glenn, Jr.'s Friendship 7.

The school buses, tour buses, charter buses begin lining up along Jefferson Drive on the Mall side of the Smithsonian Institution's National Air and Space Museum (NASM) in Washington, D.C., a half hour before the doors to the huge pink marble and glass sheathed building even open. Early visitors push eagerly forward, crowd against the immense Milestones of Flight gallery windows, shield their eyes and press their foreheads to the glass to peer inside. They point, glance away for but an instant to talk excitedly among themselves, then quickly turn back to look inside again.

Neither the visitor's age nor his nationality seems to matter, his eyes always reflect the excitement, the anticipation, the wonder of a child; and it is this joyful expectancy one sees on the Museum visitor's face that provides the most dramatic evidence of the National Air and Space Museum's enormous popularity—a popularity more scientifically measured, perhaps, by the fact that only 24 days after it had opened, the Museum had its one millionth visitor, its 50 millionth in January, 1982, and its 100 millionth in December, 1986. Within a decade of its opening, the National Air and Space Museum has become the most visited museum in the world. And by the start of the next century the equivalent of every other person in the United States will have passed through this Museum.

Each person who visits the National Air and Space Museum finds himself moved by that experience in a way he may not have anticipated, affected personally by the sudden, unexpected intimacy of his contact with history—history which, in some cases, is so recent that it is not surprising when a

Museum visitor is seen reaching hesitantly upward toward a spacecraft's heat shield as if it might be still warm to the touch. And each visitor comes away from the Museum with a renewed sense of awe and shared pride in these accomplishments, for nowhere else in the world has been gathered such overwhelming proof that some of our most elemental dreams can and do come true. But, what is so astonishing, perhaps, is that so many of these dreams come true so fast.

Only twenty-four years after Orville Wright skimmed above the Kill Devil Hill's sands for one hundred and twenty feet in twelve seconds that cold, windy, December day near Kitty Hawk in 1903, Charles A. Lindbergh flew his Ryan monoplane, the *Spirit of St. Louis*, alone for 3,610 miles in 33 hours and 30 minutes non-stop across the Atlantic.

Orville Wright was still alive on October 14, 1947, when Charles Yeager in the Bell X-1 became the first man to break the speed of sound. And less than fifteen years later on February 20, 1962, John Glenn became the first American to orbit the Earth.

In his Mercury spacecraft, *Friendship 7*, Glenn traveled 80,428 miles at 17,500 mph and within 4 hours and 55 minutes three times circled the Earth at an altitude between 101 and 162 miles. The first liquid-propellant rocket had been developed by Robert H. Goddard only thirty-six years before; the Goddard 1926 rocket had reached an altitude of 41 feet in 2.5 seconds. Three years after Glenn, astronaut Edward H. White opened the right-hand hatch of the Gemini 4 spacecraft he shared with James A. McDivitt and, connected to the spacecraft's life

Edward H. White II and James A. McDivitt's Gemini 4.

support and communications systems only by a gold-covered "umbilical cord," he climbed outside to become the first American to "walk" in space.

In July, 1969, only sixty-five years and seven months after the Wright brothers' Flyer became the world's first powered airplane to carry a man and fly under control, astronauts Neil Armstrong, Edwin Aldrin, and Michael Collins rocketed to the Moon in the Apollo 11 spacecraft and while Collins circled above in the command module *Columbia*, Armstrong and Aldrin dropped down to the surface in the lunar module and became the first men to walk upon the Moon.

The Wright Flyer, Lindbergh's *Spirit of St. Louis*, Yeager's Bell X-1, Goddard's 1926 rocket, Glenn's *Friendship 7*, White and McDivitt's Gemini 4, Armstrong, Aldrin, and Collins' Apollo 11 command module *Columbia* are all there in the National Air and Space Museum's Milestones of Flight gallery. Small wonder that those early visitors waiting for the Museum's doors to open peer so eagerly through the glass; they are looking upon treasures more fabulous than any Pharaoh's fortune. These treasures share their space with an even more exotic object older by some four billion years: a thin slice cut from a small, very hard, fine-grained piece of basalt rock picked up in December, 1972, by the astronauts of Apollo 17 from the surface of the Moon.

Ever since the Smithsonian Institution's founding in 1848 it has evidenced its interest in aerospace development. Joseph Henry, the first Secretary of the Institution, successfully persuaded President Abraham Lincoln to support balloonist Thaddeus S.C. Lowe's utilization of captive balloons by the Northern Army for military observation during the Civil War. The Institution's collection of aeronautical objects began in 1876, when a group of kites was acquired from the Chinese Imperial Commission. Samuel P. Langley, the third Secretary of the Institution (and a trained astrophysicist who established the Smithsonian

Astrophysical Observatory), was smitten by the allure of flight in his late fifties and became one of aviation's most controversial and unlucky early pioneers. He managed to produce in 1896, nevertheless, the first American heavier-than-air powered flying machine capable of making a free flight of any significant length. His machine, Langley's Aerodrome #5, containing a 1 hp steam engine, was launched from a houseboat on the Potomac River and flew about 3,000 feet at a speed of about 25 mph before landing gently in the river. The model hangs near the east wall of the Milestones of Flight gallery. When Langley attempted to build a man-carrying machine based on an enlargement of his successful model's design, he failed.

The fourth Secretary of the Institution, Charles D. Walcott, and Smithsonian Regents Alexander Graham Bell and Ernest W. Roberts, realizing the need to place American aviation on a sound scientific footing, began actively petitioning Congress for an aeronautical research and policy center in 1912. As a result of their efforts, the National Advisory Committee for Aeronautics (the forerunner of NASA) was created in 1915. The following year the Smithsonian began its long association with Robert H. Goddard, the "father" of modern rocketry. For the next twenty-nine years the Smithsonian not only published his major articles but assisted in providing funds for his research.

Congress established the National Air Museum on August 12, 1946, as a Smithsonian bureau to "memorialize the national development of aviation; collect, preserve, and display aeronautical equipment of historic interest and significance; serve as a repository for scientific equipment and data pertaining to the development of aviation; and provide educational material for the historical study of aviation." (Twenty years later that act of Congress was amended to include space flight.) Recognizing that before NASM could fulfill its mandate a new museum building

The "cockpit" of the first Wright Flyer.

would be essential, Congress in 1958
designated the present site for the
Museum—a site that has been used for
aerospace activities ranging from early
balloon ascensions to, more recently,
the colorful Kite Flying festivals. In
November, 1972, work on the new National
Air and Space Museum began. The building
officially opened to the public on July 1, 1976.

The remarkable singularity of many of its

exhibits is not, however, what makes NASM
unlike any museum in the world; it is the
methods by which the collection is made
available to the public and exhibited. The
goal of the Museum has been to present the
story of flight in all its dimensions. And
fundamentally, like other museums, NASM
uses words, images, and physical objects
to communicate with the visitor. The
basic organization of the galleries is

uncomplicated; each is devoted to a single subject or theme so that when taken in totality they do cover the entire concept of flight. But the Museum's approach to its exhibits ranges from the most simply labeled objects to the most sophisticated presentation techniques. "Our main objectives have been communication of information and feeling," Melvin B. Zisfein, NASM's former Deputy Director and director of its exhibits program, has said, "and our exhibits have been aimed at making the visitor a willing and happy participant in the process." And there is the key to the Museum's popularity. Visitors are *urged* to have fun, to stand in the cockpit of a DC-7, to walk through a Skylab Orbital Workshop, to attend an indoor aeronautical exhibition of 1913 and listen to a salespitch for a Blériot monoplane similar to the one that had made the first successful flight across the English Channel only two years before. In the Sea-Air Operations gallery the visitor stands on an aircraft carrier's "hangar deck" among actual Navy fighter planes and bombers and through a hatch to the carrier deck sees the sea rushing past. There is also "To Fly!" the filmed aerial tour of America projected on a giant, five-story-high screen. That film has now been joined by five other IMAX films including "The Dream Is Alive," an insider's view of America's space program. Still, notwithstanding the most sophisticated and dramatic exhibit techniques, the basic fare of the Museum is its mandated "aeronautical and space flight equipment of historic interest and significance."

It cannot help but be something of a shock for the Museum visitor to realize that the frail-looking canvas and wooden machine hanging overhead is the one and only Wright Flyer with which Wilbur and Orville Wright became the first men to successfully carry out controlled and powered heavier-than-air flight. All of the airplanes exhibited are genuine. The Japanese Zero, the Messerschmitt ME-109, the P-51 Mustang in the World War II gallery were flown in that war. The spacecraft are the actual craft used if they were returned from space. (If the return was not possible, NASM exhibits the real back-up vehicle or a replica made from actual flight hardware.) The Apollo 11 command module *Columbia* is the very one in which Neil Armstrong, Edwin Aldrin, and Michael Collins returned from the Moon. It seems especially fitting that Michael Collins should have become the first Director of the new National Air and Space Museum. Collins has suggested that the reason that the Museum has become the biggest tourist attraction in Washington is perhaps because so much of the collection represents specific historic events people can recall. "We are fortunate," he has said, "in that the span of history with which we work is relatively short. People find it easier to relate the exhibits to their own lives." When the visitor moves among the exhibits, he is not just looking at a collection of objects, he is reliving an experience, whether it be the first time he took a commercial airline flight, served in a war, flew his own plane, cheered Lindbergh's flight across the Atlantic, or Apollo 11's television transmissions from the surface of the Moon. At one time or another a major portion of the Museum's exhibits made world headlines. The visitor can stand next to, touch, or step inside history and share that experience with everyone else there, too.

This book is an attempt to recapture some of the National Air and Space Museum experience. Wherever possible we have supplemented the text with the personal reminiscences, eyewitness accounts, actual logs, or written records of some of the men who designed or flew in the machines on exhibit. We hope thereby to provide the Museum visitor and reader of this book with an even deeper understanding and appreciation of the wonders of flight. And though the objects exhibited in the Museum can be admired for their uncanny sculptural beauty and the sometimes marvelous functions they have performed, one must never lose sight of the fact that they are only the footprints left behind by humanity on its long, arduous, occasionally halting but wholly satisfactory, admirable, and inevitable journey to the stars.

MILESTONES OF
FLIGHT

Milestones of Flight

Turn-of-the-century America was exuberant. Optimistic. Confident that with hard work and God's divine will nothing was impossible. "God has marked the American people as His chosen nation," Indiana Senator Albert J. Beveridge proclaimed, "to finally lead in the regeneration of the world. This is the divine mission of America, and it holds for us all the profit, all the glory, all the happiness possible to man. We are trustees of the world's progress, guardians of its righteous peace."

The promise of science and technology enthralled Americans; the most marvelous new machines and inventions had already changed their lives. There now existed electricity, the telephone, the typewriter, the sewing machine, the automobile, the self-binding harvester, locomotives of seemingly incredible power and unlimited speed.

Among its cast of out-and-out cranks and eccentrics, every small town in America had its two or three inventors, tinkerers, men who believed their better mousetrap, their superior tool, their perpetual-motion machine would—after just a little more refinement—bring them fortune and fame. For them Thomas Edison epitomized the belief that by inventing something they would turn themselves into tycoons. Two such men were Orville and Wilbur Wright, whose unstinting patience, scientific intuition, methodical experimentation, and boundless confidence and optimism enabled

them within four years to succeed in resolving and providing satisfactory solutions to the seemingly insurmountable problems that had plagued all their predecessors and prevented men from achieving winged, powered, controlled flight. This is the story of Wilbur and Orville Wright and the dawn of aviation, too.

Wilbur Wright was born in Millville, Indiana, on April 16, 1867, the year gold was discovered in Wyoming, diamond fields in South Africa, Nebraska became our 37th state, Russia sold us Alaska for $7,200,000 (roughly 2¢ per acre), the year Marie Curie was born, that Charles Baudelaire died, Mark Twain wrote "The Jumping Frog" and Ibsen *Peer Gynt*, and Dr. David Livingstone was exploring the Congo in search of the source of the Nile.

Orville Wright was born in Dayton, Ohio, four years later, on August 19, 1871, the year of The Great Fire in Chicago, that Lewis Carroll's *Through the Looking Glass* and Charles Darwin's *Descent of Man* were published, Simon Ingersoll invented the pneumatic rock drill, and James Gordon Bennett's New York *Herald* reporter Henry M. Stanley greeted a stooped and sickly white man at Ujiji, Central Africa, saying, "Dr. Livingstone, I presume?"

To set the Wright brothers in time is easy enough, but to separate the reality of their personalities from the portraits painted of them over the years by myth, malice, and

The Wright Flyer in which, on December 17, 1903, one of man's oldest dreams was realized when Orville Wright achieved the world's first successful controlled, powered, manned, heavier-than-air flight. His epoch-making 120-foot journey lasted just 12 seconds, but aviation was born.

misunderstanding is more difficult. C.H. Gibbs-Smith, the Smithsonian's first Lindbergh Professor of Aerospace History, has long fought attempts to depict the Wrights as a couple of bright, local boys who "with wire, spit, sticks, canvas, and spare parts from their bicycle shop put together an airplane."

The Wrights, he has said, "were among the best educated men in the United States in the 19th Century." Although neither Wilbur nor Orville bothered to graduate from high school one has only to dip into their correspondence or attempt to follow their technical and theoretical experiments to realize the high quality of education they achieved. Orville Wright often spoke of their upbringing as having given them "exceptional advantages" but he was not referring to wealth. Their father, Milton Wright, was a Bishop in the United Brethren Church whose salary was never more than $900 a year. And although there was an additional slight income from a family farm in Indiana, money for luxuries was scarce. "All the money anyone needs," Bishop Wright used to say, "is just enough to keep from being a burden to others." The "exceptional advantages" Orville spoke of were those of an environment created in their home by their dignified and broadminded father, who never set rigid rules of conduct for his children, but rather encouraged them to be independent thinkers, to pursue intellectual interests, to investigate whatever aroused their curiosity, and to have confidence in their reasoning. His children respected and obeyed their father not out of fear but because they so admired him. In fact, it was out of deference to their father that neither Wilbur nor Orville would permit exhibition flights of their aircraft on Sundays—the day that would have been most profitable.

If Wilbur and Orville inherited their mechanical aptitude from any member of their family it would probably have come from their mother, Susan Koerner Wright, the daughter of a German-born wagon maker. According to her children, she could mend anything. If they could not afford a necessary appliance, she would adapt some other household device to do the needed task. Mrs. Wright died when Wilbur was twenty-two and Orville eighteen; their older brothers Reuchlin and Lorin had already married and moved into homes of their own. Wilbur, Orville, and their younger sister Katharine remained in their Dayton home with their father and formed the nucleus of the extraordinarily close-knit Wright family.

In the 1880s and 1890s bicycling swept Europe and the United States. Bicycle clubs sprang up everywhere. The League of American Wheelmen, which was organized in 1880, had become a major agitator in the movement for good roads. The new European "safety" bicycle with its approximately same-size front and rear wheels, sprocket chain-drive connecting its pedals to the rear wheel and pneumatic tires had come to the Wrights' attention and in 1892 Wilbur and Orville became caught up in the fad. Orville even tried some track racing. The Wrights were always mechanically minded. Lorin had invented an improvement on a hay-baling machine, Wilbur had designed and constructed a practical device for folding paper to diminish the tediousness of fulfilling a contract he had to fold the entire weekly issue of an eight-page church paper. Wilbur and Orville had both built a front porch onto their Dayton home and did all the lathe work for the posts themselves and then remodeled the interior of the house as well. What had started as an interest in making wood engravings developed within a few years into a printing-press business, and when Orville was seventeen he built a press big enough and fast enough to print a newspaper. The brothers were so handy with machinery that, inevitably, they began repairing other young men's bicycles and eventually opened their own bicycle shop, where in addition to doing repairs they sold some of the established brands like the

Coventry Cross, Halladay-Temple, Warwick, Reading, Smalley, Envoy, Fleetwing, and some they made themselves. When the perfecting press came into use in Dayton, Wilbur and Orville's small neighborhood paper, a four-page, five-column paper called *The Evening Item*, could not compete and it was converted back from a daily into a weekly, although their printing business remained about the same.

The Wrights traced their flying machine interest back to a toy Pénaud-type helicopter made of cork, bamboo, and thin paper powered by a rubber band, given them by their father in 1878, when Wilbur was eleven and Orville seven. Wilbur had almost immediately attempted to improve its design and made several models, each larger than the preceding. To their dismay the boys discovered that the larger the machine they built, the less well it flew and that, in fact, it would not fly at all if its size was much greater than the Pénaud toy. Disappointed by this line of experimentation, the boys turned to kite flying, though they never ceased to be fascinated by anything they read about flying machines. There was, of course,

the "popular" literature of the day—Jules Verne in his 1865 *From the Earth to the Moon* had launched three men into space within a conical projection fired from the Florida coast and five years later in *Around the Moon* rescued his astronauts from permanent oblivion by bringing them back to a safe ocean landing on earth. But the Wrights were never particularly interested in fiction. When, in 1895, they read of the gliding experiments being conducted by Otto Lilienthal, Germany's first and foremost contributor to the conquest of the air, they sought every piece of information they could learn about him. Between 1891 and 1896 Lilienthal had made over 2,000 glides—some of them of several hundred feet—down a large hill he had constructed near Berlin. His early gliders were monoplanes with fixed tails. The pilot's head and shoulders were above the cambered wings, his hips and legs dangled below. What limited directional control Lilienthal achieved he managed by shifting his hips and weight from side to side or back and forth. Photographs and published reports of Lilienthal's experiments fascinated the Wrights. He had effectively

Glider pioneer Otto Lilienthal achieved control by shifting his weight.

29

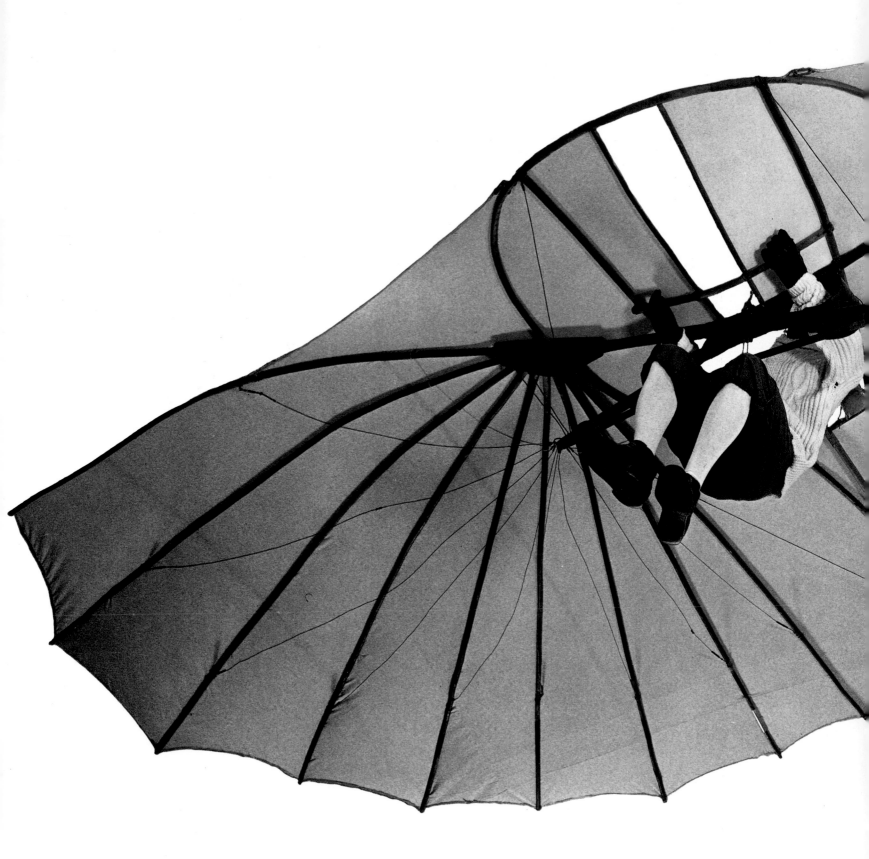

An original Lilienthal Standard glider similar to the type in which its inventor lost his life.

demonstrated that air *could* support a man in winged flight.

Ever since 1891 Lilienthal had been designing and constructing gliders with the hope that when a suitable means of propulsion was developed, it could be added to his wings. In 1896 he had built a glider with flapping wing tips powered by a small compressed carbonic acid gas motor. Unfortunately, before Lilienthal had an opportunity to test it, he was killed in one of his standard gliders when a sudden gust of wind forced the glider upward into a stall. The craft crashed to earth and broke the German aviation pioneer's back. He died the next day. One of Lilienthal's gliders can be seen at the entrance to the Flight Testing gallery on the first floor. Lilienthal's death made the Wrights even more eager to learn not only everything he had accomplished, but what progress others as well had made toward achieving human flight. Since little had been published about the subject they were unable to satisfy their curiosity. In 1899, aware of the aeronautical studies and experiments being conducted by Dr. Samuel P. Langley, the third Secretary of the Smithsonian Institution, Wilbur Wright wrote the Smithsonian:

> I have been interested in the problem of mechanical and human flight ever since as a boy I constructed a number of bats [Wilbur's name for helicopter models] of various sizes after the style of Cayley's and Pénaud's machines. My observations since have only convinced me more firmly that human flight is possible and practicable. It is only a question of knowledge and skill just as in all acrobatic feats...I believe that simple flight at least is possible to man and that the experiments and investigations of a large number of independent workers will result in the accumulation of information and knowledge and skill which will finally lead to accomplished flight.
>
> ...I am about to begin a systematic study of the subject in preparation for practical work to which I expect to devote what time I can spare from my regular business. I am an enthusiast, but not a crank in the sense that I have some pet theories as to the proper construction of a flying machine. I wish to avail myself of all that is already known and then if possible add my mite to help on the future worker who will attain final success.
>
> —*Miracle at Kitty Hawk: The Letters of Wilbur and Orville Wright*, ed. by Fred C. Kelly

The Smithsonian responded by suggesting that the Wrights read Octave Chanute's *Progress in Flying Machines*, Professor Langley's *Experiments in Aerodynamics*, and the Aeronautical Annuals of 1895–97 containing reprints of accounts of experiments going back to Leonardo da Vinci. The Smithsonian also sent pamphlets containing material extracted from the Institution's own reports including Mouillard's *Empire of the Air*, Langley's *Story of Experiments in Mechanical Flight*, and Lilienthal's paper on *The Problem of Flying and Practical Experiments in Soaring*. The Smithsonian's materials arrived in June, 1899. The Wrights read them eagerly and were surprised to learn how much time and money had already been spent attempting to solve the problem of human flight; they were perhaps even more astonished by the caliber of the men who had tried.

"Contrary to our previous impression," Wilbur subsequently said, "we found that men of the very highest standing in the professions of science and invention had attempted the problem. Among them were such men as Leonardo da Vinci, the greatest universal genius the world has ever known; Sir George Cayley, one of the first men to suggest the idea of the explosion motor; Professor Langley, Secretary and head of the Smithsonian Institution; Dr. Bell, inventor of the telephone; Sir Hiram Maxim, inventor of the automatic gun; Mr. O. Chanute, the past president of the American Society of Civil Engineers; Mr. Chas. Parsons, the inventor of the steam turbine; Mr. Thomas A. Edison; Herr Lilienthal and a host of others...."

They discovered that the period from 1889 through 1897 had been one of exceptional competiton between such distinguished pioneer scientists and inventors as Langley, Lilienthal, Chanute, and Maxim, each hoping to become the first to accomplish successfully a manned, powered flight. One by one these men had been beaten and discontinued their efforts. Rather than being discouraged by their predecessors' failures, the Wrights plunged ahead and a mere three months after they had received the pamphlets from the Smithsonian they had constructed their first aircraft, a biplane kite with a five-foot wingspan and a fixed horizontal tailplane. This craft incorporated what Gibbs-Smith describes as "their first decisive discovery and first decisive invention": wing warping.

"The Wrights had observed that gliding and soaring birds," Professor Gibbs-Smith writes, "evidenced especially by their local 'expert' the buzzard 'regain their lateral balance when partly overturned by a gust of wind, by a torsion of the tips of the wings. If the rear edge of the right wing is twisted upward and the left downward the bird... instantly begins to turn, a line from its head to its tail being the axis. It thus regains its level even if thrown on its beam's end, so to speak, as I have frequently seen them,'* and [the Wrights] decided to apply this bird practice to aeroplane wings. At first they thought of variable incidence wings but abandoned this idea for structural reasons; then they hit upon the idea of helicoidal twisting of the wings ('warping') after toying with a long cardboard box."

In spite of the Wrights' eagerness to fly, they were determined to proceed methodically. Before they would attempt any experiments with powered machines they intended to master controlled flight in gliders. Although with their first biplane kite the Wrights hoped to test their method of wing warping, they also constructed it in such a way that the wings could be shifted forward and backward in relation to one

another in order to control the center of pressure and the tailplane would then act automatically as an elevator. Once they determined that the wing warping worked, the Wrights decided to build their first full-size glider. Anyone attempting to attain sustained powered flight, the Wrights realized, was confronted with three problems: the first was the design and construction of a device capable of remaining with a man in the air; the second was finding how to control the device once it was up in the air; the third problem was that once one had a device that could remain with a man in the air and under control, one still had to provide a safe, practical propulsion system to keep him up there. They had Otto Lilienthal's research into the efficiency of various wing shapes to work with but, like Lilienthal, they knew that control could come only through practice. The Wrights' first full-size glider was completed in September, 1900. The No. 1 Glider was a biplane weighing about 52 pounds, with a wingspan of a little over 17 feet and a total lifting area of but 165 square feet. The Wrights had decided to place their horizontal elevator in front of the wing; there was no tail unit, and the wing-warping cables could be worked either by the operator or from the ground. An 18-inch-wide space at the center of the lower wing section was left uncovered so that the pilot could lie prone there with his feet extending over the rear spar.

Aware that their first full-size glider was in essence a big kite and that a strong wind would be required to keep it aloft, the Wrights wrote the Weather Service in Washington to learn where suitable winds might be found. They were given several locations with strong winds in the far west, but the closest site to Dayton was to the east at a desolate, isolated spot along North Carolina's coast: Kitty Hawk.

After exchanging letters with Joseph J. Dosher at the Kitty Hawk weather station and William J. Tate, the former postmaster and reputed to be the best educated man in that area (Tate wrote Wilbur: "I would say

*Prof. Gibbs-Smith is quoting here from a letter written by Orville to Octave Chanute on May 13, 1900.

At Kitty Hawk, the Wrights' No. 1 Glider (1900) was flown primarily as a kite.

that you could find here nearly any type of ground you could wish; you could, for instance, get a stretch of sandy land one mile by five with a bare hill in center 80 feet high, not a tree or bush anywhere to break the evenness of the wind current. This in my opinion would be a fine place; our winds are always steady, generally from 10 to 20 miles velocity per hour. We have Telegraph communication and daily mails. Climate healthy, you could find a good place to pitch tents & get board in private family provided there were not too many in your party; would advise you to come any time from Sept 15 to Oct 15. Don't wait until November. The autumn generally gets a little rough by November."), Wilbur and Orville decided Kitty Hawk would be the best spot. Wilbur Wright wrote to his father:

September 3, 1900

I am intending to start in a few days for a trip to the coast of North Carolina...for the purpose of making some experiments with a flying machine. It is my belief that flight is possible and, while I am taking up the investigation for pleasure rather than profit, I think there is a slight possibility of achieving fame and fortune from it. It is almost the only great problem which has not been pursued by a multitude of investigators, and therefore carried to a point where further progress is very difficult. I am certain I can reach a point much in advance of any previous workers in this field even if complete success is not attained just at present. At any rate, I shall have an outing of several weeks and see a part of the world I have never before visited....

The Wright No. 1 Glider had two important basic control elements: the first was a "horizontal rudder" (or elevator) placed about 30 inches in front of the leading edge of the lower wing. The Wrights believed that by placing the elevator in front rather than behind the wings they would provide themselves with a safer fore-and-aft balance control, especially should a sudden downdraft force the craft to drop. The second element was, of course, the wing warping. However, due to the small lifting area of the glider's wings, they had to fly it chiefly as a kite.

The Wrights had been guided by Lilienthal's tables relating to air pressures on wing surfaces in their determination of the amount of wing area needed. According to these tables, the Wright glider with its small lifting area of but 165 square feet should have supported a man in winds of but 17 to 21 miles per hour. The Wrights were disappointed to discover that much greater winds were necessary. They abandoned their plan to acquire hours of practice aboard their glider while flying it as a kite and were forced to fly it primarily unmanned. By the end of that season's experiments the Wrights had flown their No. 1 Glider manned just ten minutes as a kite, and had acquired only two

minutes of actual gliding time. Still, the Wrights were not discouraged; they had proved that their wing-warping method for achieving lateral control worked more successfully than any method previously attempted, and their front elevator had also provided the most successful means of vertical control. Before they left Kitty Hawk they had already decided to build a much larger glider, one capable of carrying a man in the sort of winds they had encountered at the beach. More determined than ever to fly, the Wrights nevertheless continued to proceed as deliberately, cautiously, and methodically as they had before.

The Wrights' No. 2 Glider was brought to Kill Devil Hill, four miles south of Kitty Hawk, in July, 1901. (The Wrights never did use Kitty Hawk except for their tests with their No. 1 Glider, preferring instead the Kill Devil Hill site.) Glider No. 2 was a larger machine than anyone had attempted to fly before and too large to be controlled simply by an operator shifting his weight. Its 22-foot wingspan provided a lifting area of 290 square feet. Since the glider had been designed to fly in winds of at least 17 mph and the winds on July 27th, the day the

The Wright 1903 Flyer after an attempted flight on December 14, 1903.

Wrights were ready to test it, never reached more than 14 mph, they carried the glider to the top of Kill Devil Hill for its first trial. They made five or six brief "tuning-up" flights, then made a glide that lasted 19 seconds and covered 315 feet. Although several of their flights on the first day exceeded the best of those made in the No. 1 Glider the previous year, the Wrights realized that in some ways their new glider was not as good, and that in particular it could not glide at a slope nearly as level as the 1900 machine had achieved. The Wrights became suspicious of the accuracy of Lilienthal's (and others') calculations concerning the center of air pressure on curved surfaces. Wilbur noted in his diary:

July 30, 1901

The most discouraging features of our experiments so far are these: the lift is not much over one third that indicated by the Lilienthal tables. As we had expected to devote a major portion of our time to experimenting in an 18-mile wind without much motion of the machine, we find that our hopes of obtaining actual practice in the air are decreased to about one fifth of what we hoped, as now it is necessary to glide in order to get a sustaining speed. Five minutes' practice in free flight is a good day's record. We have not yet reached so good an average as this even.

After mentioning that the machine had suffered no injury after some forty landings and that they had obtained a "free flight of over 300 feet at an angle of 1 in 6," Wilbur noted:

We have experimented safely with a machine of over 300 square feet surface [counting the elevator surfaces] in winds as high as 18 miles per hour. Previous experimenters had pronounced a machine of such size impracticable to construct and impossible to manage. It is true that we have found this machine less manageable than our smaller machine of last year, but we are not sure that the increased size is responsible for it. The trouble seems rather in the travel of the center of pressure. The lateral balance of the machine seems all that could be desired....

Although during the majority of their

flights the Wrights' method of wing warping to achieve lateral control worked properly, they discovered that at other unexplained and alarming moments the wing warping had no effect whatsoever; worse, the glider would suddenly spin around, side-slip, and crash. Obviously they had not yet achieved complete control of their equilibrium and the brothers wondered whether some sort of vertical fin placed to the rear of the wings might not help. The Wrights had proven with their No. 2 Glider that a large machine could be controlled as easily as a small one, provided that the means of control was the movement of the wing and elevator surfaces and not simply the shifting of the operator's body. The brothers had, they believed, broken all previous records for distance in their glider, but the absolute faith the Wrights had had in the scientific data compiled by their predecessors, Lilienthal especially, was badly shaken. It was becoming obvious that the center of pressure on a curved or cambered surface did not—as all the scientific books taught—travel in the same direction as it would on a plane surface, and that, in fact, the center of pressure appeared to reverse once a certain angle of attack was achieved. If there were no accurate scientific data upon which the Wrights could count then the task they had set for themselves was even more formidable than they had believed. Wilbur was so discouraged that on their way back to Dayton he said, "Not within a thousand years will man ever fly." Orville, attempting to cheer Wilbur up, argued that the unreliability of their predecessors' work did not mean that flight was impossible, it meant only that more knowledge and effort were necessary.

Throughout this early period the Wrights had corresponded with Octave Chanute, the author of *Progress in Flying Machines*, and had invited him to Kill Devil Hill to witness some of their glides. Chanute had been so impressed that he refused to let the Wrights become disheartened. He was so encouraging, in fact, that it has been said that without Chanute's insistence the Wrights might have discontinued their experiments. In August, 1901, upon their return from North Carolina, Chanute invited Wilbur to address the Western Society of Engineers on the subject of his gliding experiments. Wilbur was nagged into accepting the invitation by his sister Katharine who thought it would give him a good opportunity to "get acquainted with some scientific men," as she wrote her father, "and it may do him some good. We don't hear anything but flying machine and engine from morning till night. I'll be glad when school begins so I can escape." Octave Chanute wrote Wilbur:

September 5, 1901

The secretary of the Society, and the Publication committee are greatly pleased that you consent to giving the Western Society of Engineers a talk on the 18th [of September]. May they make it "Ladies night"?

Wilbur wrote back:

September 6, 1901

I must caution you not to make my address a prominent feature of the program as you will understand that I make no pretense of being a public speaker. For a title, "Late Gliding Experiments" will do. As to the presence of ladies, it is not my province to dictate, moreover I will already be as badly scared as it is possible for man to be, so that the presence of ladies will make little difference to me, provided I am not expected to appear in full dress.

Katharine Wright wrote her father:

September 11, 1901

The boys are still working in the machine shop. A week from today is "Ullam's" [Wilbur's] speech at Chicago. We asked him whether it was to be witty or scientific and he said he thought it would be pathetic...!

—*Miracle at Kitty Hawk: The Letters of Wilbur and Orville Wright*, ed. by Fred C. Kelly

In the first draft of his speech Wilbur unflinchingly stated that all the existing scientific tables regarding the effect of air pressure on airplane surfaces contained serious errors. But then he began to worry that it might be presumptuous of him to so publicly dispute the work of eminent scientists whose tables and data had acquired a credence simply by virtue of their having been published and accepted as fact for so long unless he could be absolutely certain he was right. Out of an old starch box Orville constructed a small wind tunnel and one day's tests were enough to show them that Wilbur's criticism could be supported by facts. Nevertheless the Wrights refrained from including Wilbur's strongest criticism of the available figures until further wind-tunnel tests provided more detailed support.

Wilbur strongly questioned the accuracy of available figures when he finally did give his speech, but in the published text he omitted his most severe attacks. His speech did, however, reflect Wilbur's conviction that scientific experiment had to be coupled with practical experience and that risks were involved:

The person who merely watches the flight of a bird gathers the impression that the bird has nothing to think of but the flapping of its wings. As a matter of fact this is a very small part of its mental labor. To even mention all the things the bird must constantly keep in mind in order to fly securely through the air would take a considerable part of the evening....The bird has learned [his] art ...so thoroughly that its skill is not apparent to our sight. We only learn to appreciate it when we try to imitate it. Now, there are two ways of learning to ride a fractious horse: one is to get on him and learn by actual practice how each motion and trick may be best met; the other is to sit on a fence and watch the beast a while, and then retire to the house and at leisure figure out the best way of overcoming his jumps and kicks. The latter system is the safest; but the former, on the whole, turns out the larger proportion of good riders. It is very much the same in learning to ride a flying machine; if you are looking for perfect safety, you will do well to sit on a fence and watch the birds; but if you really wish to learn, you must mount a machine and become acquainted with its tricks by actual trial.

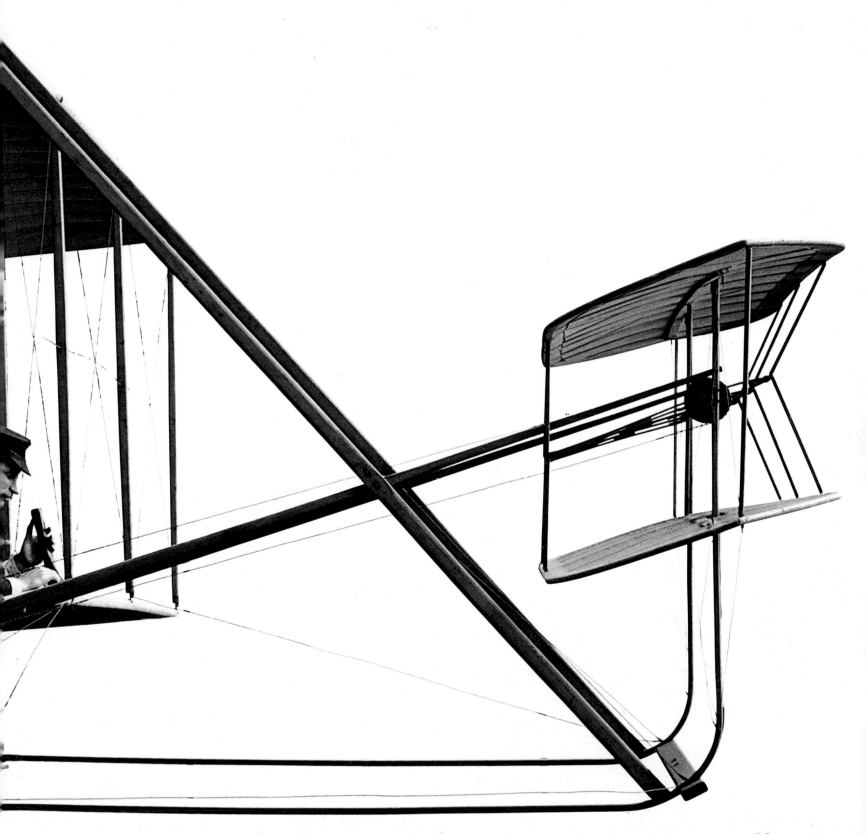

The pilot lay prone in the Wright Flyer (1903).

Although the Wrights were still not sure they could ever satisfactorily resolve the problems of manned, controlled, and powered flight and were hesitant to build another glider, their scientific curiosity, their passion for accuracy, compelled them to continue their study of the problems of air pressure. They constructed a larger, more sophisticated wind tunnel: a wooden box 16 inches square inside and 6 feet long. The Wrights still did not have electricity and derived their wind from a fan driven by a one-cylinder gasoline engine they had previously made. During that autumn and early winter of 1901, a span of just over two months, the Wrights tested a little over two hundred different types of wing surfaces in their wind tunnel. They tried monoplane, biplane, and triplane configurations in addition to models with two wings placed on the same level, one behind the other, as Langley had done. The Wrights were forced to discontinue their tests a little before Christmas since they were, after all,

> ...still in the bicycle business, still obliged to give thought to their means for earning a living, and with no idea that this scientific research could ever be financially profitable. In those few weeks, however, they had accomplished something of almost incalculable importance. They had not

only made the first wind-tunnel in which miniature wings were accurately tested, but were the first men in all the world to compile tables of figures from which one might design an airplane that could fly. Even today, in wind-tunnels used in various aeronautical laboratories, equipped with the most elaborate and delicate instruments modern science can provide, the refinements obtained over the Wrights' figures for the same shapes of surfaces are surprisingly small. But it is doubtful if the difficulties and full value of the Wrights' scientific researches within their bicycle shop are yet appreciated. The world knows they were the first to build a machine capable of sustained flight and the first actually to fly; but it is not fully aware of all the tedious, grueling scientific work they had to do before flight was possible. Important as was the system of control with which the Wrights' name has been connected, it would not have given them success without their wind-tunnel work which enabled them to design a machine that would lift itself.

—*The Wright Brothers*, by Fred C. Kelly

The Wrights finally decided they would have to test their newly acquired knowledge with a third glider at Kill Devil Hill. Their No. 3 Glider was ready for its first trial on September 29, 1902. Although the No. 3 Glider's lifting area was not that much greater than that of No. 2 (305 square feet *vs.* 290 square feet) the wingspan had been increased from the 22 feet of No. 2 to the 32

The modified No. 3 Glider (1902) being launched at Kill Devil Hill with Orville piloting.

By replacing the No. 3 Glider's fixed vertical double tailfin with the single movable rear rudder shown here, the Wrights eliminated their problems with tail spins.

feet of No. 3; this meant, of course, that the No. 3 Glider's wings were longer and narrower than those of No. 2. Wing-warping controls were linked to the "cradle" in which the operator's hips rested and to warp the wings the pilot simply shifted his hips from side to side. The most visible change was the addition of a tail consisting of two fixed vertical fins. The Wrights hoped that the tail, containing just less than a 12-square-foot total area, would counter-effect the problems they had experienced with the No. 2. Then, when the glider had been banked, the wing presenting the greater angle to the wind (and, therefore, the most resistance) had begun to slow in relation to the other wing and the machine would spin in.

The first trials with the No. 3 Glider were encouraging. The Wrights, that September and October of 1902, made over a thousand glides—the longest was 622½ feet, and the greatest duration was 22½ seconds—and a number of them were made against a 36 mile per hour wind, stiffer than any previous glider experimenter had attempted. Still, about once in every fifty glides the machine would behave in a puzzling manner: in spite of all the warp the

pilot could give to the wing tips the machine would turn up sideways in a bank and slip down to the ground. And yet at the next trial, under seemingly identical conditions, nothing would happen. The Wrights would again begin to relax but then, suddenly, the same mysterious behavior would occur, one wing would dip down, dig into the sand, the glider would spin in with what the brothers began to refer to as a "well-digging" movement. Nothing like this had occurred with their previous gliders and so the brothers deduced that whatever was happening was being caused by the addition of the tail. (Unknown to the Wrights, they had also achieved the dubious distinction of inventing the "tail-spin.") But even though the Wrights knew that the problem was caused by the tail, they had no idea why it was happening. And then one night Orville had had too many cups of coffee to sleep well and he lay awake tossing about and figured out the problem:

He was so sure he was right and that it was a basic discovery, as indeed it was, that he wanted credit for it. Sometimes when he told Wilbur something, Wilbur would act as if he already knew it. So at breakfast the next morning he [Orville] winked at his brother Lorin, who was visiting them, to get his attention as a witness, and then asked Wilbur if he could explain the machine's peculiar behavior that had puzzled them. Wilbur shook his head.

Orville then gave this explanation: When the machine became tilted laterally it began to slide sidewise while advancing, just as a sled slides downhill or a ball rolls down an inclined plane, the speed increasing in an accelerated ratio. If the tilt happened to be a little worse than usual, or if the operator were a little slow in getting the balance corrected, the machine slid sidewise so fast that this movement caused the vertical vanes [the tail fins] to strike the wind on the side toward the low wing instead of on the side toward the high wing, as it was expected to do. In this state of affairs the vertical vanes did not counteract the turning of the machine about a vertical axis, caused by the difference of resistance of the warped wings on the right and left sides; on the contrary, the vanes assisted in the turning movement, and the result was worse than if there were no fixed vertical tail.

If his explanation was sound, as Orville felt sure it was, then, he said, it would be necessary to make the vertical tail movable to permit the operator to bring pressure to bear on the side toward the higher wing.

—*Miracle at Kitty Hawk*

Wilbur realized that Orville's explanation was probably correct and immediately suggested that the mechanism controlling the wing warping and the mechanism which would move the tail should both work in conjunction; that way the pilot would only have to operate the front elevator and the wing-warping controls instead of a third control device for the tail. When the Wrights returned to work on the No. 3 Glider they attached the wires controlling the rudder to those warping the wings; further, they changed the two vertical surfaces on the rudder to one single vertical fin.

The Wrights had now solved most of the problems of securing stability and control in flight. With the movable tail the Wrights had created a practical glider so efficient it broke all existing records: it was the largest machine (32 foot 1 inch wingspan), capable of the longest time in the air (26 seconds) and covering the greatest distance (622½

feet) with the smallest angle of descent (5 degrees)—an angle less than could be attained by any of the hawks they had so carefully observed—and it had been flown in higher winds than anyone had hitherto attempted (36 mph). The 1902 flights, Orville subsequently wrote, "demonstrated the efficiency of our system of control for both longitudinal and lateral stability. They also demonstrated that our tables of air pressure which we made in our wind tunnel would enable us to calculate in advance the performance of a machine."

Because of the triumphant success of their No. 3 Glider, the Wrights applied for a patent in March, 1903. (The patent, granted in 1906, covered the principle of increasing the angle of incidence at one wing tip with the simultaneous decreasing of the angle of incidence of the other, the practical technique of wing warping, and the simultaneous use of warping and rudder to effect proper lateral control.) The Wrights were also convinced they could now build a successful *powered* aircraft.

There were two monumental obstacles the Wrights still had to overcome: they had to acquire a suitable engine, one capable of producing at least 8 horsepower that would weigh no more than 200 pounds; and second, they would need propellers. It is important to note that the Wrights never considered simply installing a motor on one of their gliders; rather they designed an entirely new machine incorporating the proven successful features of their gliders. It should also be understood that the Wrights deliberately designed their machines to be unstable. Their experience with the No. 3 Glider had shown them the importance of having their machines respond immediately and sensitively to the pilot's controls.

In an attempt to surmount the first obstacle the Wrights, in December, 1902, wrote to a number of automobile companies and gasoline-engine manufacturers asking whether they had a motor that fulfilled their requirements. None did—or if they did, they were hesitant to provide one for anything so lunatic as powering a "flying machine"—and so Wilbur and Orville, with the help of Charles Taylor, their bicycle assistant, simply designed and built one themselves. An even more formidable task was designing the propellers.

The Wrights had not anticipated the almost total lack of scientific knowledge about the workings and design of propellers that existed; after all, the screw propeller had been in use on ships for nearly a century. The Wrights knew that it was easy for a marine propeller to obtain at least a 50 percent efficiency, and they believed they would need only to study books on marine engineering to understand and learn the theory behind the operation of ship's propellers, then substitute the action of air pressure for water pressure and they would have it. When they finished their reading they realized that all the formulas on the action of propellers in water had been arrived at through trial and error rather than through any scientific theory. When marine engineers discovered that a particular propeller would not move a particular boat fast enough through the water they would substitute a propeller of a different size or different pitch until they arrived at a satisfactory solution. And yet they could neither design a propeller on paper, nor anticipate its performance in the water. Rough estimates might do for a motorboat since a propeller operating at less than the desired efficiency would still move the boat forward a little. But unless a propeller on a flying machine provided the necessary thrust, the machine would not fly.

The Wrights realized that a propeller was:

...simply an airfoil traveling in a spiral course. As they could calculate the effect of an airfoil traveling in a straight course, why should they not be able to calculate the effect in a spiral course? At first thought that did not seem too difficult, but they soon found that they had let themselves into a tough job. Since nothing about a propeller, or the medium in which it acts, would be standing still, it was not easy to

find even a point from which to make a start. The more they studied it, the more complex the problem became. "The thrust depends upon the speed and the angle at which the blade strikes the air; the angle at which the blade strikes the air depends upon the speed at which the propeller is turning, the speed the machine is traveling forward, and the speed at which the air is slipping backward; the slip of the air backward depends upon the thrust exerted by the propeller, and the amount of air acted upon." It was not exactly as simple as some of the problems in the school arithmetic—to determine how many sheep a man had or how many leaps a hound must make to overtake a hare.
—*The Wright Brothers*, by Fred C. Kelly

Wilbur and Orville argued between themselves throughout their attempts to work out a theory to explain the action of screw propellers, and it was precisely this habit of arguing technical points that made it possible for them to accomplish so much in so short a time. Neither of them would automatically defer to the other. But in their arguments about propellers they were dealing with so many unknowns that, as Orville later reported, "Often after an hour or so of heated argument, we would discover that we were as far from agreement as when we started, but that each had changed to the other's original position."

It took months for the Wrights to reduce their theories into formulas, to learn enough about how propellers did work to be able to calculate how specific designs would work. Wilbur wrote Octave Chanute:

July 18, 1903

The papers on screws [propellers], by various writers, do not seem to me of very much value. The chapter in the French book by André which is devoted to screws seems about as good as anything, but the final conclusion is that very little is known of the action of screws in motion forward. The action of screws not moving forward presents a very different case, and experiments based on such conditions are not applicable to the conditions met in practical flying...We think we have a method of figuring a screw in action but of course it is all mere theory as yet. We will know more about its correctness when we have had a chance to try it.

The 1896 Langley Aerodrome #5—the first
American powered, heavier-than-air machine
to make a flight of significant length.

Not until late September were the Wrights prepared to leave for the North Carolina coast. Their previous year's camp at Kill Devil Hill had been blown off its foundations by a storm; they repaired the damage and built a second shed to house both the No. 3 Glider (1902) and the new powered machine with which they confidently expected to fly. Just as the new shed was nearing completion, a storm came up. Winds of 40 mph increased during the night until by the next day gusts of 75 mph were blowing through the camp. Orville was forced to climb onto the roof to make repairs and while he was there the gale was so strong that it whipped his coattails over his head, pinning his arms so that he couldn't move. Wilbur hurried up onto the roof and freed him, but the wind remained too strong for either of them to pound in the nails accurately.

During this time, Samuel Pierpont Langley, the distinguished Secretary of the Smithsonian Institution, was preparing to launch his "aerodrome" from a houseboat on Washington's Potomac River. Langley's "aerodrome" was a gasoline-engine-powered aircraft with two main wings with a span of 48 feet 5 inches placed in tandem, one behind the other, with a stabilizer and tail behind. Seven years earlier, on May 6, 1896, Langley had achieved the distinction of launching the first American heavier-than-air flying machine capable of making a free flight of any significant length. The machine, Langley's Aerodrome #5, was a model, unmanned of course, and it flew about 3,000 feet at about 25 mph. It was made of steel and aluminum with wings of wood covered by silk; its wingspan was 13 feet 8 inches, it was 13 feet 2 inches long, it weighed 25 pounds and was powered by a 1 horsepower steam engine. In November, 1896, another steam-powered model, Langley's Aerodrome #6, made a similar flight. Prior to attempting to build a machine capable of carrying a man, Langley had constructed a ¼-scale model of the "aerodrome" he planned. This model, too,

October 7, 1903: Samuel P. Langley with optimistic pilot Charles M. Manly.

flew successfully. On October 7, 1903, with his gifted assistant Charles M. Manly at the controls (and with a compass optimistically sewn onto the knee of Manly's trousers), Langley prepared for what with reasonable confidence he believed would be the first man-carrying powered heavier-than-air flying machine's flight in history. At noon that day the 53-horsepower, gasoline-powered five-cylinder radial engine was started and the twin propellers flung back the air. Langley's tandem-winged "aerodrome" shuddered upon its catapult atop the Potomac River houseboat like a beast straining to be free and then, according to the Washington *Post* reporter sent to cover the flight, this is what happened:

...A few yards from the houseboat were the boats of the reporters, who for three months had been stationed at Widewater [waiting for the flight to be attempted]. The newspapermen waved their hands. Manly looked down and smiled. Then his face hardened as he braced himself for the flight, which might have in store for him fame or death. The propeller wheels, a foot from his head, whirred around him one thousand times to the minute. A man forward fired two sky rockets. There came an answering 'toot, toot,' from the tugs. A mechanic stooped, cut the cable holding the catapult; there was a roaring, grinding noise—and the Langley airship tumbled over the edge of the houseboat and disappeared in the river, sixteen feet below. It simply slid into the water like a handful of mortar...

—Washington *Post*, Oct. 8, 1903

On October 16, Wilbur wrote Octave Chanute:

We are expecting the most interesting results of any of our seasons of experiments, and are sure that, barring exasperating little accidents or some mishap, we will have done something before we break camp.

I see that Langley has had his fling, and failed. It seems to be our turn to throw now, and I wonder what our luck will be.

—*Miracle at Kitty Hawk*

And Orville wrote his sister, Katharine, shortly thereafter:

November 1, 1903

I suppose you have read in the papers the account of the failure of Langley's big machine. He started from a point 60 feet in the air and landed 300 feet away, which is a drop of 1 foot for every 5 forward. We are able, from this same height, to make from 400 to 600 feet without any motor at all, so that I think his surfaces must be very inefficient. They found they had no control of the machine whatever, though the wind blew but 5 miles an hour at the time of the test. That is the point where we have a great advantage. We have been in the air hundreds and hundreds of times, and have pretty well worked out the problem of control. We find it much more difficult to manage the machine when trying to soar in one spot than when traveling rapidly forward. We expect no trouble from our big machine at all in this respect. Of course we are going to thoroughly test the control of it on the hills before attaching the motor. We are highly pleased with our progress so far this year....I have been putting in about an hour every night down here in studying German and am getting along pretty well.

When Langley's aerodrome was catapult-launched

from atop a Potomac River houseboat it

"simply slid into the water like a handful of mortar..."

47

Edward Wilbur. NC-4. 1976. Oil on canvas, 60 × 40". Gift of Stuart M. Speiser

On December 8th, Langley prepared for another attempt. His "aerodrome" had been pulled out of the Potomac River and repaired. Convinced that his previous failure had been due to some part of his machine's structure becoming fouled in the catapult rather than there being any problem with the design of his craft itself, he ignored the catcalls and criticism of the press and prepared to catapult-launch his "aerodrome" again. There exists a photograph of the result: the "aerodrome" with pilot Manly sitting stoic and erect in the cockpit has just been launched. The nose of the aircraft is pointing straight up; everything aft of Manly and the spinning propellers is falling apart. The forward set of tandem wings appears reasonably solid still, but the second set has crumpled upward like hands clasping in prayer. One can make out broken spars, tearing canvas, bits of debris dropping into the water; the edge of the houseboat remains in view. Manly, unhurt, was once again pulled dripping out of the river; Langley's spirit, however, was crushed by the resulting brutal attacks of the press and by what one writer has referred to as "the curious sadism with which crowds had greeted aeronautical failures since the days of the Montgolfiers."

During this period the Wrights were still patiently and methodically trying to resolve

problems they were having with their propeller shafts and chain sprockets. By Saturday afternoon, December 13th, five days after Langley's failure, the machine was ready but the wind was too light for them to be able to take off from their track on level ground and it was too late in the day for them to carry the machine to one of the nearby hills where the launching track could be placed on an incline steep enough for the craft to attain airspeed. The next day, of course, was Sunday and in deference to their father the Wrights would not attempt a flight. But on Monday, December 14th, the Wrights prepared to fly.

Like Langley's machine the Wrights' "Flyer" (the name they gave all their powered machines) had twin propellers. The propellers' counter-rotation, the Wrights hoped, would neutralize any gyroscopic torque; in addition, two propellers would react against a greater quantity of air. Like their earlier gliders, the first Wright Flyer was a biplane with a skid undercarriage; but it was a considerably larger machine: its wingspan had been increased to 40 feet 4 inches with a wing area of 510 square feet. The Wrights had redesigned the elevator so that they had a "biplane" elevator in front, and the vertical surfaces of the rudder in the rear were controlled through cables linked to the operator's hip cradle.

The Wrights did not use any weights or derrick device to launch; instead, the craft's undercarriage skids rested on a 6-foot-long plank which, in turn, was laid across a smaller piece of wood to which had been attached two modified bicycle wheel hubs one in front of the other. These two hubs had ball bearings and flanges to prevent them from slipping off the 60-foot-long launching track. The track was nothing more than a series of two-by-four boards laid on their edges end to end with their upper surface covered by a thin sheet of metal.

On Monday, the wind was again too light to launch from level ground and the Wrights knew they would have to transport their Flyer to some lower slopes about a quarter-mile away.

December 8, 1903: Langley's second unsuccessful launch.

The Wrights had invited anyone in the area to come to witness their first attempts at flight, but because it was difficult for the brothers to communicate in advance the exact time of the test they had arranged to put out a signal on one of the sheds that could be seen from the Kill Devil Life-Saving Station a little more than a mile away. Shortly after the Wrights had set out the signal, they were joined by John T. Daniels, Robert Westcott, Thomas Beacham, W.S. Dough, and an "Uncle Benny" O'Neill. All of them helped the Wrights carry the Flyer to the hill.

Since obviously both Orville and Wilbur were eager to make the first flight (and since they were both equally experienced pilots by now), they decided a coin toss would determine which of them would go. Wilbur won.

The Flyer's engine was started and permitted to warm up. Then Wilbur climbed onto the machine, and settled himself into the prone position with his hips in the "cradle" to control the rudder and warp. Orville took up his position at a wing tip to help balance the machine as it gathered speed down the track. Wilbur glanced at Orville, nodded that he was ready, then he slipped the restraining wire, and the Wright Flyer took off down the rail so fast Orville couldn't keep up.

Wilbur wrote his family that night:

We gave machine first trial today with only partial success. The wind was only about 5 miles an hour, so we anticipated difficulty in getting speed enough on our short track (60 ft.) to lift. We took to the hill and after tossing for first whack, which I won, got ready for the start. The wind was a little to one side, and the track was not exactly straight down hill, which caused the start to be more difficult than it would otherwise have been. However, the real trouble was an error in judgment in turning up too suddenly after leaving the track, and as the machine had barely speed enough for support already, this slowed it down so much that before I could correct the error, the machine began to come down, though turned up at a big angle.

Toward the end it began to speed up again but it was too late and it struck the ground while moving a little to one side, due to wind and a rather bad start. A few sticks in the front rudder were broken, which will take a day or two to repair probably. It was a nice easy landing for the operator. The machinery all worked in entirely satisfactory manner, and seems reliable. The power is ample, and but for a trifling error due to lack of experience with this machine and this method of starting, the machine would undoubtedly have flown beautifully.

There is now no question of final success. The strength of the machine is all right, the trouble in the front rudder being easily remedied. We anticipate no further trouble in landings. Will probably have made another trial before you receive this unless weather is unfavorable.

It took almost two days to repair the damage to the Flyer's rudder, and the machine was not ready until too late in the afternoon of December 16 to fly. A strong, cold northerly wind blew in overnight and by the next morning puddles of water about their camp were covered with a thin skim of ice. The wind continued at about 25 mph during the early morning, and the Wrights retired indoors to wait for the wind to die down and for themselves to warm up. But when, at ten o'clock, the wind had still not diminished, the Wrights made up their minds to take the machine out anyway and attempt a flight. They hung out the signal for the men to see from the Life-Saving Station, then laid out the track. They thought that if they set the Flyer so that it would take off into the strong wind they would be able to achieve the launch from the level ground about the camp. Although they were aware of the risk in flying in so strong a wind, they believed that the potential for danger would be compensated for by the slower speed in landing. The track was on a smooth, level piece of ground about 100 feet west of the new shed. Because it was so cold, they had to duck into their camp building frequently to warm themselves by the stove. Just before they were ready, John T. Daniels, W.S. Dough, A.D. Etheridge, W.C. Brinkley, and Johnny Moore arrived. Because Wilbur had won the first toss, it was now Orville's turn to attempt the flight. Before climbing onto the

Flyer, Orville set his camera on a tripod and aimed its lens at a specific point about two-thirds down the 60-foot track, the spot being where Orville fully expected the machine to have become airborne. He asked J.T. Daniels of the Life-Saving Station to be the photographer. "When I turn the wings to a flying angle," Orville told him, "I'll leave the track and should be about two feet off the ground when directly in front of the camera. That's the time to press the button."* And then, satisfied that Daniels understood him, Orville joined Wilbur and they started the Flyer's engine. Let Orville's diary tell the story:

*The photograph later developed in Dayton turned out precisely balanced just as Orville had expected and is, in its own way, as perfect an example of the Wrights' careful and methodical preparation having led to success as the flight itself. In the photograph one can see the elevator straining upward, the Flyer piloted by Orville skimming about two feet above the track and Wilbur, no longer needing to balance the wing tip, falling away from their machine with a mixture of wonder and awe at the triumph they had achieved.

After running the engine and propellers a few minutes to get them in working order, I got on the machine at 10:35 for the first trial. The wind according to our anemometer at this time was blowing a little over 20 miles (corrected) 27 miles according to the Government anemometer at Kitty Hawk. On slipping the rope the machine started off increasing in speed to probably 7 or 8 miles. The machine lifted from the track just as it was entering on the fourth rail. Mr. Daniels took a picture just as it left the tracks.

I found the control of the front rudder [elevator] quite difficult on account of its being balanced too near the center and thus had a tendency to turn itself when started so that the rudder was turned too far on one side and then too far on the other. As a result the machine would rise suddenly to about 10 feet and then as suddenly, on turning the rudder, dart for the ground. A sudden dart when out about 100 feet from the end of the track ended the flight. Time about 12 seconds (not known exactly as watch was not promptly stopped). The flight lever for throwing off the engine was broken, and the skid under the rudder cracked.

It is not surprising that whoever was holding the stopwatch failed to promptly mark the time—he was witnessing the first time in history that a man-carrying machine had lifted itself into the air under its own power, had sailed forward without slowing, and had landed on ground as high as that from which it had taken off.

With the help of the men from the Life-Saving Station the Wrights carried their flying machine back to the launching track.

Orville's diary continues:

After repairs, at 20 minutes after 11 o'clock Will made the second trial. The course was about like mine, up and down but a little longer...over the ground though about the same in time. Distance not measured but about 175 feet. Wind speed not quite so strong.

With the aid of the station men present, we picked the machine up and carried it back to the starting ways. At about 20 minutes till 12 o'clock I made the third trial. When out about the same distance as Will's, I met with a strong gust from the left which raised the left wing and sidled the machine off to the right in a lively manner. I immediately turned the rudder to bring the machine down and then worked the end control.

Much to our surprise, on reaching the ground the left wing struck first, showing the lateral control of this machine much more effective than on any of our former ones. At the time of its sidling it had raised to a height of probably 12 to 14 feet.

At just 12 o'clock Will started on the fourth and last trip. The machine started off with its ups and downs as it had before, but by the time he had gone three or four hundred feet he had it under much better control, and was traveling on a fairly even course. It proceeded in this manner till it reached a small hummock out about 800 feet from the starting ways, when it began its pitching again and suddenly darted into the ground. The front rudder frame was badly broken up, but the main frame suffered none at all. The distance over the ground was 852 feet in 59 seconds....

After removing the front rudder, we carried the machine back to camp. We set the machine down a few feet west of the building, and while standing about discussing the last flight, a sudden gust of wind struck the machine and started to turn it over. All rushed to stop it. Will, who was near the end, ran to the front, but too late to do any good. Mr. Daniels and myself seized spars at the rear, but to no purpose. The machine gradually turned over on us.

Mr. Daniels, having had no experience in handling a machine of this kind, hung on to it from the inside, and as a result was knocked down and turned over and over with it as it went. His escape was miraculous, as he was in with the engine and chains. The engine legs were all broken off, the chain guides badly bent, a number of uprights, and nearly all the rear ends of the ribs were broken. One spar only was broken....

All possibility of further flights with the machine for that year had ended.

Curiously, despite the historic significance of what the brothers had accomplished, there does not appear to have been any sense of excitement at their camp that day, least of all shown by Wilbur and Orville themselves. Perhaps it was because they had done nothing more than they had set out to do, and were not in the slightest surprised. As for the other men, perhaps they did not fully understand the importance of the event they had witnessed, and perhaps there was something about Wilbur and Orville Wright that dampened enthusiastic responses. Certainly their "flying costume" was not the sort to encourage flamboyant celebration. They had arrived at their Kill Devil Hill camp with a supply of stiff, white collars, enough to last their stay, and each of them was wearing a stiff collar, necktie, and business suit when they made their flights. The Wrights dressed as did the average businessman of their day and for them to have worn any special "flying clothes" would have been out of character. (No one had ever seen them wear a sweater or a flannel shirt even in their bicycle shop in Dayton, where they were considered so well-dressed that they were thought to be rich.)

After surveying the damage done to their Flyer, Orville and Wilbur went into their camp building, prepared and ate their lunch, and then after washing their dishes and

Orville (left) and Wilbur Wright in their flying clothes.

putting them away, they walked the four or so miles to the Kitty Hawk Weather Station, where they could send a telegram to their father. Since so few persons had occasion to send telegrams in this near-desolate section of North Carolina, people were permitted to use the weather station's government wire to connect with Norfolk, where the message would be relayed to one of the telegraph companies. While Wilbur was examining the machine that recorded the wind velocity, Orville sent the following message: SUCCESS FOUR FLIGHTS THURSDAY MORNING ALL AGAINST TWENTY-ONE MILE WIND STARTED FROM LEVEL WITH ENGINE POWER ALONE AVERAGE SPEED THROUGH AIR THIRTY-ONE MILES LONGEST 59 SECONDS INFORM PRESS HOME CHRISTMAS. ORVILLE WRIGHT. After passing the message to Joseph J. Dosher, the weather bureau operator, Orville walked over to where Wilbur was still looking at the wind-velocity recorder. Almost immediately Dosher got the Norfolk operator on the line and after giving him the telegraph message, Dosher turned to the Wrights and said, "The operator in Norfolk wants to know if it's all right to give the news to a reporter friend." The Wrights refused permission; they wanted the news to be released first from Dayton. Dosher told the Norfolk operator of the Wrights' decision, but the operator told his friend, H.P. Moore of the Norfolk *Virginian-Pilot*, anyway. Moore attempted to get confirmation and more details about the flights, but was unable to speak to anyone who had witnessed them. Undeterred, he wrote a story that appeared the next day under a banner headline, "Flying Machine Soars 3 Miles in Teeth of High Wind Over Sand Hills and Waves at Kitty Hawk on Carolina Coast" and the subhead: "No Balloon Attached to Aid It."

When years later reporter Moore met Orville Wright and asked him what he had thought of the story, Orville replied, "It was an amazing piece of work. Though ninety-nine percent wrong, it did contain one fact that was correct. There *had* been a flight."

There had been a flight.

The Alaskan frontier had been settled; the

Spirit of St. Louis *above the Apollo 11 command module.*

The nose of the Spirit of St. Louis *wears the flags of the United States, France, Belgium, England, Mexico and other Latin American countries, and the islands of the West Indies that Lindbergh and his Ryan aircraft visited.*

Russian Social Democratic Party had split into the Mensheviks and the Bolsheviks led by Lenin and Trotsky; in London the "Entente Cordiale" had been established; Henry James had published *The Ambassadors*, George Bernard Shaw *Man and Superman*, and Jack London *The Call of the Wild*; Whistler and Gauguin had died; *The Great Train Robbery* was shown in theaters; Marie Curie was awarded the Nobel Prize in physics for her discovery of radium; Richard Steiff designed the first "Teddy bear," named after President Roosevelt; France had held its first Tour de France bicycle race; Henry Ford with capital of $100,000 founded the Ford Motor Company; and the first coast-to-coast crossing of the American continent by an automobile had taken 65 days.

There had been a flight.

And the world would never be the same.

Stand in the center of the vast Milestones of Flight gallery and look again at that fragile, austere, antique Wright Flyer skimming overhead, then look down at the sweep second hand on your wristwatch and count off twelve seconds—twelve engine-popping, chain-rattling seconds of uneven darting flight and that is all the time Wilbur and Orville Wright needed to change the destinies of man. Look up again at the Flyer and pay your respects, but then shift your gaze to the small, silver, high-winged monoplane above the Wright brothers' plane and to its right. It is the *Spirit of St. Louis* and on May 20–21, 1927, for 33 hours, 30 minutes, and 29.8 seconds it made the most famous flight in the world.

On May 22, 1919, Raymond Orteig, a New York hotel owner, offered a $25,000 prize "to the first aviator who shall cross the Atlantic in a land or water aircraft (heavier-than-air) from Paris or the shores of France to New York, or from New York to Paris or the shores of France, without stop." For the next seven years the prize went unclaimed because there simply did not exist an aircraft engine reliable enough to permit a flight of that duration and distance. But then came the

The young Lindbergh in front of the Spirit of St. Louis' *dependable Wright Whirlwind 220-hp engine.*

220-horsepower, aircooled Wright Whirlwind engine designed by Charles L. Lawrance for the Wright Aeronautical Company. On May 9, 1926, a Fokker tri-motor equipped with three of these new Wright engines and commanded by Richard E. Byrd and piloted by Floyd Bennett became the first aircraft to fly over the North Pole. The flight lasted sixteen hours.

It was an era of record-breaking attempts. In 1913 the London *Daily Mail* had offered £10,000 for the first non-stop flight between the British Isles and the New World. In May, 1919, one of the United States Navy's new Curtiss flying boats, the NC-4, had successfully flown from Newfoundland via the Azores and Portugal to England. One month later, on June 14, 1919, Captain John Alcock and Lieutenant Arthur Whitten Brown had flown a remodeled twin-engine Vickers Vimy bomber from Newfoundland 16 hours and 27 minutes across the Atlantic through appalling weather and had survived a crash the next morning in an Irish bog, thus winning the *Daily Mail*'s prize. A similar plane had won Captain Ross Smith the £10,000 prize offered by the Australian government to any Australian pilot capable of flying the 11,000 miles between London

The "NYP" below the "Ryan" painted on the
Spirit of St. Louis' *tail stood for New York–to–Paris.*

and Australia in a British-built plane within 30 days. Smith and his crew made the journey in 27 days and 20 hours, with but 52 hours to spare.

In the Pioneers of Flight gallery of the Museum there stands the huge single-engine Fokker T-2, the machine in which on May 2–3, 1923, Lieutenants John Macready and Oakley Kelly made the first successful non-stop flight across the United States. Near it is the *Chicago*, one of the two Douglas World Cruisers to have completed the first round-the-world flight that took them 26,000 miles and over five months from April 6 to September 28, 1924. Speed, endurance and distance records were being smashed left and right, but no one had been able to make the 3,300 mile non-stop Paris–New York flight required to win the Orteig Prize.

Charles A. Lindbergh was twenty-four years old and flying the mail in an old DH-4 to Chicago one night when suddenly it occurred to him to try. He had more than four years of aviation behind him, close to two thousand hours in the air, he had barnstormed over half the then-48 states, had flown the mail through the worst of weather, had learned the basics of navigation as a flying cadet at Brooks and Kelly fields. As Lindbergh sat in the open cockpit of that mailplane, considering this vision "born," he later wrote "of a night and altitude and moonlight" of crossing the Atlantic, he was a little overwhelmed by the magnitude of the undertaking he was contemplating. He had the flying experience, but not the money. So with $2,000 of his own savings he approached a group of St. Louis businessmen and managed to

convince them that a flight *was* possible, and they agreed to raise an additional $13,000 toward the purchase of a monoplane equipped with one of the new Wright engines. Although they questioned the safety of attempting the flight in a single-engine plane, Lindbergh convinced them that the more engines a plane had, the more chance there was of engine failure. In addition, he argued, the more engines the more expensive the plane.

Lindbergh was by no means alone in attempting the Paris—New York flight. René Fonck, France's leading World War I ace (75 victories), had attempted the flight in a Sikorsky tri-motor. The heavily loaded airplane's landing gear had given way on take-off from Long Island's Roosevelt Field and two of his crew members had perished in the fiery crash. Fokker was building a

tri-motor for Commander Byrd's attempt, and the Columbia Aircraft Company was preparing one of Giuseppe Bellanca's stunning Wright-powered monoplanes. Lt. Comdr. Noel Davis and Lt. Stanton Wooster of the U.S. Navy were readying the "American Legion," a large Keystone Pathfinder tri-motor biplane for their attempt. French war hero Charles Nungesser and François Coli, on the other side of the Atlantic, were preparing to attempt an east-west crossing in a 450 hp single-engine French-built Levasseur biplane.

Lindbergh and his backers could not afford a Fokker; the Wright-Bellanca could be flown only by a pilot of the Columbia Aircraft Company's choice. Other aircraft manufacturers wanted too much money or could not build the plane. So Lindbergh turned to the small, relatively unknown Ryan

60

Airlines, Inc., of San Diego, California. On February 24, 1927, Lindbergh wired his backers: BELIEVE RYAN CAPABLE OF BUILDING PLANE WITH SUFFICIENT PERFORMANCE STOP COST COMPLETE WITH WHIRLWIND ENGINE AND STANDARD INSTRUMENTS IS TEN THOUSAND FIVE HUNDRED EIGHTY DOLLARS STOP DELIVERY WITHIN SIXTY DAYS STOP RECOMMEND CLOSING DEAL. LINDBERGH

Sixty days after the deal with Ryan is negotiated, the *Spirit of St. Louis* is flying; but so are the others. The race for the Orteig Prize is becoming close. What's more, due to a technicality and the Orteig Prize committee's desire for more information about the Ryan monoplane than Lindbergh is able to provide at the time, the required sixty-day rule between a pilot's entry application and his flight means that for Lindbergh to be eligible he cannot take off until the end of May.

On April 16, 1927, Commander Byrd's big Fokker, the *America*, crashes on landing at New York's Teterboro Airport; the plane overturns and pilot Floyd Bennett is seriously injured.

On April 24th, the Wright-Bellanca loses a wheel of its landing gear during a test flight and is damaged on landing.

On April 26th, Lt. Comdr. Davis and Lt. Wooster are killed on the last of their trial flights when their Keystone Pathfinder biplane crashes into a marsh short of landing at Langley Field.

On May 8th, Nungesser and Coli take off from Le Bourget for New York—and disappear.

Newspaper editors realize the New York—Paris flight is becoming the story of the year. Interest is intense: four men are now dead, Nungesser and Coli are missing, three others are injured.

On May 10th, Lindbergh flies his new *Spirit of St. Louis* from San Diego to St. Louis non-stop in 14 hours and 25 minutes—a new record. And despite some worries caused by the Wright engine's shuddering reaction to the cold and altitude crossing the Rockies, Lindbergh is pleased with the performance of his plane. He declines all dinner invitations forwarded by his backers in St. Louis so that he might leave the next morning for New York, where the Bellanca *Columbia* is already waiting for the weather to clear. The next day, just an hour after Lindbergh's *Spirit of St. Louis* arrives at New York's Curtiss Field, Byrd's now-repaired *America* comes in to land. During the next week squabbles break out among the crew of the Wright-Bellanca; there are arguments over who is to go, whether to take a radio, what route is the best. The Bellanca's owner, Charles Levine, tells his pilot-navigator Bertaud that he is not wanted on the flight. Bertaud sues, and gets a court injunction to prevent the Bellanca from taking off without him. Meanwhile, Byrd is putting the *America* through test after test.

Thursday, May 19th: Lindbergh has been waiting now in New York for one week for the weather to break. "The sky is overcast. A light rain is falling. Dense fog shrouds the coasts of Nova Scotia and Newfoundland, and a storm area is developing west of France," Lindbergh wrote in his autobiography, *The Spirit of St. Louis*. "It may be days, it may be—I feel depressed at the thought—another week or two before I can take off."

That night Dick Blythe, the Wright Aeronautical Company's representative, invites Lindbergh to watch a Broadway musical—*Rio Rita*—from backstage.

On the way to the theater, Blythe, at Lindbergh's request, stops for one more telephone call to the weather bureau.

With the forecasts we've had, pavements shiny wet, and the tops of skyscrapers lost in haze, it's probably a waste of time to call for a final check on weather; still, I'm not going to miss any chance. We park our car at the curb, and wait while Blythe goes into an office building to phone the Bureau. When he comes out I know by his face and gait that he has news for me.

'Weather over the ocean is clearing,' he announces. 'It's a sudden change.' Unmindful of the rain, he stoops to outline the situation to us through the car window. The low-pressure area over Newfoundland is receding, and a big high is pushing in behind it. 'Of course conditions aren't good all along your route, Blythe continues. 'They say it may take another day or two for that.'

But there's a chance I'll be able to take off at daybreak. Thoughts of theater and stage vanish. The time has come, at last, for action.

—*The Spirit of St. Louis*, by Charles A. Lindbergh

The car is turned around and they start immediately for the airfield. Upon his arrival Lindbergh is surprised that there is no activity whatsoever at either the hangar where Byrd's Fokker, the *America*, is kept or at the Bellanca. Lindbergh assumes they are waiting for confirmation of the weather's improvement and he knows that if he waits for proof of good weather all the way to Paris, he could be the last to leave.

It is almost midnight before Lindbergh returns to his hotel room. The *Spirit of St. Louis* will be ready; all he will need do is take off from Curtiss and fly to Roosevelt Field, then fill the tanks and use the longer, smoother Roosevelt Field runway for his take-off to Paris. If he is to take off at dawn he will get only 2½ hours' sleep but he knows from his mailpilot experience that even minutes can help. He posts a friend outside his hotel room door to see that he is not disturbed. Lindbergh is furious with himself for not having gone to bed earlier, for having accepted the invitation to the theater, but he realizes he could not have foreseen the break in the weather. Just as he is falling asleep there is a knock at the door. The man he has posted as guard comes in.

"Slim," the man asks, "what am I going to do when you're gone?"

The stupid question wakes Lindbergh up; he cannot get back to sleep. He lies there wishing for a moment that the weather would turn bad again, that he could get a full night's rest, that the weather would sock in for a full week so that enough time would elapse for there to be no question about his eligibility for the $25,000 Orteig Prize.

From left to right: Lindbergh, Byrd, and Chamberlain smile for the press, but each was competing for the $25,000 Orteig Prize.

At 1:40, without having slept at all, Lindbergh begins to dress and he arrives at Curtiss Field a little before 3:00 on the morning of May 20, 1927. As he stands looking at the light rain falling through the haze, he is told that a message had been left for him back at the hotel saying he could sleep until dawn. A way had been found to tow the *Spirit of St. Louis* to Roosevelt Field; he wouldn't have to fly it. Lindbergh asks what the latest weather reports are and is told that in spite of the fog at Roosevelt Field, there are reports that the fog is lifting between New York and Newfoundland, the high-pressure system is still moving in over the North Atlantic and the only storms are local ones along the coast of Europe. Lindbergh orders the *Spirit of St. Louis* towed to Roosevelt Field and the fuel tanks topped.

Mechanics tie the plane's tail skid to the back of a motor truck and wrap a tarpaulin around the engine. Reporters button up their raincoats.

63

Men look out into the night and shake their heads. The truck starter grinds. My plane lurches backward through a depression in the ground. It looks awkward and clumsy. It appears completely incapable of flight— shrouded, lashed, and dripping. Escorted by motorcycle police, pressmen, aviators, and a handful of onlookers, the slow, wet trip begins.

It's more like a funeral procession than the beginning of a flight to Paris.

—*The Spirit of St. Louis*

As soon as the *Spirit of St. Louis* is positioned at the end of the runway facing into the wind, the wind maddeningly shifts 180° to blow over his tail. Conditions are inauspicious. The runway is soft, rainsoaked. Huge puddles are visible up the field. The engine is running 30 rpm too slow. The mechanic says it is the weather, but Lindbergh sees the apprehension in the mechanic's eyes. The plane's propeller has been set for cruising, not take-off. Lindbergh knows no one will think less of him if he decides not to go. The *Spirit of St. Louis* weighs 5,250 pounds—a thousand pounds more than he has ever attempted to fly it with before. The dampness of the wing surfaces not only adds weight but reduces lift. The tail wind means he will have to go that much faster before he can gain enough airspeed to leave the ground. Ahead in the misty end of the runway are the telephone wires he will have to clear. And yet, if he waits, Lindbergh knows the Bellanca will be ready, Byrd's *America*, too. But the take-off will be slow. He worries whether his engine will stand such a long ground run at full throttle without overheating.

His instinct tells him he can make it.

Lindbergh buckles his seatbelt, sets his goggles over his eyes, nods at the men on either side, and the wheel-chocks are pulled. He eases the throttle forward and the *Spirit of St. Louis* slowly creeps through the mud. Men run along on either side pushing upon the wheel struts, trying to speed the heavily laden plane along. It sloughs like an overloaded truck, its tires rutting through the mud. Gradually the speed increases and

Lindbergh's functional cockpit wicker seat was light but strong.

the men drop away.

Because of the fuel tank there is no way for him to see ahead. The periscope is of no use with the tailskid touching the ground. Lindbergh must keep the plane moving absolutely straight or one wheel will dig in, the heavy fuel load will crush the landing gear, there will be an explosion of gas. The controls seem lifeless. He has used a thousand feet of runway before the tailskid lifts and the wings begin to take up some of the load. The engine is beginning to smooth out, the propeller is taking hold of the air.

The halfway mark streaks past---seconds now to decide—close the throttle or will I get off? The wrong decision means a crash—probably in flames---I pull the stick back firmly, and---*the wheels leave the ground*. Then I'll get off! The wheels touch again...A shallow pool on the runway---water spews up from the tires---A wing drops—lifts as I shove aileron against it—the entire plane trembles from the shock---Off again—right wing low—pull it up—Ease back onto the runway—left rudder—hold to center— must keep straight—Another pool—water drumming on the fabric---The next hop's longer---I could probably stay in air; but I let the wheels touch once more—lightly, a last bow to earth, a gesture of humility before it---Best to have plenty of control with such a load, and control requires speed.

The *Spirit of St. Louis* takes herself off the next time—full flying speed—the controls taut, alive, straining—and still a thousand feet to the web of telephone wires…It'll be close, but the margin has shifted to my side. I keep the nose down, climbing slowly, each second gaining speed. If the engine can hold out for one more minute---five feet---twenty---forty---wires flash by underneath—*twenty feet to spare!*
—*The Spirit of St. Louis*

Only four hours into the flight Lindbergh finds himself fighting exhaustion, and then squalls. By the eighth hour his lack of sleep has become a serious threat: "My eyes feel dry and hard as stones. The lids pull down with pounds of weight against their muscles. Keeping them open is like holding arms outstretched without support. I try letting one eyelid close at a time while I prop the other open with my will. But the effort's too much. Sleep is winning. My whole body argues that nothing, nothing life can attain, is quite so desirable as sleep. My mind is losing resolution and control." By the tenth hour Lindbergh is flying low over the Atlantic ice fields and encounters fog. He begins climbing and at 10,500 feet he has still not broken free and is suddenly terribly cold,

colder, he knows, than he should be. He jerks his leather mittens off and thrusts his arm out the window. His palm is stung by thousands of pinpricks. He shines his flashlight out the window at the wing strut. *Ice!* He has to get out of the thunderhead, below the storm. Ice will clog the venturi tubes, his instruments will not work; ice will weigh down his plane, spoil the airflow, force him into the ocean. He must try to thread his way at night through the great cliffs of clouds that tower all around him.

During the fifteenth hour Lindbergh pushes his head out the open window to revive himself. The cold air hits his face. He lets his eyelids fall shut for five seconds, then cannot open them without forcing them up with his thumb. He lifts his brow muscles to keep his eyelids open. By the twenty-second hour he is falling asleep with his eyes open and cannot prevent it. Ghostly apparitions suddenly fill the fuselage:

These phantoms speak with human voices—friendly, vapor-like shapes, without substance, able to vanish or appear at will, to pass in and out through the walls of the fuselage as though no walls were there… First one and then another presses forward to my shoulder to speak above the engine's noise, and then draws back among the group behind. At times, voices come out of the air itself,…familiar voices, conversing and advising on my flight, discussing problems of my navigation, reassuring me, giving me messages of importance unattainable in ordinary life.
—*The Spirit of St. Louis*

Lindbergh marks on the instrument panel before him the amount of fuel consumed:

The necessity for extra fuel tanks placed in front of the instrument panel blocked Lindbergh's forward vision.

TWENTY-SIXTH HOUR
hours of fuel consumed

Nose Tank

1/4 + ⅃⊬⊤ 11

Left Wing	Center Wing	Right Wing
1/4 + 111	1/4 +	1/4 + 111

Fuselage

⅃⊬⊤ ⅃⊬⊤ 1.

He reaches into the pocket of his flying suit for a fresh handkerchief and touches something small, hard, thin he hadn't noticed before. There is a chain attached and he pulls it out; someone, unknown to him, had slipped a St. Christopher's medal into his pocket. Lindbergh is low over the ocean again, flying at little more than ten feet when he sees a porpoise, the first living creature he's seen since crossing Newfoundland fourteen hours before. He searches the water. Suddenly a black speck on the water two or three miles to his southeast catches his eye. He squeezes his lids together and looks again. It's a boat! Several small boats are dotted over the surface of the ocean!

Lindbergh banks the *Spirit of St. Louis* toward them. All sleepiness vanishes. If those are fishing boats, then the coast of Europe cannot be far away. He flies low over the first boat, drops his wing, banks, circles over the next. A man's head appears at a porthole. Gliding down to within fifty feet of the fishing boat's cabin Lindbergh closes down the throttle, leans out the window and shouts, "WHICH WAY IS IRELAND?" Three times he circles; the person at the porthole stares stupidly at the small silver plane. No one moves.

Lindbergh cannot delay, he straightens out and pushes on and into a line of squalls just a hundred feet above the ocean swells. During his twenty-eighth hour he does not dare believe his eyes. Ahead of him is land! Lindbergh has reached Ireland. He is 2½ hours ahead of schedule; he is less than six hours from Paris!

Thirty-three hours into the flight it is almost dark. He is over France when he sees a line of beacons converging with his course. Paris, where his course and the beacons intersect, will be less than a hundred miles ahead.

An hour later he is flying at 4,000 feet and sees the glow of Paris against the nighttime sky. Within an hour he will land at Le Bourget field—if he can find it.

No one Lindbergh had spoken to before departing had more than a general idea where it was. "It's a big airport," he was told, "You can't miss it. Just fly northeast from the city." But there is no airport beacon. As he reaches the city he sees a dark patch of ground where he thinks Le Bourget ought to be, then flies beyond to make sure. He turns back and circles the dark patch again. It must be Le Bourget, he thinks, but the lights seem so strange. There appear to be thousands of dim lights to one side. A factory? He descends to a lower altitude, recognizes big hangars, realizes the little lights are automobiles. It is Le Bourget! There are floodlights near the hangars; Lindbergh banks around for his final glide:

I'm too high—too fast. Drop wing—left rudder—sideslip---Careful—mustn't get anywhere near the stall. I've never landed the *Spirit of St. Louis* at night before. It would be better to come in straight. But if I don't sideslip, I'll be too high over the boundary to touch my wheels in the area of light. That would mean circling again---Still too high. I push the stick over to a steeper slip, leaving the nose well down---Below the hangar roofs now---straighten out---A short burst of the engine---Over the lighted area---Sod coming up to meet me---Deceptive high lights and shadows—Careful— easy to bounce when you're tired---Still too fast---Hold off---Hold off---But the lights are far behind---The surface dims—Texture of sod is gone---Ahead, there's nothing but night---Give her the gun and climb for another try?---The wheels touch gently—off again—No, I'll keep contact—Ease the stick forward---Back on the ground—Off—Back—the tail-skid too---Not a bad landing, but I'm beyond the light—can't see anything ahead—Like flying in fog—Ground loop?—No, still rolling too fast—might blow a tire—The field *must* be clear—Uncomfortable though, jolting into blackness—Wish I had a wing light—but too heavy on the take-off---Slower, now—slow enough to ground loop safely—left rudder—reverse it—stick over the other way---The *Spirit of St. Louis* swings around and stops rolling, resting on the solidness of earth, in the center of Le Bourget.

I start to taxi back toward the floodlights and hangars---But the entire field ahead is covered with running figures!*

—*The Spirit of St. Louis*

Ever since the first radioed reports of Lindbergh's crossing Ireland, then the Channel and being over France, masses of people have flocked to Le Bourget to watch him land. They had waited for hours and now, in their excitement, the crowds knock down the steel fences and break through two companies of soldiers and civil police guards in their eagerness to reach the small silver plane. Lindbergh barely has time to stop the engine before the first faces are at his windows. He hears them shouting "LINDBERGH! LINDBERGH!" in accents unfamiliar to his ears. He can feel the *Spirit of St. Louis* tremble as bodies crush against it. A fairing spar cracks and then another and another. Souvenir hunters tear pieces of fabric from the fuselage. When Lindbergh opens his cabin door to attempt to get help and protection for his plane he is pulled bodily out and carried over the crowd's heads. Lindbergh has become the most famous man in the world!

Today, perhaps, it is difficult for us to understand what an extraordinary feat Lindbergh achieved. Thousands of passengers cross the Atlantic non-stop every week and if they tear their eyes away from the pages of a novel or the onboard movie to look out the double-layered glass window outside at the sky, they do not consider the risks. Nor do they question for an instant that if the pilot—or more likely some flight attendant—announces that their destination is Paris that this is precisely where the plane will go.

What Lindbergh was the first to do, by an act of superb intelligence and will, millions of us accomplish regularly with the expenditure of

no more intelligence and will than is required to purchase a ticket and pack a bag. The once inimical Atlantic scarcely exists for the contemporary traveler: a glimpse of tame, pewter-colored water at the start or finish of a journey... That first New York—to—Paris flight, with its awesome risks cooly faced and outwitted by a single valorous young man had led to an ever-increasing traffic in the sky above the Atlantic and an ever-decreasing awareness of awe and risk on the part of the army of non-flyers who have followed him. His valor is hard to keep fresh in our minds when the most we are asked to face and outwit above the Atlantic is boredom
—*Lindbergh Alone*, by Brendan Gill

But on May 20–21, 1927, Lindbergh touched some responsive chord in the hearts of the world. The Government of the United States awarded him the first peacetime Congressional Medal of Honor and the first Distinguished Flying Cross. France, Belgium, England, Spain, Canada, the nations and leaders of the world vied with one another to present him with medals, commemorative plaques, scrolls, letters. George M. Cohan wrote a song about him. The Franklin and LaSalle-Cadillac automobile companies each gave him cars, Ryan an airplane. He was given life passes to baseball games, theaters, railroads. He was offered free homes, a $50,000 bonus for endorsing a cigarette, a 5-million-dollar motion picture contract, a live monkey, the most bizarre and astonishing gifts. There was something about the tininess of Lindbergh's plane against so huge and implacable a foe as the Atlantic Ocean and its storms, something about Lindbergh's shyness and modesty, his aloofness and self-confidence: he had arrived at Roosevelt Field with little fanfare, without the financial corporations the others had assembled to back their attempts. He had arrived alone and he took off alone crossing the vast Atlantic sometimes at ten feet and at other times ten thousand feet above the waves. Lindbergh did it *alone*—and from the day he landed at Le Bourget he fought in vain

---

*The official time of Charles A. Lindbergh's crossing from New York to Paris on May 20–21, 1927, in the *Spirit of St. Louis* recorded by the National Aeronautic Association under the rules and regulations of the Fédération Aéronautique Internationale was 33 hours, 30 minutes, 29.8 seconds.

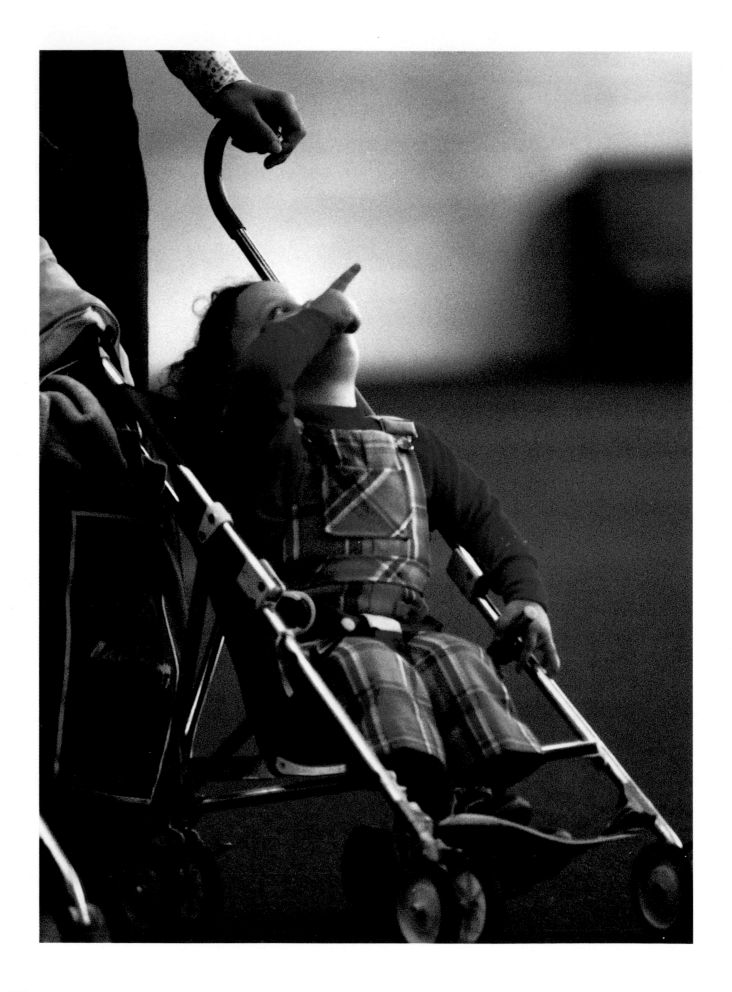

to keep this sense of self intact.

Look again at that tiny, silver plane. If Charles Lindbergh's *Spirit of St. Louis* holds a special place in the Museum visitor's heart, it is because it marks the accomplishment of an individual, an achievement that can somehow be measured and understood. There is something about Lindbergh's daring, his acceptance of chance and fate, his decision to fly without a radio or a parachute to save weight, to so load the Ryan monoplane's fuselage with fuel tanks that it was impossible for him to see straight ahead without the aid of a periscope or without swinging his plane from side to side. There is something in that plane that the visitor can relate to because Lindbergh was dealing with understandable, comprehensible risks—dangers which were simple, but whose simplicity in no way diminished the certainty of Lindbergh's valor nor the fact that the slightest error meant death.

And yet, as the visitor looks about the gallery at those other air- and spacecraft which also marked milestones in the history of flight, the "milestones" themselves become more mysterious and complex achievements, more difficult to appreciate, to measure, than a flight from point-to-point. As each new plateau of accomplishment was crossed by the individual or individuals in each machine, the flight technology demanded less and less of the individual and more and more the effort of groups, of teams of scientist-engineers, test-pilot mechanics, astronaut-scholars, physicists, and theoreticians. The machines, themselves, seem to resemble airplanes less and less and devices more and more—scientific tools designed for specific purposes, to increase understanding and knowledge. The machines' designs begin to reflect their purpose and even in cases such as the deep space probes, mechanical manifestations of pure thought itself. When the visitor, for

example, looks at that stubby-winged, bright orange and white bullet-shaped aircraft hanging high up near the gallery's west wall, he is seeing a needle-nosed Bell X-1, *Glamorous Glennis*, designed to pierce the sound barrier.

During the later stages of World War II, pilots of advanced fighter aircraft experienced strange and alarming behavior on the part of their machines when in a high-speed dive they approached the speed of sound. Their planes buffeted, wrenched about, controls sometimes reversed or became ineffective, increased drag set in; in some cases, the wings tore off, the plane broke apart and crashed. Because of these experiences there developed the idea of a sonic wall, a "sound barrier" of some kind through which no aircraft could pass unscathed. In 1944 the National Advisory Committee for Aeronautics (NACA) and the U.S. Army Air Forces initiated a cooperative program to explore this wall. For this purpose the Bell X-1 was designed and cautiously tested, its rocket-engine permitted to fire just long enough to creep the little plane's speed increment by increment up to the speed of sound. And then on October 14, 1947, Captain Charles E. "Chuck" Yeager flew the experimental *Glamorous Glennis* (named after Yeager's wife) 700 mph at 43,000 feet and thereby became the first man to fly beyond the speed of sound. Describing the flight later, Yeager reported:

With the stabilizer setting at 2° the speed was allowed to increase to approximately .98 to .99 Mach number where elevator and rudder effectiveness were regained and the airplane seemed to smooth out to normal flying characteristics. This development lent added confidence and the airplane was allowed to continue to accelerate until an indication of 1.02 on the cockpit Mach meter was obtained. At this indication the meter momentarily stopped and then jumped to 1.06 and this hesitation was assumed to be caused by the effect of shock waves on the static source. At this time the

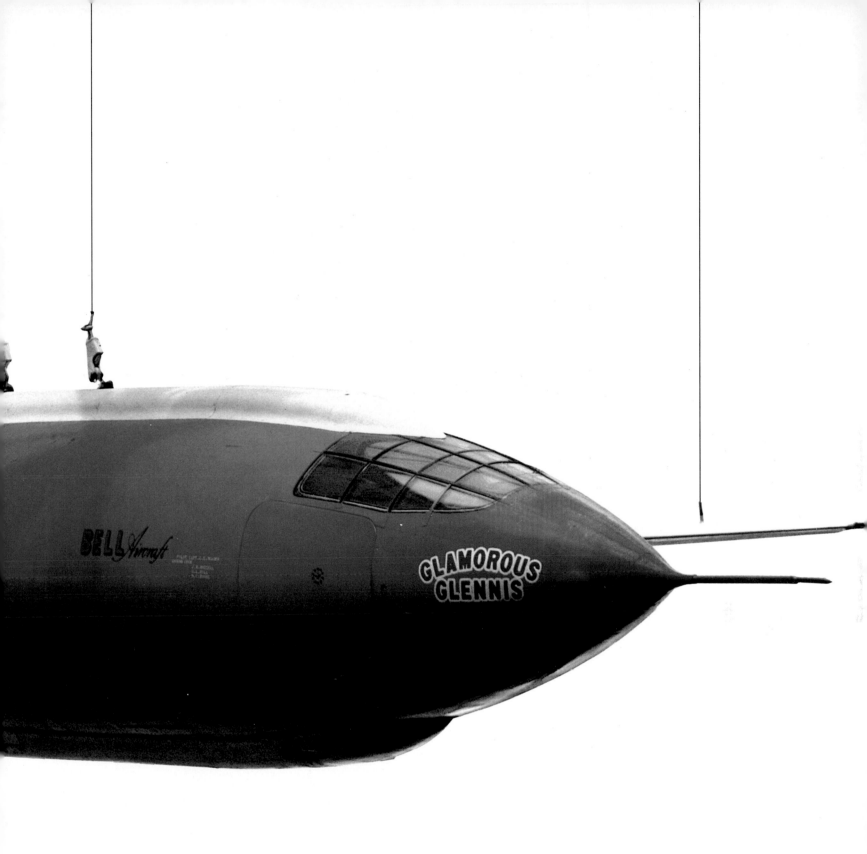

The Bell X-1, the first plane to fly beyond the speed of sound (1947).

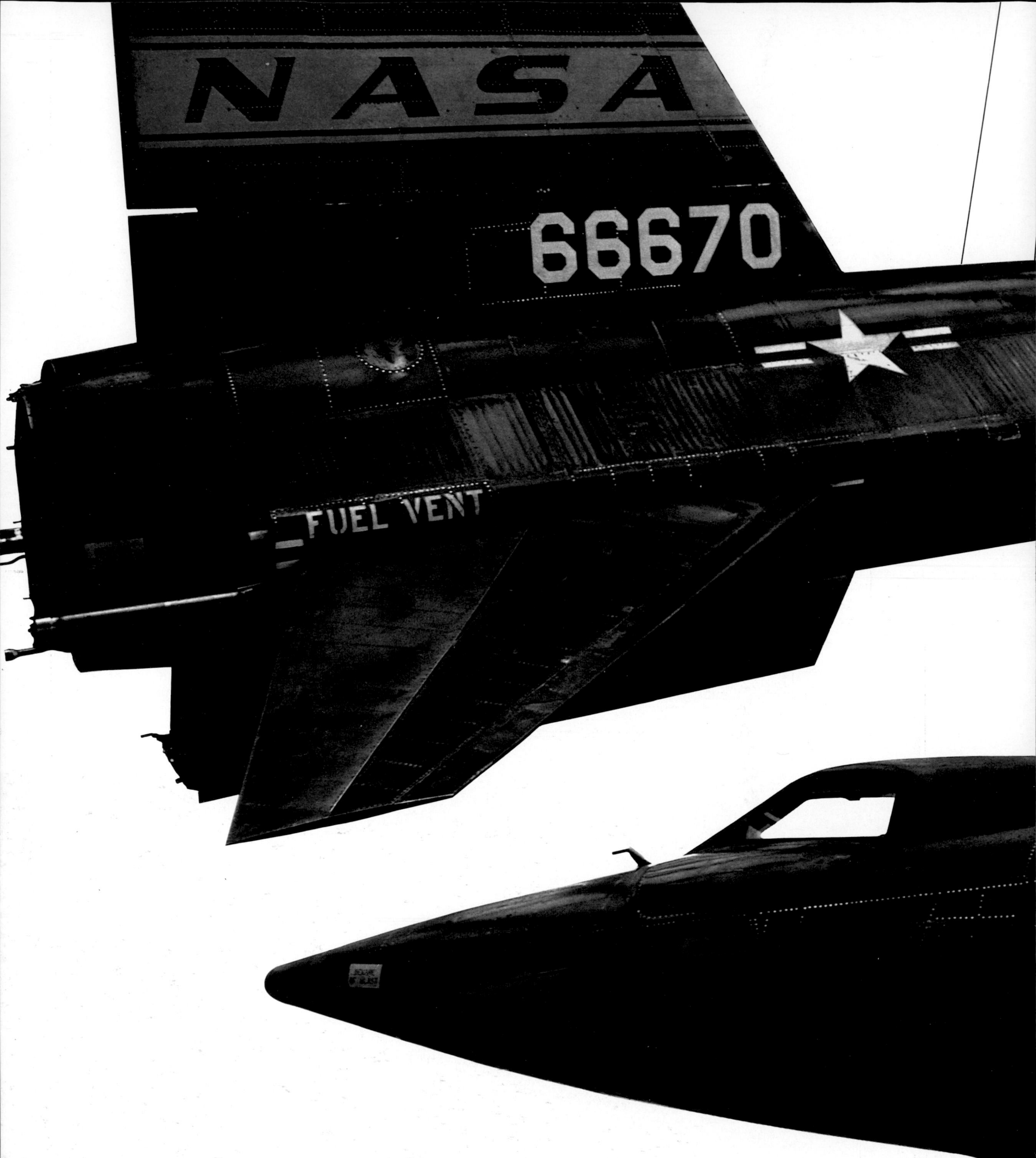
NASA
66670
FUEL VENT

This "brute of a machine" was designed to bridge the gap between manned flight in the atmosphere and manned flight in space; the rocket-powered X-15 was the first to fly four, five, and six times the speed of sound.

*Gemini 4 EVA: A mannequin in the
Milestones of Flight gallery depicts Edward
H. White II becoming the first American to
"walk in space."*

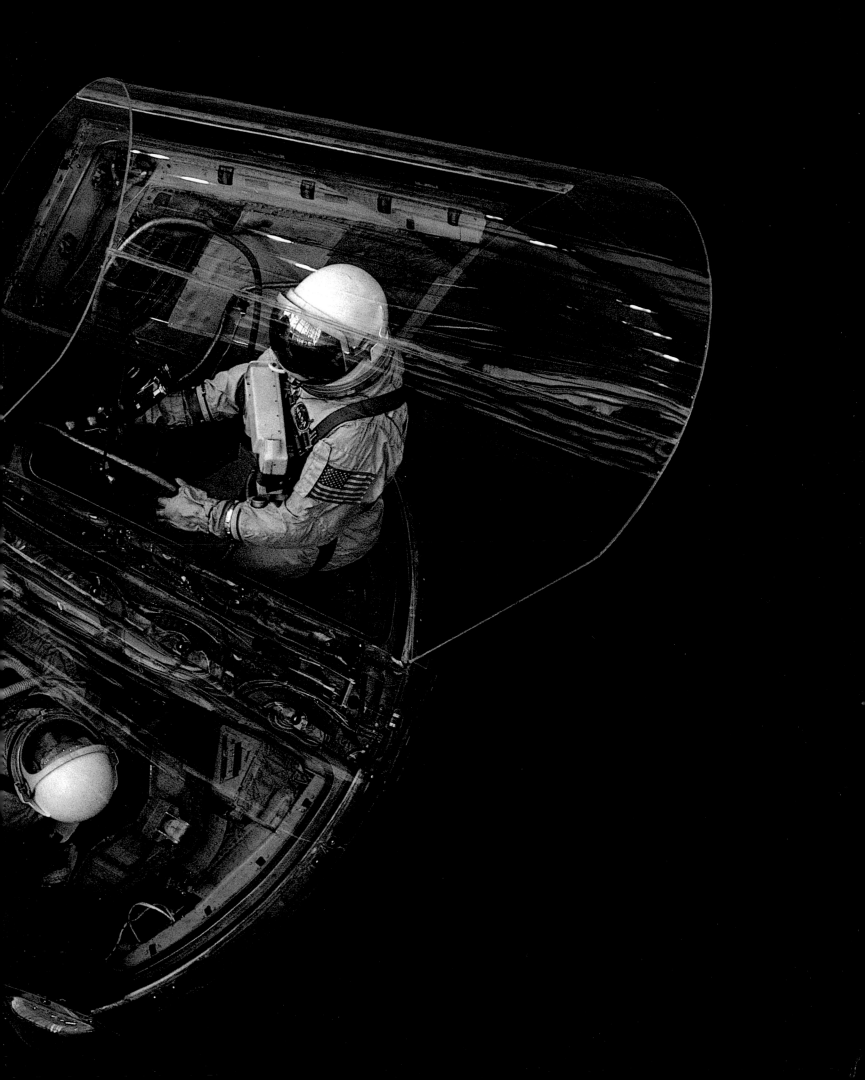

power units were cut and the airplane allowed to decelerate back to the subsonic flight condition. When decelerating through approximately .98 Mach number a single sharp impulse was experienced which can best be described by comparing it to a sharp turbulence bump.

—*Supersonic Flight: The Story of the Bell X-1 and Douglas D-558*, by Richard Hallion

The fact that nothing "spectacular" had happened belies the significance of that flight. When on August 26, 1950, General Hoyt Vandenberg as Air Force Chief of Staff presented the X-1 to Dr. Alexander Wetmore, then Secretary of the Smithsonian Institution, he stated that the aircraft "marked the end of the first great period of the air age, and the beginning of the second. In a few moments the subsonic period became history, and the supersonic period was born."

Diagonally across from the Bell X-1 is the North American X-15, that big, black, ugly brute of a machine to which the supersonic period gave birth. In 1963 this rocket-propelled research aircraft flew higher (354,000 feet), and in 1967, faster (4,534 mph) than any other airplane in history. An X-15 became the first aircraft to fly Mach 4, Mach 5, and Mach 6—four, five, and six times the speed of sound. It was designed to bridge the gap between manned flight within the atmosphere and manned flight beyond the atmosphere in space and to study hypersonic aerodynamics, control systems, winged reentry from space, and aerodynamic heating. Because of its high-speed capability, the X-15 had to withstand aerodynamic temperatures of 1,200°F.; the aircraft was therefore manufactured using a special high-strength, heat-resistant nickel alloy. The X-15 was airlaunched from a special modified Boeing B-52; the rocket-powered plane required conventional aerodynamic control surfaces to operate within the atmosphere and special "thruster" reaction control rockets located in the nose and wings to permit the pilot control of the X-15 when flying in the thin atmospheric fringe of space. The

wedge-shaped tail surfaces were needed for control at high speeds. Because the X-15's tail surface arrangement was a cruciform in shape, the lower half of the ventral fin was designed to be jettisonable prior to landing so that the rocket craft could land on a conventional two-wheel nose-landing gear and two tail-mounted landing skids. Three X-15 aircraft were built; they completed a total of 199 research flights which resulted in at least 700 technical documents, an amount equivalent to the output produced by a typical 4,000-man federal research center in the course of two years.

Beneath the *Spirit of St. Louis* rests the Mercury spacecraft *Friendship 7*. On the morning of February 20, 1962, while 135 million Americans watched on their television sets or listened in on their radios Marine Lt. Col. John Glenn was rocketed into space in that spacecraft from Cape Canaveral (briefly renamed Cape Kennedy), Florida, and became the first American to orbit the earth.

Glenn circled our planet three times at an altitude which varied between 101 and 162 miles and at a speed of 17,500 mph. The flight took 4 hours and 55 minutes and demonstrated that the human body could function in the weightlessness of space and that the spacecraft was safe for manned missions. Below are excerpts from the first radio voice messages exchanged between Astronaut Glenn and Mercury Control at Cape Canaveral on the morning that the United States truly entered the manned space exploration race.

CC stands for Cape Canaveral, P represents the pilot, John Glenn. The numerals show the flight time in hours, minutes, and seconds following the countdown and lift off.

CC [Cape Canaveral]...5, 4, 3, 2, 1, lift off!

00 00 03   P [Pilot] Roger. The clock is operating. We're under way.

00 00 07   CC Hear [you] loud and clear.

Friendship 7. the Mercury spacecraft in which
John Glenn, on February 20, 1962, became
the first American to orbit the earth.

00 00 08	P	Roger. We're programming in roll okay.
00 00 13	P	Little bumpy along about here.
00 00 48	P	Have some vibration coming up here now.
00 00 52	CC	Roger. Reading you loud and clear.
00 00 55	P	Roger. Coming into high Q [vibration] a little bit; and a contrail went by the window, or something there.
00 01 00	C	Roger.
00 01 12	P	We're smoothing out some now, getting out of the vibration area.
00 01 16	CC	Roger. You're through max. Q. Your flight path is...
00 01 19	P	Roger. Feels good, through max. Q and smoothing out real fine.
00 01 31	P	Sky looking very dark outside.
00 02 07	CC	Roger. Reading you loud and clear. Flight path looked good. Pitch, 25. Stand by...
00 02 12	P	Roger. The [escape] tower fired; could not see the tower go, I saw the smoke go by the window.
00 02 36	P	There, the tower went by right then. Have the tower in sight way out. Could see the tower go. Jettison tower is green.
00 04 08	P	Friendship Seven. Fuel 103–101 [percent], oxygen 78–100, amps 25, cabin pressure holding steady at 5.8.
00 04 20	CC	Roger. Reading you loud and clear. Seven, Cape is Go; we're standing by for you.
00 04 25	P	Roger. Cape is Go and I am Go. Capsule is in good shape...All systems are Go.
00 05 12	P	Roger. Zero G, and I feel fine. Capsule is turning around.
00 05 18	P	Oh, that view is tremendous!
00 05 21	CC	Roger. Turnaround has started.
00 05 23	P	Roger. The capsule is turning around and I can see the booster during turnaround just a couple of hundred yards behind me. It was beautiful.
00 05 30	CC	Roger. Seven. You have a Go, at least seven orbits.
00 05 35	P	Roger. Understand Go for at least seven orbits.

The vocabulary: *Max Q, Zero G, escape towers, boosters, separation*; the phrases: *we're programming in roll, all systems are go*; the *count-downs* and *lift-offs*...how swiftly we have become accustomed to the language of the space age!

The Mercury flights were the first of three steps that formed the United States' response to President John F. Kennedy's challenge a year before Glenn's flight, that "This nation should commit itself to achieving the goal, before this decade is out, of landing a man on the moon and returning him safely to Earth." The second step, the Gemini project, is represented by the Gemini 4 spacecraft located below the X-15. It was in this 7,000-pound spacecraft launched into orbit on June 3, 1965, that astronaut Edward H. White opened the right-hand hatch and "floated" out into space connected to the Gemini 4's life support and communications systems by only a gold-covered "umbilical cord." His "walk" proved that astronauts could work effectively outside their spacecraft, for example, on the lunar surface. White maneuvered by using a hand-held thruster. Other Gemini flights proved that man could stay up long enough to get to the Moon and back. This led to the Apollo program.

Directly beneath the X-15 stand two diminutive progenitors of the towering Atlas, Titan, and Saturn rockets that lifted our astronauts; they are a full-scale model of Dr. Robert H. Goddard's first liquid-propellant rocket built in 1926, and Dr. Goddard's actual final rocket built in 1941. These rockets, representing the beginning and the end of Dr. Goddard's efforts to develop high-altitude liquid-propellant rockets, mark the

Edward H. White floating in space connected to Gemini 4's life-support system only by his gold "umbilical cord."

first major breakthrough on our way to the exploration of space.

The first of Dr. Goddard's rockets to be successfully flown is the one on the metal tubular stand. It was launched from a farm in Massachusetts on March 16, 1926, and reached an altitude of 41 feet, powered by liquid oxygen and gasoline. The flight lasted but 2½ seconds, and the rocket's average speed was approximately 60 mph; but, like the first tentative flights of the Wrights' Flyer, the 1926 rocket marked the dawn of a new age. During the flight a portion of the rocket's nozzle was burned away and other parts were damaged when the rocket impacted into the ground 184 feet from its launch site; but the rocket was repaired, its pieces reassembled, and the rocket flown again on April 3, 1926.

The larger rocket was one of the last and most advanced liquid-propellant rockets tested by Dr. Goddard between 1939 and 1941. The rocket on display did not fly because of a malfunction which caused the engine to shut down shortly after ignition, but the significance of this rocket is that it incorporated most of the basic principles and elements used later in all long-range rockets and space boosters.

High up near the ceiling is a small silver ball with "feelers" extending out from it; it is a model of *Sputnik 1*, the first man-made object to be placed in orbit around the earth. *Sputnik* (meaning traveling companion) was launched, on October 4, 1957, atop a 96-foot-tall Soviet military rocket whose 1,124,440 pounds of thrust boosted the 184-pound satellite into orbit. For 22 days *Sputnik 1* transmitted internal and external temperature information and provided important orbital data concerning atmospheric and electronic densities at high altitudes. This model, on loan from the USSR Academy of Sciences, commemorates certainly one of the most dramatic milestones. The actual satellite, of course, burned up on reentry into the earth's atmosphere on January 4, 1958.

On November 3, 1957, the Soviets further stunned the world by placing *Sputnik 2* into orbit. This artificial satellite weighed 1,121 pounds and carried Laika, a dog, into orbit. On December 6, 1957, Vanguard Test Vehicle 3, carrying the first American earth satellite, a small grapefruit-sized device, exploded on its launch pad and the United States' prestige reached a new low. The little satellite continued to beep pathetically as the giant rocket consumed itself in the fire. In France café waiters the next morning approached American tourists and sardonically asked if they would like a "grapefruit" for breakfast.

Finally, on January 31, 1958, a four-stage Jupiter-C rocket designed, built and launched by the Army Ballistic Missile Agency team headed by Wernher von Braun, successfully placed *Explorer 1*, the first American satellite, in orbit. The back-up copy hangs near the *Sputnik 1* model above the balcony overlooking the Milestones gallery. *Explorer 1* measured three phenomena: cosmic ray and radiation levels (thus providing data which led to the discovery of the Van Allen belts, the belts of radiation surrounding the earth named after James Van Allen of the University of Iowa, one of the team who prepared the satellite's instrumentation), the temperature within the vehicle (essential knowledge in the design of future manned spacecraft), and the frequency of collision with micro-meteorites. The American *Explorer 1* satellite weighed 30.8 pounds including its fourth stage.

Sputnik 1 and *Explorer 1* marked the opening stages of man's space adventure. By January, 1975, 1,734 payloads had been fired into space; 1,007 had been destroyed reentering the earth's atmosphere, but 684 were still in orbit and 43 were rushing out into deep space. Among them is Mariner 2, a replica of which hangs in the Milestones gallery. On December 14, 1962, Mariner 2 was launched, making a 109-day trip that took it 36 million miles from earth to within 21,600 miles of Venus, thus completing the first successful mission to another planet. In the Milestones of Flight gallery is also the Pioneer 10 (the prototype of which hangs

At right, Robert H. Goddard's—and the world's—first successful liquid-propellant rocket (March 16, 1926) and, left, one of Goddard's last and most advanced rockets (1941).

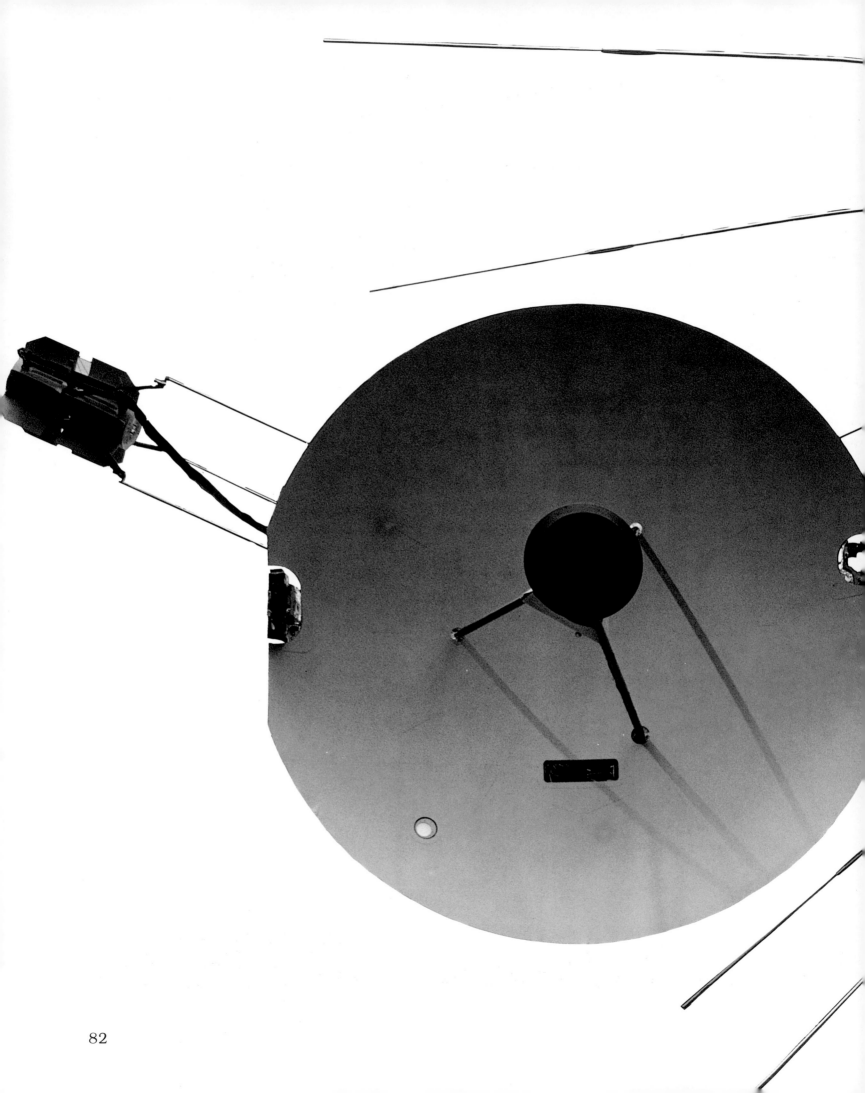

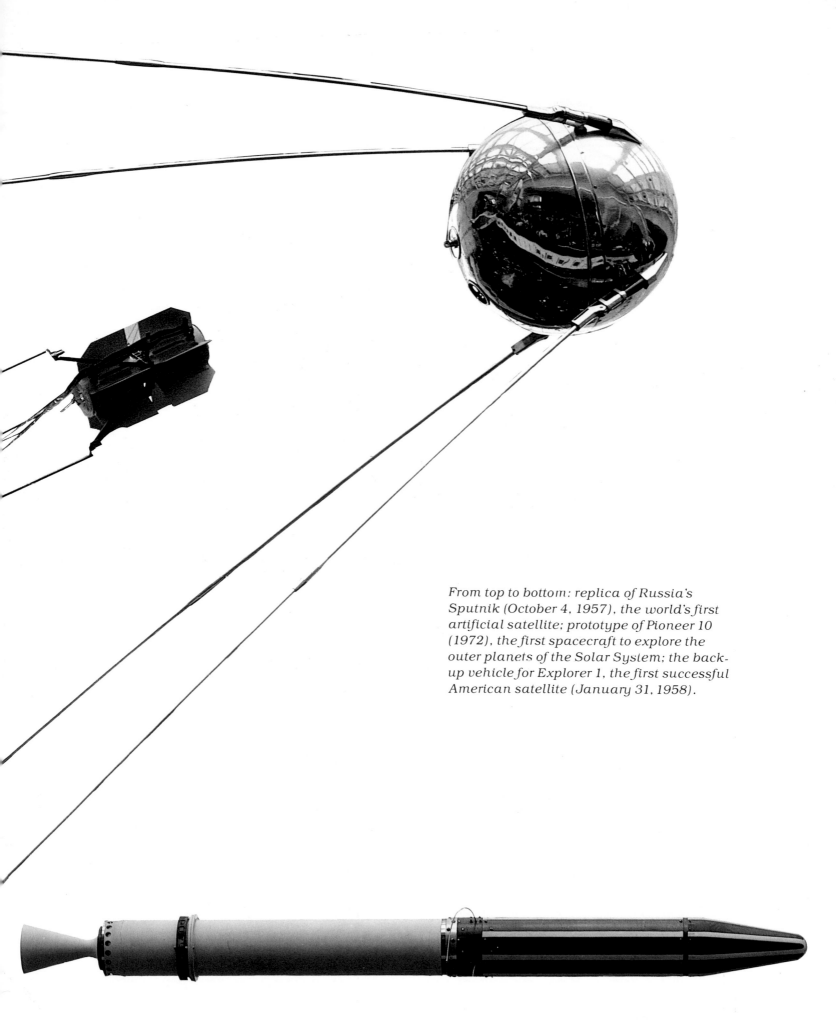

From top to bottom: replica of Russia's Sputnik (October 4, 1957), the world's first artificial satellite; prototype of Pioneer 10 (1972), the first spacecraft to explore the outer planets of the Solar System; the back-up vehicle for Explorer 1, the first successful American satellite (January 31, 1958).

near the delicate silk and metal tubing
Langley Aerodrome #5), which was launched
on March 3, 1972, and was the first
spacecraft to explore the outer planets of the
Solar System. On December 3, 1973, the
Pioneer 10 spacecraft passed within 82,000
miles of Jupiter and transmitted scientific
information and photographs of the planet's
surface 515 million miles back to earth. In
1990, Pioneer 10 will pass the farthest point
of Pluto's orbit and continue on out through
the Solar System. It carries a plaque
designed to inform any intelligent extra-
terrestrial life about the spacecraft and
where it came from. This spacecraft was
launched with the highest speed ever
achieved for a man-made object: 31,122
mph. It seems somehow especially
appropriate that the Pioneer 10 spacecraft
which was launched unmanned to explore
the farthest reaches of our Solar System
should hang so close to the Langley
Aerodrome #5, which was launched
unmanned also, on May 6, 1896, and was
the first American heavier-than-air flying
machine to make a free flight of any
significant length. It flew about 3,000 feet at
25 mph from a houseboat on the Potomac
River for about a minute and a half in several
broad circles before landing gently on the
river.

As we complete our circle of the Milestones
of Flight gallery, our eyes quite naturally
return to the 1903 Wright Flyer and the
large, conical spacecraft beneath it. The
spacecraft is the Apollo 11 command module
in which astronauts Neil Armstrong, Edwin
Aldrin, and Michael Collins flew to the Moon
and back on July 16 through July 24, 1969,
just six months before the decade expired,
thereby fulfilling President John F.
Kennedy's challenge. Apollo 11 was the first
of six lunar-landing missions (Apollos 11, 12,
14, 15, 16, 17) plus one mission abort (Apollo
13) launched between July, 1969, and
December, 1972. It was during Apollo 11 that
Neil Armstrong and Edwin E. Aldrin, Jr.,
became the first men to walk on the Moon.
The scarred heat shield and exterior are

Goddard rockets (1926-1941).

testimony to the enormous heat generated
upon the command module's return through
the earth's atmosphere when temperatures
rose to over 5,000° F. during the 25,000+
mph reentry speed. The command module
Columbia also brought back to earth the
first samples of lunar rocks and soil.

Only sixty-six years and about twenty feet
separate the Apollo 11 command module
from the Wright Flyer; this juxtaposition of
two such epochal air- and spacecraft is not
only indicative of the extraordinary
technological advances flight has achieved in
so short a time, but it is also an entirely
appropriate expression of what the
Smithsonian's National Air and Space
Museum truly is.

The Apollo 11 command module Columbia, *which brought Neil Armstrong, Edwin Aldrin, and Michael Collins back from man's first walk on the Moon.*

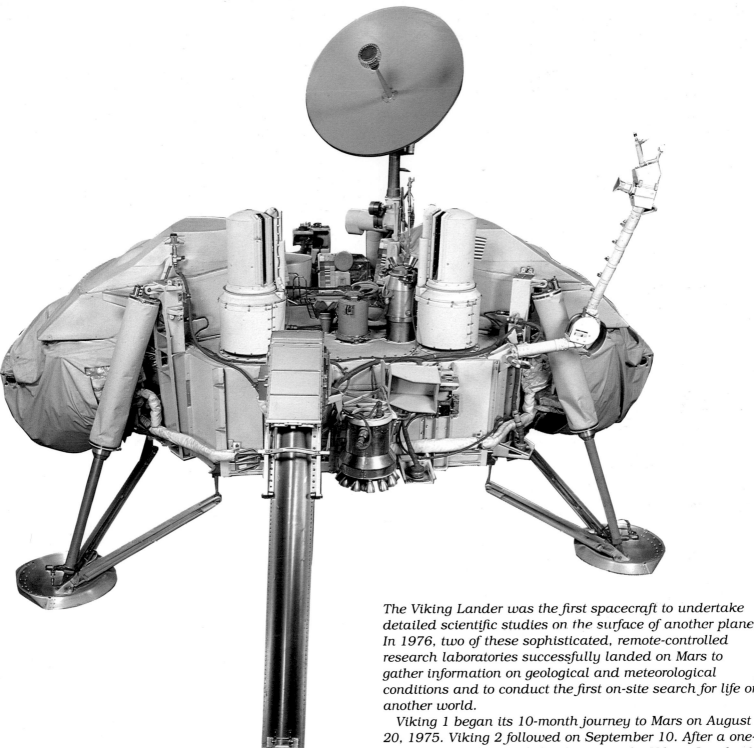

The Viking Lander was the first spacecraft to undertake detailed scientific studies on the surface of another planet. In 1976, two of these sophisticated, remote-controlled research laboratories successfully landed on Mars to gather information on geological and meteorological conditions and to conduct the first on-site search for life on another world.

Viking 1 began its 10-month journey to Mars on August 20, 1975. Viking 2 followed on September 10. After a one-month search for a safe landing site, the Viking Lander 1 touched down on the Martian surface on July 20, 1976; Viking Lander 2 made a safe descent on September 3, 1976.

The Viking Landers conducted a variety of experiments on the surface relating to Martian meteorology and geology. Soil gathered by the scoop was analyzed on board the Lander, providing new information on its chemical characteristics. One centuries-old question was answered when it was discovered that the red color of Mars is caused by its soil's iron-rich minerals. The Viking biological experiments did not detect any positive signs of life on Mars.

The Museum's Viking Lander is a Proof Test Capsule used in ground tests on Earth before and during the flight.

A child's first tentative touch of a piece of the Moon.

HALL OF AIR TRANSPORTATION

PIONEERS OF FLIGHT

Hall of Air Transportation

Visitors to the National Air and Space Museum almost automatically turn from the Milestones of Flight gallery to the Hall of Air Transportation. Perhaps it is the irresistible attraction of seeing how something as huge as a DC-3—at 17,500 pounds, the heaviest aircraft supported by the Museum structure—can be hung from the ceiling as effortlessly as if it were nothing more than a young boy's model airplane suspended on a fishing leader.*

Each of the seven aircraft on display in this hall played a significant role in the development of air transportation, a field which commenced in this country only a little more than ten years after the Wright brothers' first powered flight, when A.C. Phiel, the former mayor of St. Petersburg, Florida, paid $400 for the privilege of becoming, on January 1, 1914, the St. Petersburg–Tampa Air Boat Line's first passenger. The airline's wooden-hulled Benoist Type XIV biplane flying boat made two eighteen-mile round-trip flights per day between Tampa and St. Petersburg, and the twenty-three-minute one-way trip began regularly scheduled airline service in the United States. The Benoist XIV's top speed

was 63 mph, and when three months later the contract had ended, 1,204 passengers had paid $5.00 each for the one-way flight in the plane's open cockpit. The St. Petersburg–Tampa Air Boat Line lost only eight days due to weather or mechanical failure during its ninety days of operation and in addition to maintaining a regular schedule, the company pioneered in some advanced airline concepts: excess baggage weight or passenger weight (any passenger weighing more than 200 pounds) was charged at the rate of $5.00 per hundred pounds.

Commercial flying in this country, however, floundered about until May 20, 1927, when Charles A. Lindbergh took off from New York City's Roosevelt Field. His landing 33 hours and 30 minutes later in Paris had an extraordinary impact on American aviation. It was as though some vast psychological barrier had been crossed. Within the following year the applications for private pilot's licenses jumped from 1,800 to 5,500. In 1928 the nation's still-fledgling airlines doubled their route-miles flown, tripled the amount of mail carried, and quadrupled the number of passengers. From the moment Lindbergh landed at Le Bourget outside of Paris people realized that the airplane had suddenly come of age: airplanes could actually span continents and oceans, cross over mountains and arctic wastes. They could deliver passengers, mail, and merchandise with what America loved most of all: speed. The key factor in explaining the sudden enormous expansion of American commercial flying is that big business had come to realize that big money could be

---

*The answer lies in the genius of the Museum building's design: aircraft are hung from the triangular steel trusses that span the ceiling between the upright pylons. The trusses, which are triangular in cross section as well, provide enormous strength while creating a sense of openness and space. This technique, similar to that used to strengthen aircraft structure, makes it possible to suspend several loads along an entire span or to concentrate one major weight in one place.

made—but only if commercial aviation were run on a businesslike basis. That meant mergers and subsidies.

In 1927 Juan Trippe had received the mail contract between Key West, Florida, and Cuba. In 1928 he won additional foreign mail contracts to the Canal Zone and Puerto Rico. This meant that his young Pan American Airways would be guaranteed a backing of at least two and a half million dollars a year in mail-contract revenues. By 1931 he had locked up South America and successfully prevented foreign airlines from dominating that continent. By this time, too, three major airline companies had emerged out of the tangle of competitive small carriers in the United States: United Airlines, American Airlines, and TWA.

In 1929 Transcontinental Air Transport (T.A.T.)—which had become known as "The Lindbergh Line" when Lindbergh agreed to the use of his name—offered a unique transcontinental service requiring their passengers to split the trip between Ford Tri-Motors and pullman-car trains. The Fords flew only in daylight across safe flying country; the passengers would then board the train for the nighttime passage over the mountains. Although the trip could be made in relative safety, it was not a commercial success since the 48-hour journey across the country was not all that competitive with the railroads alone. The next year T.A.T. merged with Western Air Express and became Transcontinental and Western Air, or TWA.

That same year United Airlines, which flew the Chicago and West Coast route, bought National Air Transport, whose routes were between Chicago and New York. In the south a gaggle of small carriers combined to become American Airways.

In March, 1929, these three big systems came under the penetrating eye of Walter F. Brown, Postmaster General of the new Hoover Administration. Like his predecessors, Brown felt that the Post Office Department should encourage commercial aviation in the interests of national defense. But he also felt that rapid expansion would never take place so long as government subsidies made it more profitable for the airlines to carry mail than to carry passengers.

Consequently, he sought an amendment to the original Kelly Act that would eliminate the old pound-per-mile rate and pay operators according to how much cargo space they made available. Brown figured that this would encourage the airlines to place orders for larger airplanes; then, if mail did not fill the extra space, the operator would carry passengers rather than fly half-empty. The new proposal, passed as the McNary-Watres Bill, also attempted to reward progressive operators by providing for extra payments to airlines using multi-engine planes equipped with the latest navigational aids. The whole point of the new law, Brown said, was to develop aviation in the broad sense and to stimulate manufacturers "who would compete with each other and bring their aeronautical industry up to the point where it could finally sustain itself."

This approach meant, inevitably, that when it came to bestowing mail contracts Brown would tend to disregard the small, struggling, independent airlines in favor of the larger companies with better financing and more experienced personnel. After a series of meetings, later known sardonically as the "Spoils Conference," the big operators did walk off with most of the contracts.

—*The American Heritage History of Flight*,
by Arthur Gordon

United Airlines won the northern transcontinental route, TWA the central route, and American Airlines the southern. The size of the country, itself, had become a factor advantageous to the growth of air travel. Aircraft companies competed with each other to come up with the fastest, safest, and most economical planes. And it is this competition that resulted in some of the revolutionary designs one sees in the aircraft exhibited in this hall.

Let's begin on the floor of the Hall of Air Transportation with the rotating beacon light next to the Douglas M-2 mailplane with the Western Airlines markings. Before sophisticated radio navigation rendered beacon lights obsolete, pilots relied upon lights such as the one exhibited that were spaced approximately ten miles apart along the routes flown between major cities. In

A 1920s Douglas M-2 mailplane beneath a rotating beacon navigation aid.

The Museum's Pitcairn Mailwing
in Eastern Air Transport markings.

1946 there were 2,112 airways beacons on 124 air routes in the United States. At night under fair weather conditions a pilot could see one or more of these lights at any given time. Pilots learned the following mnemonic to help them remember the code flashed by the various airways beacon lights: **W**(.--)hen **U**(..-)ndertaking **V**(...-)ery **H**(....)ard **R**(.-.)outes **K**(-.-)eep **D**(-..)irections **B**(-...)y **G**(--.)ood **M**(--)ethods. Starting at the origin of the airway, the successive rotating beacon lights would flash the International Morse Code for the letters W, U, V, H, R, K, D, B, G, and M; this sequence would start again with the eleventh and twenty-first light, etc. The beacon on exhibit in the Museum once stood on a 60-foot-high tower on White Water Hill in California; it blinks two long and one short flash, the code letter "G," indicating it was the ninth beacon from the origin of the airway, Los Angeles.

The Douglas M-2 mailplane exhibited next to the beacon was one of a family of large biplanes built in the mid- and late-1920s by the Douglas Aircraft Company as a replacement for the venerable DH-4s used previously to fly the mails. The M-2 was a fast, sturdy, dependable aircraft for its time. The Museum's Douglas M-2 was originally delivered to the Post Office Department in 1926; Western Air Express acquired it in 1927 and flew the plane approximately 914 hours before it crashed on January 23, 1930. The plane was sold and, after passing through several owners' hands, was repurchased by Western Air Express' successor, Western Airlines, in 1940. The M-2 was restored to flying condition in 1976 in time to celebrate Western Airlines' Fiftieth Anniversary, which was the record for the longest continuous airline service in the United States.

Above the Douglas M-2 hangs the sporty Pitcairn PA-5 Mailwing, which was designed specifically for the shorter mail routes in the eastern United States. The aircraft's efficiency and economical operating costs stemmed from three factors: a clean, lightweight airframe (utilizing wooden wings and easily fabricated square tubing in the fuselage—both fabric covered), the remarkably reliable Wright Whirlwind engine (the same type of engine used by Lindbergh's *Spirit of St. Louis*), and the Pitcairn-designed airfoil and pronounced dihedral in the lower wing which permitted a relatively high speed, stability, and load-carrying capability.

The Mailwing's reputation was established on the New York—to—Atlanta mail-contract run flown by the Pitcairn Aircraft Company starting May 1, 1928, where by following the newly lighted airways between New York, Philadelphia, Baltimore, Washington, Richmond, and Atlanta the little Pitcairn Mailwings were able to make the 760-mile journey at night in seven hours—one-third the length of time it took by train. On December 1, 1928, Pitcairn Aircraft took over the Atlanta—Miami mail route, thereby creating the basic airline route upon which its successor, Eastern Air Transport (later Eastern Air Lines), would establish so successful a service. The Museum's Mailwing survived both a crash and use as a crop duster before being repurchased by Eastern Airlines employees, who restored it and presented it to Captain Edward V. Rickenbacker, who was the Eastern Airlines Company's president for so many years.

Higher up against the hall's windows in Panagra markings hangs the somewhat later Fairchild FC-2 developed by Sherman Mills Fairchild as an efficient camera plane. Fairchild, an important designer, builder, and user of aerial cameras in the early 1920s, needed an airplane that provided a wide field of view for the pilot and photographer, stability for high-altitude work, and an ability to operate out of small, rough fields. The Fairchild FC-2 which resulted could carry five persons including the pilot, and it became a great success as a light transport in the rugged Canadian bush, the jungles and mountains of South America, as well as in the United States, where its duties were expanded to include airmail delivery, passenger flights, and freight hauling. This

A favorite with bush pilots, this Fairchild
FC-2's wings could be folded within two
minutes for easier storage and
transportability. When the wings were
unfolded the pilot was reassured that they
were properly fastened because a large
padlock hung down in clear view.

The lovely 1930s Northrop Alpha flew fresh flowers from California to New York.

airplane made the first scheduled passenger flight in Peru, flying from Lima to Talara on September 13, 1928, for Pan American-Grace Airways. On January 15, 1929, Richard E. Byrd's Fairchild FC-2W2, the *Stars and Stripes*, was the first airplane to fly in Antarctica; the aircraft was left in the ice at the end of the first expedition, later recovered, and flown again.

The most unique feature of the Fairchild FC-2 was that two men could fold its 44-foot wings into a 13-foot unit within two minutes for easier storage and transportability. Unfolding took about the same amount of time. The pilot was reassured that his wings were locked in place by a large Yale padlock which would hang down in clear view.

The silver monoplane with the TWA markings hanging between the Pitcairn Mailwing and the Fairchild FC-2 is the Northrop Alpha, one of the most beautiful airplanes in the Museum's collection.

Designed by the legendary John K. "Jack" Northrop (who also designed Amelia Earhart's Lockheed Vega in the Pioneers of Flight gallery), the Alpha represents a transition in air transport design. Four passengers were enclosed in the snug, comfortable heated cabin while the pilot sat above and behind them in the open cockpit exposed to the elements in the traditional mailplane manner. The Alpha's advanced all-metal design included the Northrop multicellular cantilever wing, a semi-monocoque fuselage, and streamlined NACA cowling. The shaped fairings around the fixed landing gear provided drag reduction without the weight and complexity of a retractable undercarriage. Introduced in 1930, the Alpha was designed to be a swift, dependable airplane that could carry passengers and freight out of small airfields. With the coming of the newer, larger twin-engined Boeing and Douglas transports in

1933 and 1934, the Northrop Alphas were soon relegated to freight-carrying roles in which they flew coast-to-coast in 23 hours, bringing from California such exotic commodities as fresh-cut gardenias, silkworms, and serums to New York. Although the Alpha was in use for only a few years, its major importance lay in the advanced design concepts engineered by Northrop which were of fundamental importance in the construction of the Douglas DC-2 and DC-3. The Museum's Alpha was first owned by the Department of Commerce; its restoration was undertaken by TWA, for later exhibit in the National Air and Space Museum.

Perhaps no aircraft has been the subject of more legends, admiration, and misconceptions than the Ford Tri-Motor, one of which hangs just in front of the DC-3. Affectionately known as the "Tin Goose," the Ford Tri-Motor was the largest civil aircraft in America when it started passenger service on August 2, 1926. William Stout, an extremely gifted and brilliant designer, had originally created a high-winged, three-engine, all-metal aircraft for Henry Ford that turned out to be so ugly, so awkward in the air, and so impractical that it was grounded. Ford then turned to his chief engineer William Mayo, the racing-boat designer Harold Hicks, Otto Koppen, John Lee, and James McDonnell (who later created the McDonnell Aircraft Corporation). "It was a committee," wrote Ernest K. Gann, "utterly dedicated to a single purpose—creating an aircraft that would be good enough to please Henry Ford and thereby save their collective necks. To that desire they gave all of their talents and took aid where and in any way they could find it." Gann continues:

One clandestine effort was typical of their desperation. The first contemporary multi-motor aircraft was the Fokker F-VII, a three-engined, high-wing monoplane which had set many admirable long-distance records. By chance, Admiral Byrd, who had chosen a Fokker for his arctic expeditions, dropped in to Dearborn where his welcome was most gracious, and his Fokker hangared for the night.

It was a long night for the committee who spent most of it bending copper tubing into templates as they measured the shape of the Fokker's wing every foot of its length. Only a very few people were ever told of the night-long measuring party, certainly not Henry Ford. But those who did know were not surprised to discover a remarkable similarity between the metal airfoil of the Ford 4 A-T's wing and a certain aircraft of Dutch design.

—*Ernest K. Gann's Flying Circus*,
by Ernest K. Gann

The Tri-Motor's all-metal construction, the implied massive strength of its thick wings, the prestige of the Ford name all combined with the pure lucky timing of its debut. This happily coincided with the world-wide love affair with aviation resulting from Lindbergh's same year New York–to–Paris flight and made the Ford Tri-Motor an immediate success with the public.

Inertial starters activated the Ford's three engines. A mechanic or the co-pilot would insert a crank in each engine and, on signal, begin cranking. Slowly at first the heavy flywheel geared to the engine's crankshaft would turn and then as the speed increased and the flywheel began to whine at what sounded like full speed,

the cranker pulled a cable which engaged a spring-loaded clutch thereby transmitting the energy of the spinning wheel to turn the engine. The squealing sound was almost identical to that displeasure emitted by a resentful pig kicked in anger. If the oil was not too cold the propeller would turn at least three or four revolutions. Usually this was enough to start the engine, but if the pilot was not alert and missed the moment of truth with mixture and throttle, the crankee was obliged to start his labors all over again. Winding the cranks was very hard work and at times when the engines proved unaccountably balky there were impolite exchanges of opinion between cockpit and cranker.

...life with and aboard the tri-motored Fords was far from ideal....Ford passenger cabins were always too hot or too cold and decibel level assured them a top place among the world's noisiest aircraft. Immediately on boarding, passengers were offered chewing gum which would allegedly ease the pressure changes on

The legendary Ford Tri-Motor's all-metal construction, suggesting great strength and safety, heartened the fledgling commercial aviation industry's early passengers.

their eardrums during climb and descent, but it was just as much to encourage a cud-chewing state of nerves. They were also offered cotton which wise passengers stuffed in their ears so they would be able to hear ordinary conversation once they were again on the ground....

In the first Fords there were no seat belts. Only hand grips were provided to stabilize passengers, and summertime flying could become a purgatory. While they bounced around in low-altitude turbulence the passengers muttered about "air pockets" and a high percentage became airsick. Even with a few windows open the cabin atmosphere developed a sourness which only time and scrubbing could remove.

If the passengers retreated to the lavatory they found little comfort at any season. In winter the

expedition became a trial-by-refrigeration since the toilet consisted of an ordinary seat with cover. Once the cover was raised for whatever purpose there was revealed a bombardier's direct view of the passing landscape several thousand feet below, and the chill factor in the compartment instantly discouraged any loitering.

Even in smooth air flying a Ford became a chore if only because it was so difficult to keep in trim. The man who could coax a Ford into flying hands-off for even a few minutes was temporarily in luck and probably did not have any passengers. Even a normal bank in a Ford was an experiment in muscular coordination mixed with a practiced eye for anticipation since whatever physical input was directed to the

controls a relatively long time passed before anything happened. To stop or reverse as desired *any* maneuver required a keen sense of anticipatory delay. In rough air these delays and willfulness were compounded, and just keeping the Ford straight and level became a workout. In a thunderstorm or line squall the pilots sometimes wondered who was in charge of affairs.

In the devious way of legends the Ford has somehow emerged as a stable aircraft. In the sense that they were always controllable and therefore safe the recognition is true, but no more is deserved. Legend also has it that Fords could be landed in any small cow pasture, which was not so, particularly if loaded. The ability lies somewhere in between, depending on many

factors, including the hunger status of the pilot or airline. Takeoff performance was actually more remarkable than landing speeds, which were much higher than most historians seem inclined to acknowledge. Perhaps they do not realize that behind the more spectacular short-field stops made by Fords which actually touched down between 65 and 70 miles per hour, there was a grunting copilot pulling for all he was worth on the long [Johnson] bar which extended from the floor upward between the two cockpit seats. The bar activated the hydraulic brakes which in their own obstinate way gave airline mechanics perpetual trouble.

—Ernest K. Gann's Flying Circus,
by Ernest K. Gann

Ford Tri-Motor passengers sat on wicker seats chosen for their strength, lightness, and comfort. The first models of this plane carried twelve passengers, later models as many as fourteen at 110 mph. The most distinctive aspect of the aircraft was its rugged all-metal construction readily identified by its corrugated aluminum skin. Its only rival was the tri-motored Fokker to whose design the Ford owed so much. But when one of these Fokkers crashed, killing Notre Dame's famous football coach, Knute Rockne, and blame for the crash was placed on the structural failure of the Fokker's wooden construction, the Ford Tri-Motor emerged as the unchallenged transport of that era.

The Museum's Ford began its career on

April 12, 1929, with Southwest Air Fast Express (SAFE), a company bought out the following year by American Airlines, which flew the plane until retiring it in 1935. In 1936 the Ford was sold to TACA International Airlines and flew in Nicaragua for several years. From 1946 to 1954 this Ford hauled passengers and cargo in Mexico, and then was sold to a Montana crop-dusting company which also used it to fly a cargo route in Alaska before returning the Ford to Mexico, where it was sold again. By the time the Ford was reacquired by American Airlines it had ended up parked beside a small airfield in Oaxaca, where someone had installed a woodburning stove inside it, stuck a chimney through its roof, and was using it

The Boeing 247D, which entered service in 1933, was the world's first truly modern airliner.

When United Airlines' huge order for these 247Ds monopolized Boeing's production lines, competing airlines were forced to turn to other aircraft companies. Douglas' response—the DC-1—was a design that rendered Boeing's airplane immediately obsolete.

as living quarters. American Airlines restored the Ford and flew it on publicity tours until donating it to the National Air and Space Museum.

The Boeing 247D, the first truly modern airliner, was a pioneer whose clean lines readily distinguished it from the tri-motors and biplanes it had immediately rendered obsolete. When it entered service in 1933 its design features included low-wing all-metal construction necessitating

a padded step in the middle of the passenger aisle beneath which the main wing spar was located. This inconvenience was accepted as a visible proof of the great structural strength of the plane. Its retractable landing gear was clean and simple; the wheels protruded slightly, thereby providing some protection should a wheels-up landing be necessary. And its 550 hp Pratt and Whitney Wasp engines were supercharged to provide a top speed

of 200 mph in later models.

The Boeing 247D on display in the Museum placed third overall and second in the transport category in the 1934 London–to–Melbourne MacRobertson International Air Derby during which it was piloted by Colonel Roscoe Turner and Clyde Pangborn. It made the 11,300-mile trip in 92 hours, 55 minutes, and 30 seconds elapsed time. NASM's 247D subsequently was used as a regular commercial transport by

United Air Lines, which had ordered the first sixty 247s in 1932. This order fully tied up Boeing's production lines and effectively denied the use of that plane by any other airline. On May 22, 1933, when the new Boeing 247D went into service on the cross-country run, it made the flight from San Francisco to New York in 19½ hours—knocking eight and a half hours off the previous 27-hour flying time. The original 247 had a cruising speed of 170

mph compared to the 110 mph speed of the Ford Tri-Motor then in general use. The tactic of preventing other airlines from purchasing the Boeing 247 worked to a disadvantage for United Air Lines since it forced TWA to go to Douglas for an aircraft that could compete.

The all-metal skin of the Boeing 247 was anodized aluminum, which gave the airplane its gray color; but because the Museum exhibit's anodized skin was badly weather-worn, it was painted gray to protect it—except for the cowlings and vertical surfaces which, less worn, retain the original anodized finish. Museum visitors may notice that the airplane on display has two different markings; the left side is painted as it was when Roscoe Turner flew it in the London–to–Melbourne race, the right side carries the paint scheme it wore as a part of United Air Lines.

Between the Ford Tri-Motor and the

Boeing 247D hangs the Douglas DC-3, the single most important aircraft in the history of air transportation. Almost every Museum visitor is familiar with it; many have flown in it as a commercial airliner or in its military C-47 version. Until 1934, airline passenger planes were either too slow or carried too few passengers to create much profit. The Boeing 247 carried only ten passengers and it was at about this time that airline companies could no longer depend upon airmail contracts for their primary source of income. When Transcontinental and Western Airways (to become TWA) ordered Douglas to develop an airplane capable of competing with the Boeing monopolized by United Air Lines, Douglas came up with the DC-1. On its maiden flight, July 1, 1933, the DC-1 exceeded all of TWA's specifications. Only one DC-1 was built. On May 18, 1934, the first production DC-2 went into service with TWA on the Columbus–Newark route.

It could carry fourteen passengers in comfort and it was fast. The DC-2 began its career by breaking the commercial New York–to–Chicago speed record four times in eight days. So many airlines ordered the DC-2 that Douglas had to carefully schedule deliveries. The beginnings of the "Great Silver Fleet" came when Eastern Air Lines ordered fourteen DC-2s for its New York–Miami route. In 1934, a KLM Royal Dutch Airlines DC-2 finished second in the London–to–Melbourne MacRobertson Derby by covering the distance in 90 hours and 17 minutes, almost three hours faster than the Boeing 247D hanging in the hall. Although the DC-2 resembled the Boeing 247 in some ways, it offered major advantages: the overall design was aerodynamically cleaner, it used the more powerful Wright R-1820 engines (875 hp), had split-trailing edge flaps, and controllable pitch propellers. The DC-3 came about when C.R. Smith, the

President of American Airlines, asked Douglas to build a larger version of the DC-2 that could be fitted with fourteen berths so that passengers could sleep on the long transcontinental flights. This version became the Douglas Sleeper Transport, the DST. The DST was not a financial success and was turned into the DC-3, which was identical except that instead of sleeping berths the aircraft was fitted with twenty-one seats.

The DC-3's success was due to the fact that it was the first airplane in the world that could make money by hauling passengers only. The flying public liked the plane. Its 180 mph cruising speed was the fastest of its day, and it carried eleven passengers more than the Boeing 247 and in greater comfort. It was considered much safer than any previous transport. Pilots liked its stability, ease of handling, and single-engine performance. The airlines liked it because it was reliable and inexpensive to operate. By 1938, 80 percent of all American commercial airline traffic was carried on DC-3s. And by the next year most of all the airline traffic in the world was carried on DC-3s.

The DC-3 displayed in the Hall of Air Transportation made its last commercial flight on October 12, 1952, when it flew from San Salvador to Miami. By then it had flown more than 56,700 hours! The DC-3 has proved itself the most popular transport airplane ever built.

Even though the engineering of the DC-3 was advanced for its day, uncertainties on the part of its designers led to the aircraft's being built much stronger than necessary. Much of its strength came from the multicellular stressed-wing construction devised by Jack Northrop, a technique employing many small formed-metal members rather than the until-then traditional rib-and-spar wing design. The redundancy of its strength made it possible for the DC-3 to operate under difficult conditions. During the war the military version of the DC-3, the C-47, was used by the thousands in every theater of the war. Affectionately nicknamed "Gooney Bird," they were flown over the Hump, over Europe, in the Pacific. After the war, many of the surplus C-47s were converted back into civilian transports. Almost every warring power used the DC-3 or a license-built derivative. The Russians used C-47s supplied by lendlease and Lisunov LI-2s made in

The 1937 Beechcraft Model 18 had the longest production span of any aircraft in history—almost 33 years.

Russia; the Japanese built a series known as an L2D. Because post-war surplus made it possible to purchase a C-47 for a low price, innumerable small airlines were created with but one or two of these converted aircraft. Over 9,123 C-47s were built during the war, and there is probably not a nation in the world today where one cannot find at least one DC-3 still flying.

On the floor of the hall rests a smaller twin-engine Beechcraft Model 18—an airplane that was first produced on January 15, 1937, and remained in production for almost thirty-three years, the longest production span of any aircraft in history. It was designed as a rugged, fast, easily maintained aircraft and was used by a large number of smaller airlines.

Its greatest impact on general aviation lay in its use for many years as the most prestigious of the executive transports. During most of its more than thirty-two-year production it was considered the best small twin-engine transport in the world.

The ultimate development in large piston-engine airliners was the Douglas DC-7. American Airlines introduced this model on its New York–to–Los Angeles run on November 29, 1953, and it became the first airliner to offer non-stop service from either coast. The DC-7 cruised at 360 mph and was the fastest aircraft in service; as a result, eighteen different airlines ordered these planes. The nose section of the DC-7 displayed in the Museum is from the American Airlines flagship *Vermont*, which

Museum visitors enter the cockpit of American Airlines' DC-7 flagship Vermont. *On November 29, 1953, a DC-7 became the first airliner to offer non-stop transcontinental service from either coast.*

carried about 130,000 passengers in its almost 13,500 hours aloft. With the advent of the faster, turbine-engine-powered Boeing 707s, the piston-engine airliner became obsolete.

Should the Museum visitor wonder why the NASM does not exhibit a Boeing 747 "Jumbo Jet," the answer is simple: the aircraft is so huge, it wouldn't fit.

Because the airways crossing America are under constant surveillance by the U.S. Air Traffic Control System, operated by the Federal Aviation Administration (FAA), air travel is the safest form of transportation in the United States today. The Air Traffic Control System is a vast network of facilities located in all 50 states, Puerto Rico, and in such places as Guam and American Samoa.

Interior view of the DC-7's cockpit.

The FAA staffs and maintains 450 airport control towers, 22 en route air traffic control centers, and more than 300 flight service stations. The 22 centers log more than 32 million operations a year; 16 of these centers record more than one million each, and four over two million.

While the average size of aircraft increases on the densely travelled routes to keep pace with the traffic volume, the number of feeder services and small airlines will also grow. Increased air traffic volumes have demanded a staff of highly skilled air traffic controllers and flight service specialists. As time goes on, the complexities of our air transport system will multiply, continuing to present a constant challenge to those responsible for air traffic control.

116

Pioneers of Flight

After passing through the Hall of Air Transportation, Museum visitors will often climb up to the balcony overlooking the DC-3 and the Ford Tri-Motor hanging just out of reach, then continue along beneath the Wright 1909 Military Flyer to the balcony above the Milestones of Flight gallery where they can look through the open door of Lindbergh's *Spirit of St. Louis* into the cockpit at the panel whose instruments, Lindbergh wrote, during the long night of flight "stare at me with cold, ghostlike eyes." But then the visitor will turn back, drawn by the seven very special aircraft spanning the years 1911 through 1977 that are a part of the Pioneers of Flight gallery. In this gallery the visitor also finds documents and newsreel clips commemorating the Fiftieth Anniversary of Lindbergh's New York – to – Paris flight and a wall display honoring the many varied and important contributions to flight safety and research made possible by the Guggenheim family. And although the visitor cannot help being drawn to the larger planes in this exhibit, no one who spends any time in this gallery emerges without a very special affection for that frail, canvas-covered Wright EX biplane, the *Vin Fiz*. It was in this machine that Calbraith Perry Rodgers completed the first transcontinental flight while attempting to win William Randolph Hearst's $50,000 prize offered to that person who completed the coast-to-coast trip in thirty days.

Cal Rodgers, a tall, wiry, cigar-chomping, motorcycle-racing, daredevil descendant of two Naval heroes, had less than sixty hours total flying experience when, on September 17, 1911, he took off from Sheepshead Bay, Brooklyn, in the *Vin Fiz* on the first leg of his trip. He flew to Middleton, New York, where he landed and spent the night. The next morning he took off again and flew directly into a tree, damaging his aircraft so badly it took three days for it to be repaired. (During a demonstration flight at Fort Myer, a Wright 1909 Military Flyer hit a tree, but it took only four *hours* for its damage to be repaired.) Rodgers' attempt at the Hearst prize was sponsored by the Armour Company which, at that time, was manufacturing a grape-flavored soft drink called "Vin Fiz." Rodgers was paid $5.00 for every mile he flew with the Armour Company's soft drink's name emblazoned on his wing. When the *Vin Fiz* was repaired, Rodgers pushed stubbornly west, followed and occasionally led by a special three-car train carrying his wife, his mother, mechanics, some Armour representatives, and $4,000 worth of spare parts. Rodgers navigated using "the iron compass" (railroad tracks) whenever possible, but because he followed the wrong switch he once ended up in Scranton, Pennsylvania, rather than, as he had intended, Elmira, New York. During his leapfrog journey across the country Rodgers made sixty-nine stops—not all of them intentional. He crashed nineteen times, once suffered an in-flight run-in with an eagle, and his *Vin Fiz* was wrecked so many times that when he finally arrived in California enough spare parts had been used to completely build four planes, and the only pieces that remained of the original *Vin Fiz* in which he had taken

off from Sheepshead Bay were the rudder and two struts from the wing. Rodgers, himself, finished the flight with his leg in a cast and an ugly scar on his forehead. Cal Rodgers did not win the Hearst prize; he had, in fact, reached only Oklahoma when the thirty-day requirement ran out. But with the Armour Company's backing and his own pertinacity, Rodgers continued on from Oklahoma to Fort Worth and El Paso (he made twenty-three stops in Texas alone), then west to Tucson, and on to Pasadena, where 20,000 well-wishers celebrated his successful completion of the first trans-continental flight by draping him in an American flag even though the crossing had taken him forty-nine days. The in-flight time for Rodgers was 82 hours and 2 minutes at an average speed of 52 mph. It beat driving; don't forget in 1903 when the first automobile traversed the United States it had taken sixty-five days. Four months after Rodgers reached Pasadena he was dead; during an exhibition flight he had crashed into the Pacific.

The Wright EX Vin Fiz *in which in 1911 cigar-smoking, ex-motorcycle racer Calbraith Perry Rodgers crashed nineteen times, made 69 stops, and lost the Hearst $50,000 prize en route to completing the first trans-continental flight across the United States.*

The huge high-wing monoplane that dominates the Pioneers of Flight gallery is the Fokker T-2 in which United States Army Lieutenants Oakley G. Kelly and John A. Macready took off from Long Island's Roosevelt Field on May 2, 1923, and not quite twenty-seven hours later landed at Rockwell Field in San Diego, California, thus completing the first successful U.S. *non-stop* coast-to-coast flight.

Kelly piloted the plane on take-off and Macready was given the honor of landing. The pilots exchanged places five times during the flight—an extraordinary maneuver which required that one man fly the plane from the near-blind rear controls within the fuselage (there was vision only to the sides) while the other abandoned the open cockpit, folded down the back of his seat, then opened a small triangular panel, then slithered through the opening and on back. The pilot in the rear would then snake his way forward to the open cockpit and once he was seated in the cold and drizzle he would shake the wheel

119

ARMY AIR SERVICE
NON STOP
COAST TO COAST

In this giant Fokker T-2, Army
Lieutenants Kelly and Macready completed
the first successful non-stop U.S. coast-to-
coast flight in 1923.

to show he could now take over command. The Fokker T-2's first two attempts to cross the country had started in the west but because of the weight of the fuel-laden aircraft they could not gain enough altitude to climb over the mountains. The third and successful attempt was made by flying west from a take-off in the east. Here Lieutenant Macready discusses the successful flight:

We took off from Long Island at about 11:30 in the morning. It took us about a mile, maybe a mile and a half, to get off, and our wheels were still on the ground when we came to the drop-off from Roosevelt Field to Mitchell Field. We went over the ledge and down. Kelly was taking it off and I was behind. I was wondering "Is it going to touch the ground? If it does, it's no use." But it maintained its flight and we just got over the hangars ahead. We had to keep from stalling and get just as much climb out of it as we could. Then we were over the water, near Staten Island, getting up around one hundred feet, maybe even a little more than that.

We got across the Alleghenies all right, but it began to get dark around Dayton, and it began to drizzle. You couldn't see anything but the lights of the cars on the highways, and we could only keep our compass from getting entirely out of control by checking it with the highways....We went over by St. Louis...and then we were over some mountains. As long as you can see some outside fixed point you can balance your plane and everything's O.K. But if you've got nothing, you can't tell if you're upside down or how you are. I was flying then and there'd be times when you wouldn't get any outside point at all in the drizzle. We were over the Ozarks and I knew that if that kept up, I was just going to lose control of that plane. Then there'd be a light, some mountain cabin. Then you'd find you're cocked up on the side, but you're still going ahead. When you see that light, you can get yourself adjusted again. In Kansas we came out from under the storm. Then we could see the section lines and we could keep our course.

The impression that you get when you're flying across and you see no lights is that you're alone, and of course, we didn't know exactly where we were....When we got out into Arizona the mountains were getting higher. We couldn't get over them so we had to deflect our course. We were just going over treetops, not very high, I'd say one hundred feet, something like that. Then we saw a place...a break in the mountains, and

we came through...and then pretty soon I saw a little town. You can work out a pattern by the way the railroads come in. Then you check it with your map. The town was nothing, but we were located. For the first time during the night we were located, the first time since St. Louis.

Then we were over Tehachapis and then down over San Diego. Well, as I said, we had the impression that nobody was interested. Then we came gliding toward North Island, and we came pretty close over the U.S. Grant Hotel. We saw a bunch of people on top of the hotel waving sheets and so forth. We went on and landed at Rockwell Field, and, of course, we were pretty tired when we got there.

—Oral History Collection of Columbia University

Nestled under the giant T-2's wing is another significant Army plane, a Douglas World Cruiser, one of two that completed the first successful around-the-world flight. By the end of World War I, the airplane had achieved a level of development whereby it had become a practical means for aerial exploration. By 1923 several long-range and endurance flights had succeeded and the goal of achieving an around-the-world flight was inevitable. Several pilots had tried or were preparing an attempt at this goal. The British had failed in 1922 and again in 1924. The Italians had failed then, too. But then on April 6, 1924, some of the United States Air Services' best pilots took off in four Douglas World Cruisers (modified Douglas DT torpedo planes) and on September 28, five months and 27,553 miles later, two of the planes completed the circle and returned once more to Seattle.

The flight's route spanned major oceans—the Pacific for the first time—and encountered weather extremes ranging from arctic to tropical. Although much of the support and safety facilities were provided by the United States Navy and the Coast Guard, conditions were often primitive. Fuel was rowed out to the World Cruisers in 55-gallon drums and had to be hand-pumped into the aircraft. The Douglas World Cruiser #1, the *Seattle*, piloted by Major Frederick L. Martin with Sergeant Alva Harvey as crew flew into heavy fog and low ceilings after taking off from Kanatak, Alaska, and struck a mountain slope and crashed. The men fortunately were not seriously injured and their ten-day "walk-out" to civilization is a story almost as remarkable as the flight itself. The Douglas World Cruiser #3, *Boston*, was lost en route to Iceland when engine trouble necessitated a forced landing in the ocean. Although its crew was rescued, the aircraft had to be abandoned in the rough seas.

Lieutenant Leslie P. Arnold, one of the four pilots successfully completing the circumnavigation of the globe, recalls the experience:

> The local papers intimated that there was not much hope that we would complete the journey, but as we circled over Seattle, I thought that there was at least a fifty-fifty chance.... The airplane was a biplane and had a metal framework covered with linen; there was a single Liberty engine of 400 horsepower. The maximum ceiling was around 8,000 feet so we had to fly under things instead of over them. There was just a small glass windshield in front of us, probably six inches high and eighteen inches in width. It was no protection at all unless you got way up under. It was nothing to have your shoulders wet through. They got iced up in no time at all.
>
> Nobody ever touched the planes save us. If an engine had to be changed, we changed it. We washed them, babied them, put them to bed. We did everything but sleep with them. After a flight we had a routine procedure which we followed. If the water connection had to be taped or shellacked, we did that—so we knew it wasn't going to fall down on the next hop. Everybody was a mechanic and everybody was a flier.
>
> ...We had a forced landing in the jungles of Indochina. We were flying down the coast when a spray of water hit us. We knew that something had gone wrong in the engine. It began to heat up too much. We landed in the mouth of a river to discover that one of the cylinders had cracked. The other boys landed alongside. We had to get a new engine, so they took off and went to Tourane; Lowell Smith and I stayed in the airplane, sleeping on the wings. Finally, out of the darkness, came a Swedish voice. It was Erik Nelson [another one of the fliers]. He had made arrangements through Standard Oil and the local tribes, and we were towed up the river to the capital of the French kingdom of Annam. It was an interesting ride, with the natives in their sampans towing us. The chief was riding in his sampan with a parasol and with his favorite wives fanning him.
>
> The trip from Iceland to Greenland was the most frightening. We were two planes by that time. Everything was fine for about one hundred miles when we ran into fog and drizzle. We kept circling lower and lower until we were just over the water. There we saw broken ice and drifting icebergs, but we were past the point of no return and we had to keep going. So we plowed through this. The icebergs and the fog are both a dirty gray color, and it was very difficult flying seventy-five or one hundred feet off the water. We would veer here and there; when you saw one ahead there was no way to go except up in the fog. Finally we turned to the west and found the shore of Greenland. Then we had one hundred miles of clear weather and again ran into fog. We decided to stay on top this time, and kept our position by the mountain peaks. At last we arrived at a certain peak and thought the harbor should be under us. So we came down through the fog and landed in a nice little harbor. The second plane had not come in, however, and we went about our routine maintenance work, never saying a word. We felt too bad to say anything. But after about half an hour we heard the unmistakable noise of a Liberty engine. Then we had one of the finest parties Greenland had ever seen.
>
> Flying on from Greenland to Labrador, we got halfway across when two of our gas pumps failed—we had to pump the gas by hand. It got tiresome and my arm began to get numb. So I thought the only way to do it was to use a combination of my belt and a handkerchief around my neck like a sling and to pull it with my other hand. I did that for close to four hours. It was tiresome, but it beat the hell out of swimming.

—Oral History Collection of Columbia University

Certainly one of the most beautiful planes in the Air and Space Museum's collection is the little black-fuselaged biplane with floats that hangs suspended near the back wall and the *Vin Fiz*. It is the Curtiss R3C-2, flown on October 25, 1925, by United States Army Lieutenant James H. Doolittle in the course of winning the Schneider Trophy Race (limited to seaplanes) at a speed of 232.57 mph. The following day Doolittle set a world speed record over a straight course with the

A Douglas World Cruiser which in 1924 successfully completed the first round-the-world flight. Four of these modified torpedo planes took off from Seattle on April 6, 1924— five months and 27,553 miles later two of them safely returned.

*In this handsome, sporty Curtiss R3C-2,
U.S. Army Lieutenant Jimmy Doolittle won
the 1925 Schneider Trophy Race (limited to
seaplanes) with a speed of 232.57 mph.*

same plane at a top speed of 245.7 mph. As the Curtiss R3C-1 with a fixed wheeled landing gear, the racer had won the Pulitzer Race earlier that same year. The R3C-2 was the last biplane to win the Schneider Trophy. From 1926 on monoplanes dominated the races and the Supermarines, designed by R.J. Mitchell, whose innovations led to the great British fighters like the Spitfire of World War II, beat all other competitors.

The other float plane in this gallery is the John K. Northrop-inspired Lockheed Sirius in which Charles A. Lindbergh and his wife Anne Morrow Lindbergh made two major flights in 1931 and 1933 to survey possible overseas airline routes in the early days of international air travel. The Lindbergh's 1931 "North-to-the-Orient" flight—the first

east-to-west flight by way of the north—took them from Maine across Canada to Alaska and Siberia, over the Kurile Islands to Tokyo and finally to Lotus Lake near Nanking, China, and demonstrated the feasibility of using the "Great Circle" route to reach the Far East.

Charles A. Lindbergh had described the flight as one with "no start or finish, no diplomatic or commercial significance, and no records to be sought"; but for the first half of their journey the Lindberghs flew where no airplane had flown before. The Sirius was equipped with Edo pontoons since the majority of their trip would be over water, and they wore electrically heated flying suits against the arctic cold. As one might anticipate of any journey across the ice packs to the Seward Peninsula, the

During a stop in Greenland, the Lindberghs' Lockheed Sirius was christened Tingmissartoq ("the man who flies like a big bird") by an Eskimo boy.

frozen Bering Straits to Siberia and down to Japan, the Lindberghs encountered terrible weather and flying conditions. Charles Lindbergh, with typical foresight, had arranged for fuel dumps at their rendezvous points along their route. But finding them could be difficult. Anne Lindbergh was the radio operator and navigator. She was responsible for keeping the outside world in touch with their progress, and it was often very trying for her to locate a radio station, tap out a Morse Code message and acknowledge the reply before she would have to haul in the antenna because her husband had found a hole in the clouds and could get low enough to confirm a reference point. Anne Lindbergh reported that the greatest compliment she had received on this flight was from another

radio operator who had congratulated her on her skill in sending and receiving messages, saying, "no man could have done better."

The Lindberghs were offered the hospitality of the British aircraft carrier *Hermes*, which was anchored in the Yangtze River off Shanghai. At night the Sirius would be lifted out of the swollen river's raging current and the Lindberghs would sleep on board—Anne Lindbergh being the first woman ever permitted to do so. But, putting the Sirius back in the water one morning, the current caught the aircraft. Both Lindberghs were in the cockpits and as the current swung the plane, tipping it, the crane's cable caught against the Sirius' wing and their plane began to roll under. Lindbergh yelled to his wife to jump. He waited just long enough to see that she had

In 1932, with this Lockheed Vega, Amelia
Earhart made the first solo flight by a woman
across the Atlantic.

In 1937, Amelia Earhart disappeared over the Pacific while attempting a round-the-world flight.

surfaced safely, then he jumped into the river too. They were rescued by a tender, but the Sirius was too damaged to continue its flight and had to be returned to the United States for repairs.

Charles and Anne Lindbergh's next journey in the Sirius, which had been overhauled and refitted with a larger engine, was in 1933. Lindbergh was serving then as a technical advisor for Pan American Airways. With Anne Lindbergh again working as the navigator and radio operator, they took off from New York in July and flew up the eastern coast of Canada to Labrador, then 650 miles across the water to Greenland. While there, a young Eskimo boy painted the word "Tingmissartoq"(meaning "the man who flies like a big bird") on the Sirius' fuselage. From Greenland the Lindberghs flew their newly christened plane to Baffin Island and back before going on to Iceland. From Iceland, the Lindberghs flew to the major cities of Europe as far east as Moscow, then down the west coast of Africa, where they crossed the South Atlantic and flew down the Amazon River before turning north through the Caribbean and on back to the United States, where on December 19 they landed in New York. They had traveled some

30,000 miles across four continents, visited twenty-one countries, and gained invaluable information toward the planning of north and south Atlantic commercial airline routes. Their flight was an incredible one for that period of aviation; they had encountered the worst possible hazards and flying conditions: blizzards, sandstorms, hurricanes, arctic cold and tropical heat. Anne Lindbergh again displayed extraordinary skill as a navigator and radio operator: in one instance a Pan Am radio operator after transmitting a 150-word coded message to her through heavy static was heard to remark, "My God, she's got it!" One additional benefit of the flight had nothing to do with aviation, but was of importance to the Lindberghs; it gave them things to worry about, plan for, and cope with that could permit them time to recover from the brutal February, 1932, kidnap-murder of their firstborn son.

That beautiful bright red Lockheed Vega—another John K. Northrop design—was used by Amelia Earhart during two historic flights in 1932: the first solo flight by a woman across the Atlantic, and the first solo flight by a woman across the United States. In 1935 she flew a similar Vega to become the first person—man *or* woman—to fly solo from Hawaii to the mainland of the United States.

Amelia Earhart had become interested in aviation in 1918 when she witnessed an exhibition of stunt flying while serving as a Red Cross nurse's aide in Canada. In June, 1928, she had become internationally famous overnight when she was the first woman to fly across the Atlantic, but she had been only a passenger and suffered the frustration of never once during the 20 hour and 40 minute flight being permitted to touch the controls. Although the world lavishly praised her courage in having made this flight, Amelia Earhart pointed out, "The bravest thing I did was to try to drop a bag of oranges and a note on the head of an ocean liner's captain—and I missed the whole ship." Determined to prove she could make the Atlantic crossing by herself, Amelia Earhart took off on May 20, 1932, from Harbor Grace, Newfoundland, and immediately ran into bad weather. Ice accumulated on her bright red Vega's wings and at one point the weight of the ice forced her into a 3,000-foot uncontrollable descent. She did finally manage to level out when the warmer air near the ocean's surface cleared the ice, but it wasn't until 14 hours and 52 minutes of having fought heavy storms and terrible fatigue that she landed in a field in Northern Ireland, 2,026 miles from her starting point. She was, of course, celebrated upon her return but instead of basking in admiration she prepared for her second flight, the first woman's solo non-stop transcontinental hop.

Amelia Earhart, in the same Vega, took off from Los Angeles, on August 24, and 19 hours and 5 minutes later landed in Newark having covered the 2,448-mile distance at an average speed of 128½ mph. The Vega was the first airplane built by the Lockheed Aircraft Company and established its tradition for excellence in design.

In 1937, Amelia Earhart and her navigator Fred Noonan lost their lives while attempting a round-the-world flight in a twin-engined Lockheed Electra. Noonan and Miss Earhart had reached New Guinea in late June of that year and their next stop was to be Howland Island, a tiny dot 2,556 miles across the South Pacific. Noonan had reported having trouble setting his chronometers, time pieces whose accuracy is essential in determining longitudinal measurements during navigation over water. But ignoring Noonan's misgivings, Amelia Earhart and her navigator took off from New Guinea on July 2 and headed west. The *Itasca*, a United States Coast Guard vessel stationed at sea near Howland Island, received radio messages from Miss Earhart and Noonan reporting strong headwinds and heavy fuel consumption. A final fragmentary message was received indicating that the Electra was off course, lost. Then no further messages were received and the radio was silent. No

trace of the aircraft or its crew has ever been found.

The *Gossamer Condor* became the newest addition to the Pioneers of Flight gallery by achieving one of mankind's oldest dreams: sustained, maneuverable man-powered flight, and thereby won the £50,000 Kremer Prize established in 1959 by British industrialist Henry Kremer and administered by the Royal Aeronautical Society of Great Britain. The open competition's rules were clearly defined:

> The course shall be a figure of eight, embracing two turning points, which shall be not less than half a mile apart.
>
> The machine shall be flown clear of and outside each turning point.
>
> The starting line, which shall also be the finishing line, shall be between the turning points and shall be approximately at right angles to the line joining the turning points.
>
> The height, both at the start and the finish, shall be not less than ten feet above the ground; otherwise there shall be no restriction in height.
>
> The machine shall be in continuous flight over the entire course.

It was also understood that all take-offs had to be from approximately level ground and in winds not exceeding ten knots.

According to legend, of course, the first attempt at man-powered flight occurred in ancient Greece when the Athenian architect Daedalus fashioned wings out of feathers and wax so that he and his son Icarus might escape from King Minos' labyrinth on Crete. The flight ended tragically when Icarus flew too close to the sun and the wax holding his feathers melted. Still, for centuries inventors have experimented with various forms of self-propelled aircraft. Some were totally absurd, some were designs that were too impractical ever to emerge off the drawing board, but others provided major advancements in the field. The most obvious problem was the power-to-weight ratio, but others were the structural efficiency of the design and the ability of the pilot to control his craft.

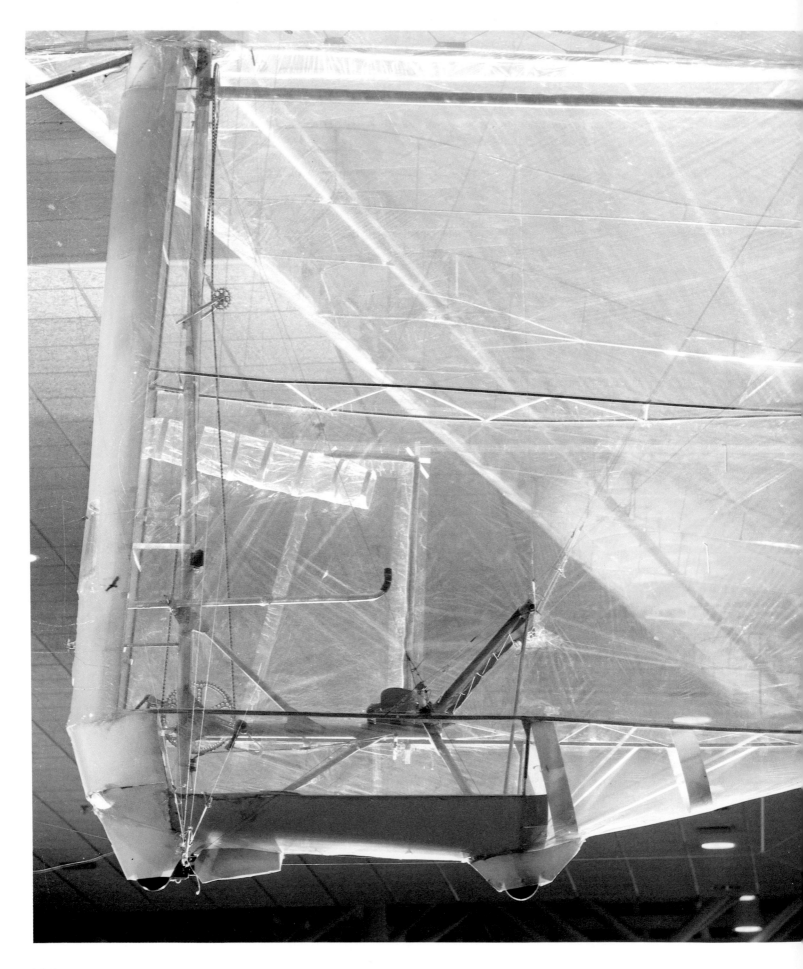

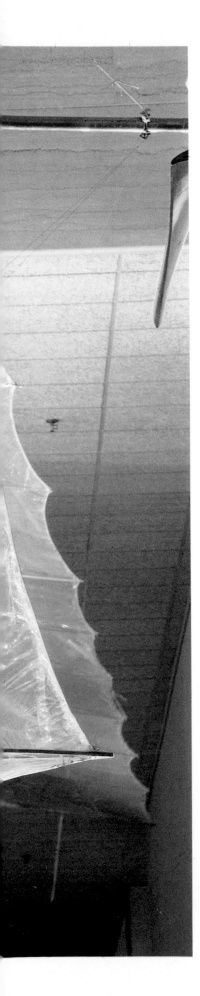

The Gossamer Condor's 96-foot wingspan and incredibly light weight of 70 pounds enabled it to become in 1977 the world's first successful sustained, maneuverable man-powered flying machine, thereby winning for its inventors a $50,000 prize.

Britain's Hawker-Siddeley engineers constructed the *Puffin*, a man-powered craft, in 1961 that the following year flew 993 yards before crashing. It was rebuilt as the *Puffin II* but crashed again and the designers abandoned the project. In 1967 a Japanese team from Nihon University constructed a series of man-powered aircraft with some success. Their *Linnett II* had a 73.17-foot wingspan, and on February 19, 1967, it flew 299 feet. The major problem with the *Linnett II* lay in the difficulty its designers had in constructing a torque shaft of sufficient strength and lightness that was still capable of driving the propeller mounted in their flying machine's tail. Portions of the ill-fated *Puffin II* re-emerged in the *Liverpuffin*, a project of Great Britain's University of Liverpool. *Liverpuffin* made a short flight in 1972. It had a wingspan of 65 feet and weighed 125 pounds. The *Liverpuffin*'s best feature was the ease with which it could be assembled and disassembled. In 1973 an English two-place man-powered aircraft called the *Toucan* flew 700 yards. In 1974 the French constructed the *Hurel Aviette*, a huge 132-foot wingspan man-powered aircraft with outrigger airfoils that flew 1,093 yards. Joseph A. Zinno of North Providence, Rhode Island, was the first American to achieve success with man-powered flight. His Zinno ZB-1 weighed 150 pounds and had a 78-foot wingspan. The ZB-1 was made of aluminum, balsa wood, and thin plastic sheeting. It was powered by bicycle pedals connected by chain to a balsa propeller located behind the cockpit. The Zinno ZB-1 flew about 26½ yards at an altitude of one foot. In 1977 the Japanese built a man-powered aircraft called the *Stork* that flew 1.3 miles in a straight line. And then that same year the *Gossamer Condor* successfully completed the world's first sustained, maneuvered man-powered flight.

Paul B. MacCready, Jr., the designer of the *Gossamer Condor*, built his 96-foot wingspan craft with thin aluminum tubes covered with plastic and braced by stainless-steel wires. The leading edges of

This Grumman Gulfhawk II, first flown in 1936, was one of the most exciting aerobatic aircraft of all time.

the wings are made with corrugated paper and styrene foam. MacCready was able to keep the weight of his machine down to an incredible 70 pounds.

At 7:30 in the morning on August 23, 1977, pilot Bryan Allen seated himself in a semi-reclining position in the "cockpit" of the *Gossamer Condor* and began pedaling with his feet. With one hand he gripped the handle that controlled both the craft's vertical and horizontal movement and with his other hand he operated the lever, located beside the seat, which was connected to the control wires that "warped" the wings. The *Gossamer Condor* rose slowly and gently off Shafter Airport in California and landed 7 minutes and 27.5 seconds later, having successfully negotiated the figure-8 course around the pylons placed one-half mile apart and the ten-foot-high hurdles at either end. The *Gossamer Condor* flew a total of 1.35 miles between take-off and landing at a speed of between 10 and 11 mph. One advantage the *Gossamer Condor* had over previous man-powered craft was the ease with which its structure could be repaired and its design modified; this enabled its designers to test the craft extensively and make alterations and repairs where needed. The *Gossamer Condor*'s flight proved that unassisted take-offs and sustained flight with turns and climbs were possible, and the designers in the field could now turn their attention to refining man-powered flight and to finding practical applications for its use.

The *Voyager* aircraft in which pilots Dick Rutan and Jeana Yeager made the first non-stop flight around the world without refueling was virtually a flying fuel tank. It had eight storage tanks on each side of the airplane and a fuel tank in the center. The 7,011.5 pounds of fuel aboard at take-off amounted to 72.3 percent of *Voyager*'s gross weight. When it landed, after 9 days, 3 minutes, and 44 seconds in the air, only 106 pounds of fuel remained.

The *Voyager* began its long take-off roll at 7:59:38 A.M. Pacific Standard Time, the tips of its long, sagging, fuel-laden wings—their span greater than that of a Boeing 727—perilously grinding along the runway. Two minutes and six seconds later, with less than 800 feet of the 15,000-foot runway remaining, *Voyager* lifted into the air and shortly thereafter pilot Dick Rutan maneuvered the aircraft to shake off both damaged wing tips. For the benefit of those who had held their breath throughout the unnerving take-off, co-pilot Jeana Yeager radioed back, "If it were easy, it would have been done before."

In order to conserve fuel, *Voyager* went most of the flight with only its rear engine providing power. Five days into the flight continuous thunderstorm activity over the continent of Africa strained the aircraft and crew. The long, thin main wing of *Voyager* was so flexible that the wing tips could deflect as much as 30 feet up and down before reaching their structural limit—but for the major portion of the flight the tips deflected upward from 3 to 5 feet in the air turbulence the vulnerable *Voyager* encountered. Rutan and Yeager were subjected to endless buffeting within the cramped, noisy confines of their narrow cockpit.

The crossing of the west coast of Africa on December 19th was an emotional moment for *Voyager*'s crew and support team; it meant that the aircraft was on the homestretch. And then, off that coast, the oil pressure suddenly fell, and it was discovered that the rear engine was running out of oil. Fortunately Yeager and Rutan were able to add lubricant and continue on before any damage to the engine occurred. The following night, over the Atlantic Ocean near Brazil, a violent thunderstorm flipped *Voyager* over onto its side in a 90° bank; Rutan was able to right the aircraft but, as he later admitted, he thought he had "lost it."

Once *Voyager* entered the Caribbean, the weather calmed down. By then, according to Dick Rutan's brother, Burt, designer of the *Voyager*, the pilots were so battered and exhausted they were incapable of accurately

The Voyager *aircraft in whose narrow, cramped, noisy cockpit pilots Dick Rutan and Jeana Yeager spent 9 days, 3 minutes, and 44 seconds in December, 1986, while completing the first non-stop flight around the world without refueling was virtually a flying fuel tank.*

relaying to him basic flight data. On December 21st, more bad weather, this time over Texas, forced *Voyager* to follow a southerly course across Central America. On December 22nd, while flying up the coast of Mexico, an electrical transfer fuel pump failed, but Rutan and Yeager were able to

bypass the problem by transferring fuel using the engine-driven mechanical pump.

And then, early in the morning on December 23rd, during the last leg of their flight, while the rear engine mechanical pump was drawing fuel from the canard tank, the strain on this small pump and an air pocket in the line starved the rear engine of fuel. *Voyager*'s rear engine stopped and the suddenly powerless aircraft began to drop toward the sea. While Rutan and Yeager struggled to make an emergency restart of the front engine, *Voyager* lost 5,000 feet of altitude. Finally the front engine burst into life and the plane fought its way back up to a safer altitude while the pilots eliminated the air block and restarted the rear engine. *Voyager* was leveled off at 3,500 feet where, at this lower altitude, the headwinds were less severe and Yeager and Rutan continued home. Later that morning, at 8:05:28 Pacific Standard Time, *Voyager* touched down at Edwards Air Force Base, its non-stop flight around the world without refueling successfully completed.

Dick Rutan and Jeana Yeager had flown around the world at speeds comparable to that at which Charles Lindbergh had crossed the Atlantic in his *Spirit of St. Louis* and done so in a cabin even more cramped and noisy than the "Lone Eagle's" Ryan NYP.

"Perhaps in some ways," editorialized the Washington *Post* the day following their landing,

> the flight of Dick Rutan and Jeana Yeager was not quite the leap into the dark that Charles Lindbergh made nearly 60 years ago when he flew solo from New York to Paris; the two who landed in California yesterday after flying around the world without refueling had the benefit of meteorological, technical and medical consultation not available to Lindbergh on his lonely flight. But in other ways, especially its length, the flight that ended yesterday was more grueling.

Two weeks later, on January 6, 1987, the two pilots squeezed back into the plane's cockpit and took off from Edwards Air Force Base for *Voyager*'s final flight to Mojave, California, and home.

BALLOONS
AND AIRSHIPS

EARLY FLIGHT

WORLD WAR I
AVIATION

GOLDEN AGE
OF FLIGHT

Balloons and Airships

Everybody *knows* hot air rises; still, visitors to the Balloons and Airships gallery seem momentarily stunned to *see* the principle in action right at the entrance, but soon they become drawn by one of the most colorful and striking exhibits in the gallery: the large facsimile of the Montgolfier balloon in which one hundred and twenty years before the Wright brothers achieved powered flight at Kill Devil Hill, two daring Frenchmen, Pilâtre de Rozier and the Marquis d'Arlandes, made the first sustained aerial flight. The Montgolfier hot-air balloon in which this ascent was made was 70 feet high and 46 feet in diameter; the balloon's covering was blue cotton cloth embroidered in gold with lions' heads, eagles, anthropomorphized suns, the ornamental devices of that era, and signs of the zodiac. The hot air that made the 1,600-pound balloon lift was furnished by the heat rising from dry straw burned on an iron grate directly beneath the open neck of the huge balloon. When Pilâtre de Rozier expressed his intention of making the first manned ascent, Louis XVI thought the plan so foolhardy and the risk so great that he proposed that two prisoners already condemned to death and not de Rozier should make that first attempt. Pilâtre de Rozier was appalled that "two vile criminals" might be granted "the first glory of rising into the sky" and when the Marquis d'Arlandes interceded with the king in de Rozier's behalf, offering as a gesture of his faith in the Montgolfier balloon's safety to accompany de Rozier on the voyage, Louis XVI relented.

The Marquis d'Arlandes, the world's first flying passenger, later wrote a friend the following account of man's first free flight in a hot-air balloon:

We went up on the 21st of November, 1783, at near two o'clock. M. Rozier on the west side of the balloon, I on the east. The machine, say the public, rose with majesty....

I was surprised at the silence and absence of movement which our departure caused among the spectators, and believed them to be astonished and perhaps awed at the strange spectacle...I was still gazing when M. Rozier cried to me—

"You are doing nothing, and the balloon is scarcely rising a fathom."

"Pardon me," I answered, as I placed a bundle of straw upon the fire and slightly stirred it. Then I turned quickly, but already we had passed out of sight of La Muette. Astonished I cast a glance toward the river. I perceived the confluence of the Oise. And naming the principal bends of the river by the places nearest them, I cried, "Passy, St. Germain, St. Denis, Sèvres!"

"If you look at the river in that fashion you will be likely to bathe in it soon," cried Rozier. "Some fire, my dear friend, some fire!"

We traveled on...dodging about the river, but not crossing it.

"The river is very difficult to cross," I remarked to my companion.

"So it seems," he answered; "but you are doing nothing. I suppose it is because you are braver than I, and don't fear a tumble."

I stirred the fire, I seized a truss of straw with my fork; I raised it and threw it in the midst of the flames. An instant afterward I felt myself lifted as it were into the heavens.

"For once we move," said I.

At the same instant I heard from the top of the balloon a sound which made me believe that it had burst. I watched, yet I saw nothing. My

A model of de Rozier's 70-foot-tall hot air Montgolfier balloon (1783).

143

companion had gone into the interior, no doubt to make some observations. As my eyes were fixed on the top of the machine I experienced a shock, and it was the only one I had yet felt. The direction of the movement was from above downward. I then said,

"What are you doing? Are you having a dance to yourself?"

"I'm not moving."

"So much the better. It is only a new current which I hope will carry us from the river," I answered.

I turned to see where we were, and found we were between the Ecole Militaire and the Invalides.

"We are getting on," said Rozier.

"Yes, we are traveling."

"Let us work, let us work," said he.

I now heard another report in the machine, which I believed was produced by the cracking of a cord. This new intimation made me carefully examine the inside of our habitation. I saw that the part that was turned toward the south was full of holes, of which some were of a considerable size.

"It must descend," I then cried.

"Why?"

"Look!" I said. At the same time I took my sponge and quietly extinguished the little fire that was burning some of the holes within my reach; but at the same moment I perceived that the bottom of the cloth was coming away from the circle which surrounded it.

"We must descend," I repeated to my companion. He looked below.

"We are upon Paris," he said.

"It does not matter," I answered. "Only look! Is there no danger? Are you holding on well?"

"Yes."

I examined from my side, and saw that we had nothing to fear. I then tried with my sponge the ropes which were within my reach. All of them held firm. Only two of the cords had broken.

I then said, "We can cross Paris."

...I then threw a bundle of straw on the fire. We rose again, and another current bore us to the left. We were now close to the ground, between two mills. As soon as we came near the earth I raised myself over the gallery, and leaning there with my two hands, I felt the balloon pressing softly against my head. I pushed it back, and leaped down to the ground. Looking around and expecting to see the balloon still distended, I was astonished to find it quite empty and flattened. On looking for Rozier I saw him in his shirt sleeves creeping from under the mass of canvas that had fallen over him. After a deal of trouble we were at last all right.

—*The Saga of Flight*, ed. by Neville Duke and Edward Lanchbery

Ten days after Pilâtre de Rozier and the Marquis d'Arlande's flight, Professor J.A.C. Charles made the first free ascension in a hydrogen balloon incorporating many of the essential features of modern balloons. Professor Charles and a companion—one of the Robert brothers who had earlier shown Charles how to construct a balloon capable of containing the hydrogen within its envelope through coating its silk covering with rubber—took off in an elaborately gilded wicker-work car suspended beneath their small candy-striped balloon from the Tuileries Gardens. Their balloon, equipped with food, extra clothing, scientific instruments, and ballast bags of sand, rose swiftly 800 feet into the air as Professor Charles and Monsieur Robert waved flags to the assembled crowds below. Professor Charles, who was able to control the ascent and descent of his balloon by dumping ballast and by opening a crude valve at the top of the balloon that permitted the hydrogen to escape, thus quickly proved the superiority of the gas or hydrogen-filled *Charlières* over the slow, fire-breathing, clumsy hot-air *Montgolfières*. In the two-hour flight the balloon was carried by the wind twenty-seven miles from Paris to Nesle where, at dusk, the two men brought themselves gently down to the ground. Monsieur Robert got out and Professor Charles decided to make one more ascent; the balloon lightened now by one less occupant, sprang up to 9,000 feet, and Charles became the first man to see the sun set twice in one day. But because of the cold and a sharp altitude-pressure pain in his ear, Professor Charles descended soon after.

The story is told that Benjamin Franklin, while visiting Paris that same year, witnessed one of the first balloon ascents. Many of those about him saw no practical value in the flights, and a French Army officer

A model of J.A.C. Charles' hydrogen-filled balloon (1783).

present with Mr. Franklin although impressed by the balloon's beauty and value as a toy asked, "But of what use is it?" Franklin, who was always quick to appreciate the potential significance of any new scientific experiment, is supposed to have replied, "Of what use is a new-born baby?"

These first balloon adventures generated the most enormous excitement—an excitement which might be difficult for us to understand today; but if one strolls through the gallery and looks at some of the snuff boxes, prints, tapestries, inlaid furniture, cabinets, china and enamels all with ballooning motifs, it may help to understand that balloons were in their time a craze comparable, say, to the Beatles.

A puppet theater in this gallery reenacts the dramatic 1785 aerial crossing of the English Channel for the first time by two of the least convivial aeronauts in history: the French pilot Jean-Pierre Blanchard and his American-born patron and passenger Dr. John Jeffries. Blanchard had become interested in flight in 1777 when he began experimenting with parachutes. In 1781 he built a "flying chariot," a man-powered ornithopter with four beating wings that, of course, did not fly. But with the success of the *Charlière* balloons he gave up heavier-than-air flight and concentrated his attention on ballooning and made several successful ascents. Dr. John Jeffries was so taken by Blanchard's triumphs that he was determined to make a flight with Blanchard across the English Channel—so determined, in fact, that as Jeffries later wrote in his *A Narrative of the Two Aerial Voyages of Dr. Jeffries*, published in London in 1786, "I agreed with M. Blanchard in consideration of my engaging to furnifh him with all the materials and labour to fill the Balloon; and to pay all the expenfes of tranfporting them to Dover." There, while waiting for more auspicious weather, Blanchard, who wanted the honor of making the first Channel crossing all to himself, made every effort to get rid of Jeffries. Blanchard secretly attached a heavy lead girdle to himself under

his coat to show, as Jeffries wrote, "from the incapacity of the Balloon, it was madnefs to attempt the experiment with two perfons, unlefs the Balloon could carry an *hundred pound* weight of *ballaft*. The pretended friends of M. Blanchard, his Countrymen, publicly circulated *fuch* reports of my having *declined* the enterprize, as occafioned my being repeatedly infulted while preparing for our experiment." The quarrel was resolved only when the Governor of Dover Castle intervened and heard Dr. Jeffries explain how he had paid all of Blanchard's London debts thereby alone making it possible for Blanchard "*to purfue this* experiment," and that Jeffries had even agreed "in case of neceffity on our paffage, *I would get out* of the Car *for his prefervation*." Furthermore, Jeffries explained to the Governor, he "was refolved to undertake the Voyage at all events, without M. Blanchard, unlefs he thought proper to accompany me, without further artifice or objection; as I was fully fatiffied of the practicability of the plan."

At one in the afternoon on January 7, 1785, Blanchard and Jeffries lifted off for the coast of France. With them in the car suspended beneath the balloon were "three facks of fand ballaft, of ten pounds each; a large parcel of pamphlets, two cork jackets, a few extra cloaths of M. Blanchard; a number of inflated bladders, with two fmall anchors or grapnels, with cords affixed, to affift our landing." Attached to the car was a *moulinet*—a hand-cranked propeller and some aerial oars. The balloon "rofe flowly and majeftically" over the edge of the Dover cliff then, as thousands watched, drifted slowly out to sea and "paffed over feveral veffels of different kinds, which faluted us with their colours, as we paffed them; and we began to overlook and have an extenfive view of the coaft of France; which enchanting views of England and France being alternately prefented to us by the rotary and femicircular motion of the Balloon and Car greatly increafed the beauty and variety of our fituation."

When Blanchard and Jeffries had reached

2,000 feet thirty minutes into their voyage, the balloon had swollen to its utmost extent and it became necessary to vent some of the gas. Twenty minutes later it became obvious they had released too much for Jeffries "found we were deſcending faſt." They cast out one and a half sacks of ballast and the balloon steadied; but with two-thirds of the Channel still to cross, their balloon began to lose altitude again, and they "were obliged to caſt out the remaining ſack and an half of ballaſt, ſacks and all; notwithstanding which, not finding that we roſe, we caſt out a parcel of pamphlets, and in a minute or two found, that we roſe again."

At two-fifteen they had to throw overboard the remaining pamphlets. At two-thirty, Jeffries "found we were again deſcending very rapidly...We immediately threw out all the little things we had with us, ſuch as biſcuits, apples, &c. and after that one of our oars or wings; but ſtill deſcending, we caſt away the other wing, and then the governail ...and unſcrewing the moulinet, I likewiſe caſt that into the ſea..." The balloon continued to lose altitude so "we cut away all the lining and ornaments, both within and on the outſide of the Car and...threw them into the ſea." Overboard next went their only bottle of cognac, the anchors and cords; "but ſtill approaching the ſea, we began to *ſtrip ourſelves*, and caſt away our cloathing, M. Blanchard firſt throwing away his *extra coat* ...after which I threw away my *only coat*; and then M. Blanchard his other coat and trowſers: We then put on...our cork jackets, and prepared for the event," a crash landing into the sea.

At last, as they neared the coast of France, their balloon again began to rise—higher in fact than it had reached before and as a result "from the loſs of our cloaths, we were almoſt benumbed with cold." The wind had increased and the two intrepid voyagers soon found themselves descending again and fast approaching a forest. Over land at last they cast away their cork life preservers but "we had now approached ſo near to the tops of the trees of the foreſt, as to diſcover that they

In 1785, Jean Pierre Blanchard and Dr. John Jeffries, two of the least convivial aeronauts in history, made the first crossing by air of the English Channel.

were very large and rough, and that we were deſcending with great velocity towards them." There was nothing left for them to throw overboard and then it suddenly occurred to Jeffries "that probably we might be able to ſupply it from within ourſelves, from the recollection that we had drunk much at breakfaſt, and not having had any evacuation...that probably an extra quantity had been ſecreted by the kidneys, which we might avail ourſelves by diſcharging. I inſtantly propoſed my idea to M. Blanchard,

147

The dirigible USS Macon dwarfs its Sparrowhawk defensive fighter preparing to begin hook-on maneuvers below.

A close-up of a Curtiss F9C-2 Sparrowhawk with the airship's trapeze gear above.

and the event fully juftified my expectation; ...however trivial or ludicrous it may feem, I have reafon to believe [the act] was of *real utility* to us, *in our then fituation*; for by cafting it away, as we were approaching fome trees of the foreft higher than the reft, it fo altered our courfe, that inftead of being forced hard againft, or into them, we paffed along near them...as enabled me to catch hold of the topmoft branches of one of them and thereby arreft the further progrefs of the Balloon." Branch by branch Jeffries and Blanchard guided the balloon to an open spot where they could let it sink to the ground. They landed with what must have been the most enormous relief about twelve miles inland from Calais and the open sea; and thus completed the first crossing of the

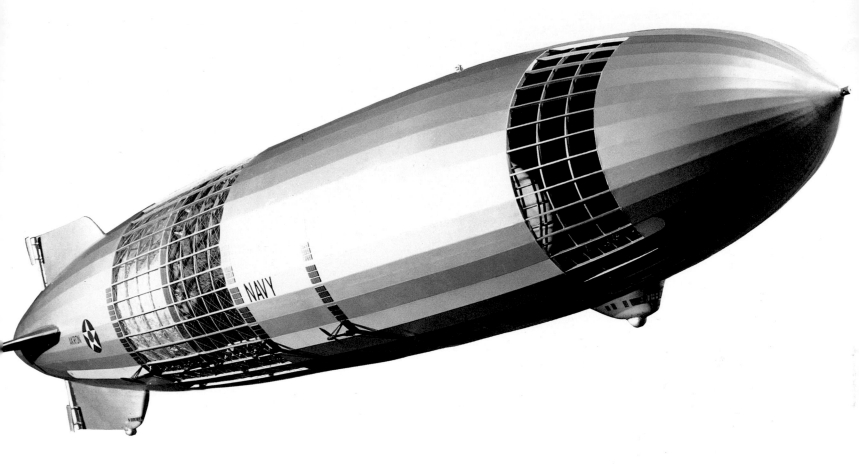

A cut-away model of a Macon-class dirigible showing its rigid frame construction.

English Channel by air. It would be almost 125 years before the Channel would be crossed by a heavier-than-air flying machine; a Frenchman would be at the controls then, too: Louis Blériot on July 25, 1909, crossed the twenty miles of open water in a monoplane of his own design in thirty-seven minutes.

As the visitor passes among the exhibits within the Balloons and Airships gallery, he begins to realize the uniqueness of lighter-than-air flight. From the free-floating *Montgolfières* and *Charlières* that led to the lovely sweep of sport ballooning now making such an extraordinary comeback, he can see balloons that are used to gather meteorological data and high-altitude research gondolas as well. A diorama depicts

the use of balloons in war from the Battle of Fleuris in 1794 to U.S. Navy blimps that made anti-submarine patrols in the Pacific during World War II. One of the most interesting of the exhibits is the tiny Sparrowhawk, a Curtiss F9C-2 scouting fighter biplane that was carried in its belly as protection by the Navy's 1930s giant Macon airship. But the exhibits that seem most to capture the gallery visitor's attention are those pertaining to the *Hindenburg*, the last passenger-carrying Zeppelin.

Toward the end of the nineteenth century enough had been learned about structures, streamlining, power plants, and control devices to construct powered balloons or airships that could be flown successfully. On October 19, 1901, Santos Dumont, a

149

Open latticework dirigible construction gave strength with minimum weight.

The Hindenburg's *place settings did not sacrifice elegance in favor of lightness.*

wealthy Brazilian heir to a coffee fortune who had emigrated to France and built the most advanced one-man, gasoline engine-powered dirigible, won the coveted Deutsch prize for having the first dirigible or airship capable of navigating from the Aéro Club at St. Cloud around the Eiffel Tower and back within thirty minutes. Although he covered the seven-mile distance in 30 minutes and 40 seconds, it was good enough to make him a hero. But it wasn't until Count von Zeppelin of Germany—an avid balloonist who recognized the enormous reconnaissance potential of dirigibles as well as the structural and weight-carrying limitations of the blimps then being built—that a sensational breakthrough in dirigible design was achieved by creating motorized airships with a rigid construction. The lifting gas bags or balloons were contained within and fastened to a rigid, aluminum-girdered external skeleton covered by fabric. The control cabin, engines, and maneuvering surfaces were distributed along and suspended from this load-carrying frame thereby not only making higher speeds possible, but eliminating as well the problems hitherto caused when winds and aerodynamic forces deformed the surfaces of the uncontained gas bags. Zeppelin's early series of rigid airships interested the German government enough for it to offer to buy one for the army if a

Zeppelin could be made that could fly from its base 435 miles to an assigned point and remain aloft for twenty-four hours. When the LZ-4 which Count von Zeppelin had constructed in an attempt to win the government contract exploded, the German people were so moved by the catastrophe that they patriotically raised one and a half million dollars for a Zeppelin fund and the Army accepted an earlier model, the LZ-3, in its stead. By 1910 aerial passenger service by Zeppelin in Germany had become a reality. Passengers comfortably seated in wicker chairs on pile rugs were served wine and cold lunches; but their travel was not without its risks. More than half of the twenty-six dirigibles built by Count von Zeppelin by 1914 had come to a violent end. Squadrons of Zeppelins bombed London during World War I, but they caused more of a nuisance than damage. British fighters would fly directly above the lethally inflammable airships and drop bombs along their lengths. After the war, passenger service was resumed.

The two most famous of the rigid airships were the *Hindenburg* and her earlier sister airship, the *Graf Zeppelin,* and it was the successful commercial exploitation of these two giant airships that kept lighter-than-air aviation alive. Passengers paid $720 apiece for a round-trip ticket between Germany and the United States and cruised in luxury at a stately 78 mph during the

Passengers on the 804-foot-long Hindenburg *paid $720 apiece for a round-trip ticket between the United States and Germany*

voyage. The *Hindenburg* made ten round trips between New York and Germany in 1936; and then on May 3, 1937, ninety-six persons boarded the Zeppelin for its first flight to New York of the 1937 season. The crossing was prolonged by headwinds and in the New York area thunderstorms delayed the landing, but finally on May 6, a little after seven P.M., the *Hindenburg* approached its mooring mast at Lakehurst, New Jersey, dropped its cables and suddenly, without warning, disaster struck. Sheets of flame enveloped the tail section; the exploding hydrogen leapt high in the air, the massive skeleton buckled, girders twisted in the raging heat; and, as horrified spectators and newscasters watched, passengers flung themselves out of the flaming airship as it crumpled like some huge, mortally wounded animal to the ground. Thirty-five of its ninety-six passengers and crew were killed, countless others were terribly burned. The spectacle of the catastrophe was so appalling that the end of the *Hindenburg* marked the end of the Zeppelin era.

The newest exhibit in the gallery is that of the *Double Eagle II.* Between 1873 and August, 1978, a number of balloons were readied in preparation for various attempts to cross the Atlantic, but only thirteen were actually launched. By 8:43 P.M. on August 11, 1978—the moment that Ben L. Abruzzo, Maxie L. Anderson, and Larry Newman in

their 11-story-high silver and black neoprene-coated, nylon cloth covered, helium-filled *Double Eagle II* balloon lifted into the dark almost windless sky over Maine's Presque Isle potato fields—five aeronauts, in all, had died attempting to be the first balloonists to cross the Atlantic.

Ben L. Abruzzo and Maxie L. Anderson's attempt in *Double Eagle* the year before had ended when a storm and reports of what Anderson called "the worst local weather I'd heard about in nearly 30 years of flying" compelled them to make a near-fatal ditching in the frigid waters off the coast of Iceland. On the first night of their second attempt, "a failure from the first attempt emerged again despite all our application to avoid it—the radios worked poorly or not at all," Abruzzo later wrote. ". . . There we were again, wishing for a long string and two tin cans. We were flying by the seat of our pants, but at least this time the weather was good."

The *Double Eagle II*'s announced plan was not just to cross the Atlantic from continent to continent, but to duplicate Lindbergh's feat by crossing from New York to Paris. As Abruzzo explained:

Unlike other balloonists who had tried it, we planned to climb to high altitude quickly and maintain it. As it turned out, for about a third of the flight we were above 15,000 feet and breathing

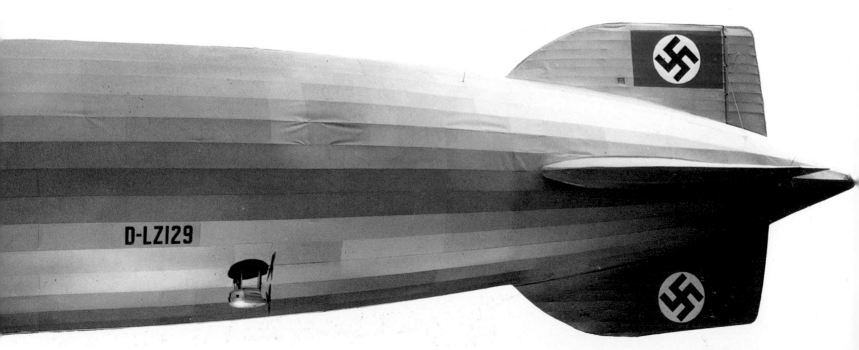

The era of the giant Zeppelins came to a catastrophic end
shortly after seven P.M. on May 6, 1937, when the Hindenburg
with 96 passengers and crew aboard exploded in flames
approaching its Lakehurst, New Jersey, mooring mast.

oxygen. At these higher altitudes we could expect stronger winds to speed us along.

Yet going to high altitude is, by itself, not enough. If you just drift with the winds, you can end up going anywhere. Our plan was to pilot the *Double Eagle II* by moving into a migratory high-pressure system of a kind fairly common in the North Atlantic between the storms of spring and autumn.

Such a high-pressure system is basically a mass of air that rotates slowly clockwise as the whole mass moves from west to east. By launching when we did . . . we climbed into the heart of the high, which was now squeezed into a ridge by two low-pressure systems, one ahead of us and one behind. If the high kept its shape and strength, we could ride to Europe with it. And, as it rotated like a giant, sluggish merry-go-round, we would follow a curving path around it and grab the brass ring at Paris.

—"'Double Eagle II' Has Landed!" by Ben L. Abruzzo with Maxie L. Anderson and Larry Newman in *National Geographic*

On *Double Eagle II*'s third day out a radio link was established through an amateur-band radio operator in England, but the balloonists had little opportunity to relax. Ninety percent of their waking hours was spent communicating, navigating, planning and executing ballast, updating their logs, and housekeeping.

At noon on their fifth day unanticipated atmospheric conditions forced *Double Eagle II* into an alarming descent. It was not just the high cirrus clouds that had moved in to screen the sun thereby cooling the balloon, but something else that, as Abruzzo later wrote, "was pushing us or sucking us right down into the very center of a perfect circle punched in the cloud deck below." Despite careful ballasting, "we continued to drop from our high of 23,500 feet, past 20,000, into the teens, past 10,000. And we kept descending." The balloon dropped 19,500 feet before bottoming out at a low of 4,000-foot altitude. Only then did the *Double Eagle II* again slowly begin to rise. That night, "The full moon shone brightly on the cloud deck below and a line of towering cumulus that was rising westward," Ben

The 11-story-high Double Eagle II *was the first balloon to successfully cross the Atlantic.*

Abruzzo wrote. "It was at such moments that I wished my family, all my friends—everybody really—could be in the gondola with me.

Not long after that lovely, silent pause, we received a transmission from air control at Shannon: '*Double Eagle II*, you are over the coast of Ireland.'" Abruzzo continued:

Strangely, perhaps, we felt no great elation. We were crushed by fatigue and, with our oxygen thrown over, plagued by headaches. Paris, our goal, was still a long, uncertain way off. . . . We had *flown* across the Atlantic, maneuvering to take advantage of favorable winds, not just drifting. Before us was the final test of our finesse as balloon pilots. . . .

That test was the notoriously rough 120-mile-long stretch across the English Channel. High over England they passed—"too high to see people on the ground," Abruzzo wrote, "[but] they were there, no

155

The gondola of Ben L. Abruzzo, Maxie L. Anderson, and Larry Newman's Double Eagle II *helium balloon.*

doubt about it. From every village and town, farm and roadside, the flash of mirrors sparkled up at us. Everybody was shining mirrors, it seemed, everybody. So were we. We signaled back down to that lush countryside sequined with flashes."

As the sun began to set, *Double Eagle II* had crossed the Channel and was floating over Le Havre. The crew was jubilant. They started gradually descending, at 100 and 200 feet per minute, and the crew searched for a place to drop their remaining heavy ballast and for a landing site. At 300 feet they dropped an empty propane tank and a battery over some plowed fields and heard

the *thump, thump* as they impacted harmlessly below. Free of that ballast *Double Eagle II* rose again to clear the town of Evreux. And then, once beyond the town, the balloon began its final descent, "with one power line to clear and a green, then a golden field to set down in. The long journey was all but ended," Abruzzo wrote.

Just before *Double Eagle II* touched down in a barley field near the town of Miserey, France, 137 hours and six minutes after it had commenced its historic leap across the Atlantic, the crew made its last radio transmission: "All aircraft in the area, *Double Eagle II* is landing."

Early Flight

At the entrance to the Museum's gallery of Early Flight a marquee proclaims:

WELCOME
TO THE
FIRST ANNUAL
AERONAUTICAL EXHIBITION
HELD
UNDER THE AUSPICES OF THE
SMITHSONIAN INSTITUTION
1913

Immediately upon entering the bunting-draped Early Flight gallery the visitor is immersed in an exhibition suggesting an aeronautical trade show of the pre-World War I era. With its fragile, antique aircraft suspended overhead, its salesbooths for the Curtiss Aeroplane Company, the American Aeroplane Supply House whose salesman is promoting a Blériot XI, and the Wright Company ("Our Product: Wright Aeroplanes"), its models of fanciful early unsuccessful aircraft designs, its displays of daring aviators from all over the world, significant aero engines, propellers, and "the latest in Flying Togs," this gallery ably captures the flavor and excitement of the first decade of manned, powered flight.

"I sometimes think," wrote Wilbur Wright in 1908, "that the desire to fly after the fashion of the birds is an ideal handed down to us by our ancestors, who, in their grueling travels across trackless lands in prehistoric times, looked enviously on the birds soaring freely through space at full speed, above all obstacles, on the infinite highway of the air."

Of course, until Wilbur and Orville Wright's epochal 12-second, 120-foot flight on December 17, 1903, to embark upon those highways of the air had been only a dream. For centuries our ancestors peopled the heavens with winged gods; they wrote and told stories of men who had tempted fate by emulating the birds, been borne aloft by gryphons, or carried to the Moon by migrating birds.

Less fanciful, but no more practical, were the grand schemes for the construction of flying machines that abounded. In the 15th century, Leonardo da Vinci, obsessed by the notion that man someday would fly, covered pages of his notebooks with sketches of wing- and paddle-powered aeronautical devices that seemingly sprang wholecloth from his fertile imagination.

But, the "Father of Aerial Navigation" and the first person to apply the methods of science to the problems of flight was Sir George Cayley, an English baronet born in 1773.

"An uninterrupted navigable ocean, that comes to the threshold of every man's door, ought not to be neglected as a source of human gratification and advantage," Cayley would write in 1816. Seventeen years earlier, in 1799, Cayley had already published, in visual form on a silver disk, his breakthrough concept of the modern airplane with separate lift, propulsion, and control systems. In 1804, Cayley had constructed the world's first model glider; 5 feet long, it featured a main plane set at a 6° angle and a moveable tail that combined

rudder and elevator in a single unit. (A small-scale model of Cayley's glider is on display in this gallery.)

By 1809–10, Cayley had published his famous "triple paper" in which he suggested an aircraft with fixed cambered or covered wings instead of the wing-flapping ornithopter devices until then universally favored. And by the time of his death in 1857, Cayley had constructed three full-scale gliders that had flown short distances with human passengers aboard.

Another model found near the entrance to this gallery is that of the *Aerial Steam Carriage.* Inspired by Sir George Cayley's example, William Samuel Henson (1805–88) designed and patented the *Aerial Steam Carriage* in 1842–43. This widely publicized high-wing, wire-braced aircraft contained an engine housed in the fuselage, separate tail-control surfaces, and screw-type propellers. With his partner John Stringfellow (1799–1883) Hensen founded the Aerial Transit Company to finance the construction of full-scale machines. Stock certificates portraying the proposed machine were sold and the thousands of lithographs that had been distributed in Europe and the U.S. to raise funds for the Aerial Transit Company venture succeeded, at least in fixing the modern configuration of the airplane in the public mind. However, the *Aerial Steam Carriage* was never built and attempts to fly even a large model were unsuccessful.

Just over 25 years later, Stringfellow entered a triplane version of the *Aerial Steam Carriage* in a competition sponsored by the Aeronautical Society of Great Britain Exposition in 1868, and although his craft—a model of which is suspended from the ceiling near the entrance to this gallery—did not receive any awards, he won a prize of £100 for the engine he designed and displayed with it. Stringfellow's steam engine was judged the lightest engine exhibited in proportion to its power; it produced one horsepower for a weight of 13 pounds. (It, too, is on display in this gallery.)

After 1860, professional engineers, inspired by Cayley and the publicity engendered by Stringfellow and Henson, began to work on the problems of flying machines. The late-19th-century experimenters attempted to construct full-scale machines with little preliminary testing of models or gliders. Sir Hiram Maxim (1840–1916), inventor of the rapid-fire gun, for example, built a huge steam-powered test bed in Kent, England, in 1893 that was designed to run back and forth on an 1,800-foot track. His enormous craft required a 3-man crew and weighed 8,000 pounds. Although in 1894 it lifted from the track, his machine was incapable of true flight. During this same period an electrical engineer in France, Clément Ader (1841–1925), developed and tested three large, powered machines, the first of which, the *Éole*, made a powered hop of 165 feet in 1890. However, since the *Éole*'s "hop" was neither sustained nor controlled, it cannot be regarded as a true flight. Other experimenters such as Russian Navy Captain Alexander Mozhaiski (1825–90) also built large machines that reportedly became airborne. In 1884, Mozhaiski's large, steam-powered machine, completed the year before, ran down a ski jump and "hopped" 65–100 feet. But, again, its flight was neither sustained nor controlled.

Meanwhile, some of the most significant aeronautical experimenters between 1860 and 1903 were working with powered models—among them Samuel P. Langley (1834–1906), the third Secretary of the Smithsonian Institution. Langley had become interested in the problem of flight in 1884. The large steam-powered "aerodromes" constructed by Langley between 1891 and 1896 were the best known flying models of the pre-Wright period. Meanwhile, in Australia, Lawrence Hargrave (1850–1915) was concentrating on kites and models. One Hargrave model, flown in 1890, had a compressed air engine powering twin ornithoptering wings on its nose and covered distances of up to 343 feet. Models of both the Hargrave ornithopter and Langley's gasoline-powered "Quarter Scale Aerodrome" hang near the entrance.

The culmination of Samuel P. Langley's work in aeronautics was in the design and construction of a full-scale, man-carrying version of the steam- and gasoline-powered aerodromes; a 52-horsepower gasoline engine driving two pusher propellers provided power for the craft which, like its smaller predecessors, was catapult-launched from the roof of a houseboat anchored in the Potomac River. The two attempts to launch the machine in October and December, 1903, ended in disastrous crashes, a full account of which can be found in the Milestones of Flight chapter.

The final foundations for the airplane were laid by the men who built and flew manned gliders in an effort to solve the problems of a successful powered aircraft.

The first American to build and fly a man-carrying glider was John Joseph Montgomery (1858–1911), who conducted tests near San Diego in 1884. Between 1884 and 1886 Montgomery constructed perhaps as many as six other gliders, but none was successful and he abandoned his aeronautical experiments until 1904 when he began flight testing the *Santa Clara*, a tandem-wing man-carrying glider designed to be launched from a balloon.

Unveiled in public flight for the first time on April 29, 1905, Montgomery's craft was steered by pilot Daniel Maloney operating foot stirrups that lowered the trailing edge of the *Santa Clara*'s wings. Montgomery's glider attracted a good deal of attention but he abandoned the design after Maloney was killed in a crash later that year and a second pilot was slightly injured in a similar craft. On October 13, 1911, Montgomery, himself, was killed in an accident during test flights in California's Evergreen Valley.

Five years earlier, the German glider pioneer Otto Lilienthal had been killed, too, testing one of his standard sailing machines similar to the 1894 Lilienthal-built *Normal-Sagelapparat* that hangs suspended from the ceiling at this gallery's entrance. Lilienthal had written, "one can get a proper insight into the practice of flying only by actual flying experiments"—and fly he did. In machines of this type, Lilienthal made glides of up to 1,150 feet. Photographs and published reports of Lilienthal's experiments fascinated the two men who were destined to achieve the final victory. Although Lilienthal's scientific data had greatly influenced the Wrights, the brothers found friendship, moral support, and a sounding board for their ideas closer to home in Octave Chanute who, by 1900, had become the most influential flying-machine experimenter in the world.

A renowned civil engineer, publicist, and a promoter of active gliding, Chanute's 1894 collection of articles on the history and theory of flight, *Progress in Flying Machines*, had helped the Wrights keep abreast with aeronautical developments (and with how little still was known). In 1897, Chanute and August M. Herring, a New York engineer, developed and flew a trussed biplane glider whose design influence can be seen in the Wrights' unpowered Flyer of 1902. Several models of Chanute's gliders are in this gallery.

Just beyond these models is the entrance to the gallery's small theater in which visitors may watch "newsreels" of early flyers: Wilbur Wright demonstrating his Flyers in Europe; a Voisin float glider; the canard-type *14-Bis*, in which Brazilian airship pilot and balloonist Alberto Santos-Dumont made the first successful heavier-

Seeking to develop a stable glider as the first step toward powered flight, German glider pilot Otto Lilienthal was the most successful and influential of the early airmen who built and flew hang gliders in the late 19th century.

than-air flight in Europe (on November 12, 1906, Santos-Dumont, standing in the cockpit, had made a wavering flight of 722 feet); Orville Wright's successful tests of the Wright 1909 Military Flyer before the military at Fort Myer; Hubert Latham's unsuccessful attempt to cross the English Channel in the lovely, delicate *Antoinette*; Louis Blériot's successful crossing; and the World's First Great Air Meet held in August, 1909, at Rheims, France.

"Sixty thousand spectators turned up every day of the week-long show," Arch Whitehouse reported in his book, *The Early Birds*, "some sitting in private boxes, others at the tables of the open-air restaurants, or crowded in the temporary stands. Many drank champagne under cool arbors, or joined the mobile throngs that gaped at the stilt-walkers, tight-rope artists, or the peep-shows. Aviation had become—in France at any rate—the world's greatest outdoor attraction."

By the summer of 1908, five years after their first flight, Wilbur Wright was in Europe preparing to show a 1907 Wright

July 1, 1909: As part of its flight tests, the Wright 1909 Military Flyer rounds a tower erected on the parade grounds of Fort Myer, Virginia.

aircraft to an audience skeptical about the controversial American brothers' claims of flight—a popular saying abroad in those days had been that the Wrights were "either flyers or liars" and in France a new word had been invented to describe them: *bluffeurs.* On Saturday, August 8, 1908, Wilbur was preparing to give his first flying exhibition on foreign soil and prove the Wrights' claims. The flight was to take place at the race track at Hunaudières—"a sandy, open terrain surrounded by pine trees near the quiet town of Le Mans about 130 miles west of Paris," Henry Serrano Villard later wrote. And it was at Hunaudières that, for several weeks, Wilbur Wright had been methodically repairing and reassembling the

1907 Flyer the Wrights had shipped to France the year before. Villard's account continues:

> . . . The impatiently awaited trial took place before a weekend throng of spectators—many surveying the scene from . . . trees, some with picnic lunches, and all keyed up in the belief at last that something momentous was going on. . . . As reported in the Paris *Herald*, the plane looked frail in relation to its bulky motor, which balked at starting on the first two or three tries. A tear on the lower wing had been noticeably repaired with a patch and glue. Wilbur Wright himself, far from dramatizing the occasion, was dressed as if for a stroll: gray suit; high starched collar; and a cap. When the weight on the catapult dropped, the plane lunged forward on its monorail "like an arrow from a cross-bow shot into the air. . . ."

THE WRIGHT
1909

Within fifty feet of the start, recounted the *Herald* next day, "the machine rose to a height of eight to ten meters, circled twice, took turns with ease at almost terrifying angles and alighted like a bird. The flying time was 1 minute, 45 seconds."

It was a stupendous vindication. No longer was the term *bluffeur* to be heard. As the voluble balloonist Surcouf put it to the members of the Aéro Club de France: "*C'est le plus grand erreur du siècle!*" ("It is the greatest error of the century!") Disbelief in the Wright claims had indeed been a colossal mistake, and the French were magnanimous in admitting it. "This is the beginning of a new phase of mechanical flight!" exclaimed Blériot. "Wright is a genius. He is the master of us all." The newspaper *Figaro* said: "It was not merely a success but a triumph. . . ."

—*Contact! The Story of the Early Birds,*
by Henry Serrano Villard

On August 20, 1908, Orville Wright, in accordance with the Army's contract specification—and partly in response to President Theodore Roosevelt's keen interest in their accomplishments—brought a 1908 Wright Flyer to Fort Myer, Virginia, to demonstrate the craft's abilities to the Army. On September 9th, he made a flight of 1 hour, 2 minutes, and 5 seconds—the first flight of over an hour. Orville had made several other impressive flights and had taken up passengers. But on September 17, 1908, carrying Army Lieutenant Thomas E. Selfridge as an official observer, Orville crashed.

Lieutenant Selfridge had been a member of Alexander Graham Bell's Aerial Experiment Association along with John A. McCurdy, Frederick W. "Casey" Baldwin, and Glenn H. Curtiss—the Wrights' adversary in their patent infringement suit instigated just two months before. Selfridge had asked Orville to take him aloft that afternoon but, as Arch Whitehouse was later to write, because of Selfridge's relationship with Curtiss:

Orville was reluctant, but decided to make this flight one that would impress the young officer. . . . This is not to imply that [Orville] made a show-off display; on the contrary he flew his biplane with conservative skill. . . . After a

beautiful take-off, Orville put the machine through three perfect circuits of the Fort Myer area, and was starting a fourth at an altitude of around 250 feet when a light tapping, or vibration, was heard. . . . The pilot thought it was coming from the chain drive, but when he glanced back he could see nothing out of the ordinary. He decided, however, to shut off the engine and make an immediate landing. He had hardly moved the controls . . . when two sharp thumps shook the machine violently, and the biplane swerved sharply to the right, indicating that something very serious was wrong. Orville then cut the engine completely, and saw that he was heading straight into a gully filled with small trees.

Instead of risking a landing there, he tried to turn to the left and get down on the parade ground. By that time the controls of the tail assembly were inoperative, so he worked on the wing-warping gear, hoping to level the machine out and fly straight on. At that point the machine suddenly . . . seemed to be heading for a nose-on crash. . . . When the machine was about 25 feet from the ground, it began to right itself, and had there been another 20 or 30 feet to spare, they might have landed safely. But there was not enough height to make a complete recovery. The biplane hit with such impact that Lieutenant Selfridge was fatally injured, dying a few hours later. . . . Orville escaped with a fractured leg and four broken ribs.

—*The Early Birds,* by Arch Whitehouse

In Europe, upon learning of Selfridge's death and his brother's injuries, Wilbur cancelled his entry in the Michelin & Commission de Aviation prize events. But once he determined that Orville was going to recover and was receiving adequate medical care, Wilbur returned to giving exhibitions—in part to fulfill a contract to demonstrate his aircraft to the Lazare-Weiller syndicate which was considering launching a European company to manufacture the Wright biplane for the European market. As a result of his demonstrations the French syndicate announced its determination to build 50 Wright airplanes in Europe—but only if Wilbur agreed to train three volunteer pilots.

It was about this time that Lord Northcliffe of the London *Daily Mail* offered a £500 prize (raised to £1,000) to anyone capable of flying across the English Channel. As incentive to Wilbur to compete, Northcliffe privately offered him a $7,500 bonus if he would try for the prize and win it. Due to his business obligations, failing weather, and his own indifference to the challenge ("a useless risk," Wright called it, and dismissed a Channel attempt adding, "nor if effected would it prove anything more than is braved by flying over dry land"), Wilbur Wright was unwilling to make the attempt; and, the following summer, the honor of making the first flight across the English Channel and the *Daily Mail*'s £1,000 prize went to France's Louis Blériot who, in his Blériot XI monoplane, powered by a 3-cylinder, 25-horsepower Anzani engine, took off from a farm near Calais at 4:35, the morning of July 25, 1909, and made the 23½-mile Channel crossing in 37 minutes.

Blériot's own account of his feat appeared in the following day's *Daily Mail*:

At 4:30 we could see all around. Daylight had come. [Blériot's business associate] M. Le Blanc endeavored to see the coast of England, but could not. A light breeze from the southwest was blowing. The air was clear. . . . I was dressed as I am at this moment, a khaki jacket lined with wool for warmth over my tweed clothes and beneath my engineer's suit of blue overalls. My close-fitting cap was fastened over my head and ears. I had neither eaten nor drunk anything since I rose. My thoughts were only upon the flight and my determination to accomplish it this morning.

4:35! All set! Le Blanc gives the signal and in an instant I am in the air, my engine making 1,200 revolutions—almost its highest speed—in order that I may get quickly over the telegraph wires along the edge of the cliff. As soon as I am over the cliff, I reduce my speed. There is now no need to force my engine.

I begin my flight, steady and sure, toward the coast of England. I have no apprehensions, no sensations, not at all. The *Escopette* [a fast French destroyer ordered to escort any attempting aircraft across the Channel] has seen me. She is driving ahead at full speed. She makes perhaps 42 kilometers (about 26 miles) an hour. What matters? I am making at least 68 kilometers (42½ miles).

Rapidly I overtake her, traveling at a height of 80 meters (about 250 feet). The moment is supreme, yet I surprise myself by feeling no exultation. Below me is the sea, the surface disturbed by the wind, which is now freshening. The motion of the waves beneath me is not pleasant. I drive on. Ten minutes have gone. I have passed the destroyer and I turn my head to see whether I am proceeding in the right direction. I am amazed. There is nothing to be seen, neither the torpedo destroyer, nor France, nor England. I am alone. I can see nothing at all—not a thing! For ten minutes I am lost. It is a strange position to be alone, unguided, without compass, in the air over the middle of the Channel.

I touch nothing. I let the airplane take its course. I care not whither it goes. For ten minutes I continue, neither rising nor falling, nor turning. And then twenty minutes after I have left the French coast, I see the green cliffs of Dover, the Castle, and away to the west the spot where I had intended to land. What can I do? It is evident that the wind has taken me out of my course. . . .

Now it is time to attend to the steering. I press the lever with my foot and turn easily toward the west, reversing the direction in which I am traveling. Now indeed I am in difficulties, for the wind here by the cliffs is much stronger, and my speed is reduced as I fight against it. Yet my beautiful airplane responds. Still steadily I fly westward hoping to cross the harbor and reach the Shakespeare cliff. . . . I see an opening in the cliff. Although I am confident that I can continue for an hour and a half, that I might indeed return to Calais, I cannot resist the opportunity to make a landing upon this green spot. Once more I turn my airplane, and describing a half circle, I enter the opening and find myself again over dry land. Avoiding the red buildings on my right, I attempt a landing; but the wind catches me and whirls me around two or three times. At once I stop my motor, and instantly the machine falls straight upon the land from a height of 20 meters (65 feet). In two or three seconds I am safe upon your shore. Soldiers in khaki run up, and a policeman. They kiss my cheeks. The conclusion of my flight overwhelms me.

—"Cross-Channel Flight" by Louis Blériot
in *The Saga of Flight*

No single flight attempted up to that point had so immediate and striking an impact as Blériot's crossing of the English Channel. Millions around the globe became convinced that aviation was a rapidly maturing

166

Swiss exhibition pilot John Domenjoz's 1914 Blériot XI.

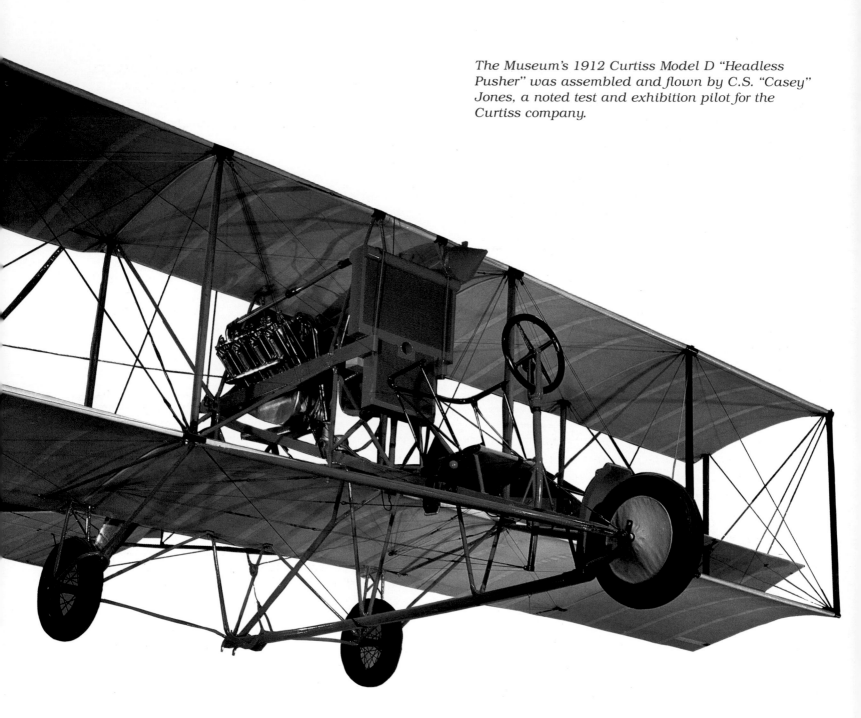

technology capable of overcoming barriers and altering mankind's notions of distance and speed. As Lord Northcliffe's *Daily Mail* editorialized:

> The British people have hitherto dwelt secure in their islands because they have attained at the price of terrible struggles and of immense sacrifices the supremacy of the sea. But locomotion is now being transferred to an element where Dreadnoughts are useless and sea power no shield against attack. . . .

On display in the Early Flight gallery is a Blériot XI similar to the one in which Louis Blériot made his Channel crossing. It was acquired by the Museum in 1950 and is an example of one of the most sophisticated of the Blériot XI designs. The Museum's Blériot was flown in Europe by John Domenjoz, a Swiss exhibition pilot of that period, who received this aircraft in July, 1914. By then nearly 900 Blériot XIs had been built.

The Wright 1909 Military Flyer, the

world's first military aircraft, was developed for the use of the U.S. Army Signal Corps and is often referred to as "Signal Corps No. 1." This airplane was constructed in 1909 in response to a specification for a 2-seat observation aircraft, and was flown by Orville Wright at Fort Myer, Virginia, that July before many dignitaries, including then-President William Howard Taft. On July 27, Orville set an endurance record of 1 hour, 12 minutes, and 40 seconds. And three days later, on July 30th, accompanied by Lt. Benjamin D. Foulois, he set an average speed record over a ten-mile course of 42.5 miles per hour. As a result of these successful trials (which were the extension of the 1908 trials), the Signal Corps purchased this aircraft from the Wrights for $30,000.

Like the Wright brothers, Glenn H. Curtiss was a bicycle builder; and, like Orville Wright, he also raced them. From bicycles he moved on to motorcycles and, in 1907, set a speed record of 136.3 miles per hour at Daytona, Florida, on a motorcycle he had developed.

During this period, at the request of Alexander Graham Bell, Curtiss was also designing and building lightweight internal combustion aero engines. Curtiss was given the title "Director of Experiments" for the Aerial Experiment Association (AEA) founded by Bell in 1907. The AEA's goal was to build a successful man-carrying airplane.

The first AEA airplane to fly was the *Red Wing* (called such because it was covered with red silk). Powered by a Curtiss-designed 40-hp V-8 engine and piloted by "Casey" Baldwin, this aircraft completed a brief flight at Lake Keuka, New York, on March 12, 1908. The AEA's second aircraft, the *White Wing*, used the same Curtiss engine. This machine first flew on May 18, 1908, and Curtiss, himself, completed his first flight in the *White Wing* on May 21st. In a subsequent accident this plane was badly damaged and the AEA turned to Glenn Curtiss for their fourth aircraft: the *June Bug*.

The *June Bug* first flew at Hammondsport, New York, on June 21st, and two weeks later, on July 4, 1908, Curtiss won the Scientific American Trophy for the first official public flight of one kilometer in a straight line:

> . . . On the Fourth of July, 1908, the June Bug made its first bid to win the trophy. With flags, fireworks, and all-day picnics to celebrate the occasion, the citizens of Hammondsport were joined by a small band of enthusiasts from the Aero Club of America. . . . They were accompanied by a similar group from Washington. . . . Thus there were witnesses aplenty to make the flight official; even the trees held spectators. None, however, were more excited than the founders of the Aerial Experiment Association. The June Bug's engine had been tuned up with meticulous care, and an attempt had been made to render the wing surfaces airtight by coating the silken fabric with varnish—an early example of the use of aircraft "dope." If anything was likely to defeat the effort, it would be the weather. . . .
>
> At five o'clock on the morning of that Glorious Fourth there were clouds and wind; at noon it was raining; but late in the afternoon conditions began to improve. About seven o'clock, in gathering dusk and against a dramatic background of green-clad hills and inky black clouds, Glenn Curtiss handily captured the trophy. With a perfectly functioning motor supporting him, he flew past the flag marking the end of the course and came down in a meadow more than a mile from the starting point. Only a forbidding row of trees and the frightening specter of making a turn in the air prevented him from going farther. The speed of this flight—the first to be officially recorded in America—was 39 m/hr.
>
> —Contact! The Story of the Early Birds, by Henry Serrano Villard

Although Curtiss' flight in the *June Bug* was the first *public* demonstration of flight in America, the Wright brothers had, of course, flown farther and earlier. Shortly thereafter, Curtiss left the AEA to develop his own aircraft, among them the famous Curtiss Model D III. This Model D was known as the "Headless Pusher" because, unlike the earlier Curtiss machines, it had no forward elevator surface. It was a highly maneuverable biplane and was powered by a

Lincoln Beachey, famous for his chilling aerial exhibitions, raced his modified Curtiss pusher "Little Looper" against automobiles, flew it under a suspension bridge near Niagara Falls, and, as he expected, died young.

variety of Curtiss engines ranging from 40 horsepower for training to twice that for exhibition flights. The aircraft could be readily dismantled and reassembled and was flown by several famous early airmen including Eugene Ely (later the first man to land on and take off from the deck of a Naval vessel), and AEA alumni John McCurdy and Glenn Curtiss in addition to "Casey" Baldwin. But perhaps the most famous of the Curtiss exhibition team was Lincoln Beachey who flew a specially reinforced and strengthened Model D known as his "Little Looper" and earned a reputation as the greatest stunt pilot in the United States, if not in the world.

"Short, clean shaven, a jaunty young man who disdained special flying clothes and flew usually in a business suit with cap turned backward," wrote aviation historian Henry Serrano Villard,

> . . . Beachey seemed a lonely and fearsomely exposed figure in the little Curtiss. Hands clasping the steering wheel, feet clamped behind the front wheel of his tricycle landing gear, he descended from the heavens in the "death dives" that were his speciality. . . .
> Beachey was widely known for a circus act that pitted his biplane against the 300-hp racing automobile of the begoggled, cigar-chewing Barney Oldfield for the "Championship of the Universe"; the aeroplane generally won. He gathered more laurels on June 27, 1911—flying over Niagara's spectacular Horseshoe Falls, under the steel arch of the International Bridge, and down the gorge in six minutes of supreme suspense for a hundred fifty thousand spectators. . . .
>
> —*Contact! The Story of the Early Birds*

The Lilienthal Glider, the Wright 1909 Military Flyer, the Blériot XI, and the Curtiss Model D "Headless Pusher" are all famous aircraft, or the creations of famous aeronautical figures. The one remaining large aircraft hanging in this gallery is—although not so recognizable—arguably as historically important. It is the Ecker Flying Boat—a copy of the well-known Curtiss flying boat of that time. And since this

Herman A. Ecker fitted his 42-foot wingspan, 26-foot-long, homemade aircraft with a boat hull and, beginning in 1913, gave flying exhibitions all along America's East Coast and in upper New York State.

gallery reproduces a 1913 trade show, it includes this example of the sort of aircraft many of that era's Americans were building in their own back yards.

The Museum's Ecker Flying Boat is one of the oldest flying boats in existence, and the only flying boat in the building. It was built in 1912 by Herman A. Ecker who, having first learned to fly, next fashioned this aircraft out of supplies obtained through manufacturers' catalogues, hardware stores and lumber yards and then set about earning money by giving demonstrations. Though Ecker, himself, was not a famous, pioneering individual, he does represent that significant segment of early American aviators who built aircraft based upon the sort of information and supplies readily obtainable from manufacturers whose displays would be on view at aeronautical exhibitions such as the one the Museum has created here.

In the early years of aviation prior to World War I, it was France, not the United States, that was the leading nation in both military and civilian aviation. Without financial encouragement to create new flying machines, and with Congress' seeming indifference to the airplane's military potential, the lag in America's development of aircraft began the moment Europe in general and France in particular witnessed the extraordinary advances the Wrights had made with their Flyers and learned how the Wrights had mastered lateral control. Four years before Wilbur Wright's 1909 European exhibitions, nine nations—France, England, the United States, Spain, Belgium, Germany, Italy, Switzerland, and Sweden— had banded together to establish the Fédération Aéronautique Internationale to regulate record and competition flights and to establish guidelines for the world aviation community. And in 1912, three years after Wilbur Wright's European tour, there were according to *Jane's All the World Aircraft* 29 nations in which aircraft were being built. By then the Wrights were so enmeshed in their patent suits (first against Glenn Curtiss, and later against the French aviator Louis Paulhan, the English aviator Claude Graham-White, the Anglo-French aviator Henri Farnum, and nearly anyone else involved in flying for profit) that American leadership in aviation was fading.

It is only fitting, therefore, that as visitors to the Early Flight gallery complete their tour and pass the exhibit of aero engines showing the development of motors designed to produce the maximum amount of horsepower with the minimum amount of weight (those lighter, more powerful motors that had made possible the enormous advances in aircraft capabilities between 1903 and 1913), they also pass the exhibits devoted to aeronautics in nations other than the United States. For by 1911 European aircraft were superior to American craft in both performance and design.

The world's first licensed woman pilot was Raymonde de la Roche who had earned her license in France on March 8, 1910. The first American woman to be licensed was the lovely Harriet Quimby who passed the test on August 1, 1911. As part of the Moisant International Aviators, an exhibition team, Quimby quickly became famous as an excellent and daring pilot although her plum-colored satin flying suit did nothing to detract from her public visibility. On April 16, 1912, Harriet Quimby became the first woman to fly an airplane across the English Channel. Less than three months later she was dead from a tragic, and still unexplained, accident while giving an exhibition flight over Boston Harbor. By the beginning of World War I, eleven women in the United States had earned their pilot's license and countless others, in those preregulatory days, were flying without one.

The Early Flight gallery charts the progress of aviation from dream to reality; but visitors to this gallery, standing amid the potted plants, the bunting-draped balcony with its ragtime band, and the rich, decorative touches of a bygone era, may find it difficult to believe that they, themselves, are not dreaming and have not stepped back to a time when the airplane was still very new and young and every flight was a grand adventure.

World War I
Aviation

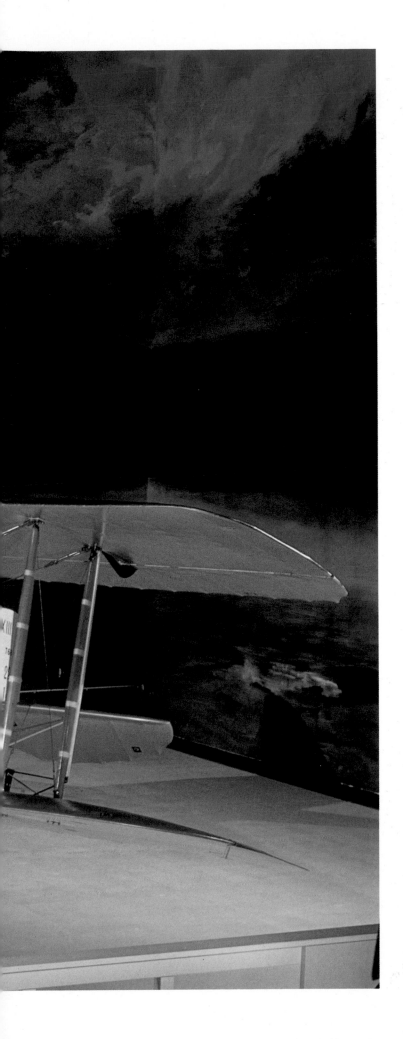

In 1914, when World War I began, the airplane was still a crude, powered "box kite" of dubious structural strength and reliability. Although there were aircraft that could fly 125 miles per hour, or travel non-stop over hundreds of miles, or attain an altitude of 19,800 feet, no one machine could accomplish all of these feats, and an aircraft capable of high performance in one area was usually too specialized to perform well in another.

Both sides initially used the aircraft for observation. A slow, stable aerial platform was needed from which an observer or pilot could study or photograph enemy movements, troop disposition, and activities on the ground. "When we started the First World War," recalled T.O.M. Sopwith, the British aircraft designer and builder, "there were no fighters. The small, rather high-performance—for their day—aircraft that we were building were really built as scouts. From scouts they developed into fighters, literally—from going up with rifles and revolvers to the day when we learned to fire through the propeller." The pressure the outbreak of the war created upon the

One of the best fighters of World War I was the SPAD *XIII, an improved version of the successful* SPAD *VII. A total of 8,472* SPAD *XIIIs were built, including 893 for the American Expeditionary Forces.*

An Albatros D.Va, one of two remaining in the world, sits in front of the Museum's recreation of a World War I forward airfield's primitive hangar tent.

designers of aircraft to come up with new machines of advanced design and greater capabilities was enormous. "Development was so fast!" Sopwith commented. "We literally thought of and designed and flew the airplanes in a space of about six or eight weeks. . . . From sketches the designers went to chalk on the wall. Until about the middle of the war there was no stressing at all. Everything was built entirely by eye. That's why there were so many structural failures. We didn't start to stress airplanes at all seriously until the [Sopwith] Camel in 1917." The rapid, alternating, and increasingly sophisticated technological advancements made by both sides in aerial tactics, weaponry, aircraft performance and design were brought about by the opposing forces' determination to deny each other the advantages of aerial observation. Therefore, by 1917 the crude "box kite" had developed into a relatively reliable aerobatic weapon capable of shooting down its opponents.

The photographs accompanying this text depict the gallery as it looked in 1976. In 1986 it was discovered that insects, inadvertently introduced into the gallery through the wooden and canvas set, were attacking the uniforms on exhibit and threatening the canvas-covered aircraft. The entire exhibit had to be dismantled and the gallery fumigated. A newly designed World War I gallery will be installed in 1990.

The Museum's aim in creating this exhibit was to reconstruct from available records and interviews the environment of the actual Allied advance airstrip near Verdun, France, at which, just two days before the signing of the Armistice that ended World War I, American pilots captured Lt. Freiherr Heinz von Beaulieu-Marconnay's brand-new Fokker D.VII, the "U-10."

The sky was sodden, patchy with fog. Rain was falling that cold November 9, 1918, morning, as it had been ever since the detachment from First Flight of the First Pursuit Group's 95th Aero Squadron had been ordered to the small forward airfield just east of the ruins of Verdun. Although the

French army continued to block the German advance on Verdun, the city's buildings had been shattered and burned by years of German artillery fire. The Meuse-Argonne-St. Mihiel offensive was in its final days. To the east of a French infantry division and elements of the American 81st Division were engaging the Germans near Moranville. Far to the north and west Pershing's army had pushed the enemy back to the Meuse line and the outskirts of Sedan. The German army was in general retreat all along the western front; a revolution was threatening in Germany itself. The detachment of First Flight had been ordered up to the advance airfield to destroy German observation balloons and any low-flying German observation aircraft attempting to sneak across the lines should a break in the weather occur. Because of the fog, heavy sporadic rains, and overcast skies of the past few days, however, German air activity had almost completely ceased.

Living conditions at the advance base were primitive. The pilots were housed in two shell-pocked, aged buildings; the unit's SPADs were tucked inside two hangars built and abandoned long before by the French. German artillery and anti-aircraft guns were dug in just behind a ridge to the east and the northeast of the airfield. The base had been hit a number of times in previous months by German artillery fired not at the airfield but at the huge 16-inch U.S. Naval guns five hundred yards south of the rudimentary runway. These naval guns were mounted on railroad trucks and wheeled out on spurs of track just to the south and west of the field when they were to be fired. Captain Alexander H. McLanahan, A.S., U.S.A., who was in command of the First Flight detachment recalls that the

two American Naval guns went off at regular intervals of 20 minutes day and night. These tremendous pieces must be heard to be appreciated. They caused such terrific displacement of the air when they exploded that our ships were gradually being shaken to pieces. It was impossible to keep the wings in line or the wires tight. Neither could the pilots sleep with

The captured German Fokker D.VII is visible through a gaping shell-hole in the pilots' building.

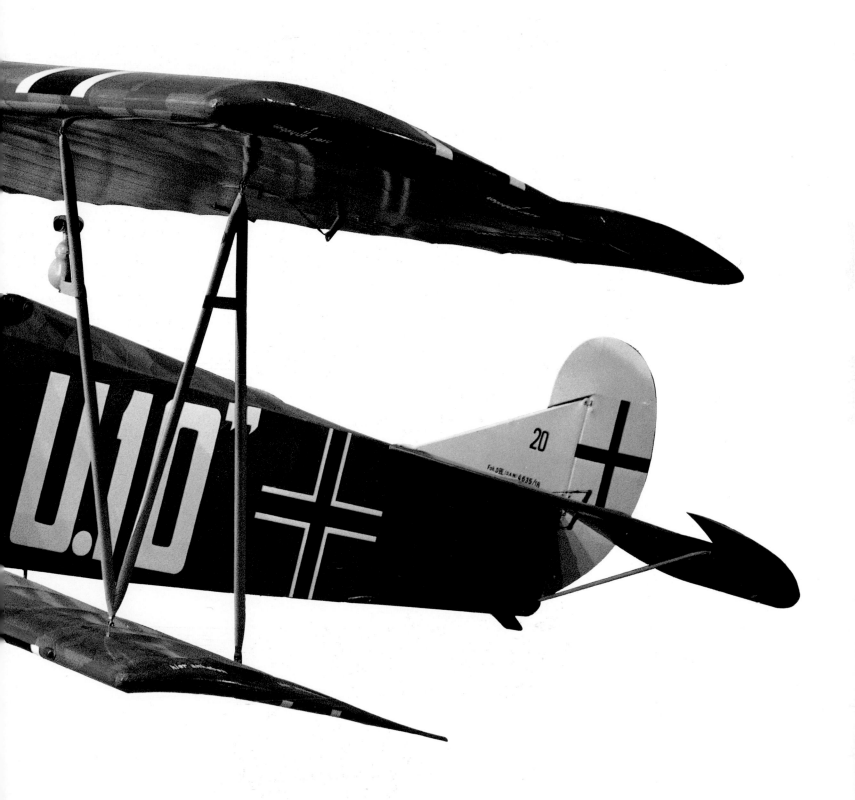

The Fokker D.VII was so respected by World War I Allied airmen that its surrender was specified in the Armistice agreements.

continual explosions which almost bounced them from their beds and shook their very teeth out. Plates of food jumped all over the table unless held down and packs of cards flew about like swallows every time the guns went off. To add to it all the German artillery was constantly trying to land shells on the American battery, many of which fell short here and there about the airdrome and barracks. Despite all, we carried on until we began to fall asleep while walking about.

Due to the sporadic artillery bombardment and continuing bad weather no flight operations had been scheduled that morning and the detachment's SPAD XIII aircraft had been withdrawn out of sight within the hangars. Detachment Commander McLanahan, Flight Commander Lt. Edward P. Curtiss, and Lt. Sumner Sewall were playing cards in their barracks when, according to McLanahan's recollection,

we suddenly heard a strange sounding plane— not like any of ours. Upon looking out the window we saw this Fokker flying very low and circling the field. We grabbed our revolvers—as in my case with the playing cards still in my left hand—and we dashed out and waved to the plane to come down. To our astonishment he did, and we were able to surround him. The three of us with our revolvers had captured him before he could set fire to his plane.

Upon questioning the German pilot, he said he was flying this brand-new Fokker from the rear to the Metz airdrome but apparently got lost. He said he had been flying on the British front and was not familiar with our territory....When the pilot found he was captured he appeared to turn philosophical and said that after all the war would soon be over.

The German pilot, Lt. Freiherr Heinz von Beaulieu-Marconnay, of the 65th Jagdstaffel was, Lt. Sumner Sewall later recalled, "welcomed with an appropriate shot of Cognac and good fellowship, after which we turned him over to the nearby artillery outfit for further processing."

Over the years since the Fokker D.VII became the possession of the Smithsonian, speculation has centered upon why von Beaulieu-Marconnay gave up so readily and what the possible significance might have

been of the "U-10" painted in such large white block letters on both sides of the Fokker's fuselage between the cockpit and the Roman cross and over the center section of the upper surface of the top wing. The answer to the first question appears to be that the German pilot might easily have mistaken the advance airdrome for one behind the German lines; von Beaulieu-Marconnay was admittedly lost, no Allied aircraft were visible on the field, and the German pilot, once down on the ground and having recognized his mistake, seemed depressed and did not appear to care where he had landed, according to Detachment Commander McLanahan and Lt. Sewall. It was subsequently learned that the pilot's older brother, Lt. Freiherr Oliver von Beaulieu-Marconnay (winner of Germany's highest decoration, *Pour le Mérite*...for his twenty-five aerial victories) had been shot down in combat that October 18th and had died of his wounds on the 30th, just twelve days before Freiherr Heinz von Beaulieu-Marconnay dropped out of that gray November sky to land at the 95th Aero Squadron's advance airbase.

The mystery of the enigmatic "U-10" was solved when a careful search into von Beaulieu-Marconnay's military records turned up the fact that prior to his service as a pilot he had served with one of Germany's most prestigious cavalry regiments: the 10th Uhlans.

The Fokker D.VII flown by von Beaulieu-Marconnay was one of the best single-seat fighter aircraft produced during the war and was said to make good pilots out of poor ones. Although the D.VII was not as fast as the SPAD or the S.E.5, it was responsive and easy to control all the way to its ceiling. Its ability to hold a steep climbing angle without stalling made it especially dangerous in the favored attacking angle of coming in from the rear and below. And when the D.VII did stall, its nose dipped forward without the plane falling off into a spin. Since the D.VII could attain and maintain a higher altitude than most Allied

machines it was able to drop down on unsuspecting aircraft with lethal results. The Fokker D.VII was so feared and respected by Allied airmen that the surrender of all this specific model of German aircraft was incorporated into the Armistice terms.

The two Allied aircraft in this gallery are both Spads—an acronym for *Société pour Appareils Deperdussin*, or, in rough translation, the Deperdussin Aeroplanes Society. Armand Deperdussin had been the founder, but he had resigned as head of the firm before the war and been replaced by Louis Blériot, the famous French aviator. Blériot had changed the name of the firm to *Société pour Aviation et ses Dérives* [Society for Aviation and its Derivatives] so its initials remained the same. The SPAD was probably the most famous aircraft of World War I and the SPAD VII, hanging upside down in the gallery as though in inverted flight, was considered a very "hot" machine. They were noted for their high rate of climb and rugged construction which enabled them to dive at high speeds without losing their wings—a not uncommon accident with other aircraft. Although the SPAD VII was a favorite with good pilots, it was a very tricky plane to handle at low speeds because of its very steep gliding angle with

General "Billy" Mitchell's two-seater SPAD XVI was filled with Scarff ring-mounted twin Lewis machine guns—but this model SPAD was never popular with pilots. The osprey insignia was Mitchell's own.

power-off, a rather technical way of saying that without its engine it dropped like a brick. Pilots landing them had to "fly them onto the ground," with their power on. Many fatalities occurred when the engine quit just after take-off or on landing approaches. The Museum's specimen carries the Indian head insignia of the famed Escadrille Lafayette, the unit with which American pilots first saw combat during the war. James Norman Hall, who with Charles Nordhoff was the author of *Mutiny on the Bounty*, served with the Escadrille. Hall, again with Nordhoff, wrote one of the best books to come out of that experience. Here in this excerpt from his *Falcons of France* he tells of his first impressions of aerial combat:

> While in training in the schools I had often tried to imagine what my first air battle would be like. I haven't a very fertile imagination, and in my mental picture of such a battle I had seen planes approaching one another more or less deliberately, their guns spitting fire, then turning to spit again. That, in fact, is what happens, except that the approach is anything but deliberate once the engagement starts. But where I had been chiefly mistaken was in thinking of them fighting at a considerable distance from each other—two, or three, or even five hundred yards. The reality was far different. At the instant when I found myself surrounded by planes, I heard unmistakably the crackle of machine-gun fire. It is curious how different this sounds in the air when one's ears are deafened by altitude, the rush of wind, and the roar of the motor. Even when quite close it is only a faint crackle, but very distinct, each explosion impinging sharply on the eardrums. I turned my head over my shoulder, to breathe the acrid smoke of tracer bullets, and just then—whang! crash!—my wind shield was shattered. I made a steep bank in time to see the black crosses of a silver-bellied Albatros turned up horizontally about twenty yards distant, as though the German pilot merely wanted to display them to convince me that he was really a German. Then, as I leveled off, glancing hastily to my right, I saw not ten metres below my altitude and flying in the same direction a craft that looked enormous, larger than three of mine. She had staggered wings, and there was no doubt about the insignia on her fishlike tail: that too was a black cross. It was a two-seater, and so close that I could clearly see the pilot and the gunner in the back seat. Body and wings were camouflaged, not in daubs after the French fashion, but in zigzag lines of brown and green. The observer, whose back was toward me, was aiming two guns mounted on a single swivel on the circular tract surrounding his cockpit. He crouched down, firing at a steep angle at someone overhead whom I could not see, his tracers stabbing through the air in thin clear lines. Apparently neither the pilot nor the rear gunner saw me. Then I had a blurred glimpse of the tri-color *cocardes* of a Spad that passed me like a flash, going in the opposite direction; and in that same instant I saw another Spad appear directly under the two-seater, nose up vertically, and seem to hang there as though suspended by an invisible wire.

> What then happened is beyond the power of any words of mine to describe. A sheet of intense flame shot up from the two-seater, lapping like water around the wings and blown back along the body of the plane. The observer dropped his guns and I could all but see the expression of horror on his face as he turned. He ducked for a second with his arms around his head in an effort to protect himself; then without a moment's hesitation he climbed on his seat and threw himself off into space. The huge plane veered up on one side, turned nose down, and disappeared beneath me. Five seconds later I was alone. There wasn't another plane to be seen.

—*Falcons of France*, by James Norman Hall
and Charles Nordhoff

No Allied pilots and, until near the end of the war, very few Germans carried parachutes. The pack parachute did exist; it had been invented by an American showman before the war, but only spotters in observation balloons were equipped with them. The reasons for not giving parachutes to pilots were both cruel and stupid. The excuses were that parachutes were not reliable enough to justify mass production of them and that if a pilot wore a parachute he might be tempted to use it instead of committing himself to the fight. Hall's account of the observer leaping to his death was not an uncommon occurrence. Most pilots were more afraid of fire than bullets and would jump from their burning aircraft, choosing the quick certainty of that death to slowly burning alive.

One of the most famous airmen of World War I was Manfred von Richthofen, the Red Baron, Germany's leading ace who shot down 80 Allied planes. In this excerpt from his autobiography he describes his battle with a British aircraft in April, 1917:

Suddenly one of the impertinent Englishmen tried to drop down upon me. I allowed him to approach me quite near, and then we started a merry quadrille. Sometimes my opponent flew on his back and sometimes he did other tricks. He was flying a two-seater fighter. I realized very soon that I was his master and that he could not escape me.

During an interval in the fighting I assured myself that we were alone. It followed that the victory would belong to him who was calmest, who shot best, and who had the cleverest brain in a moment of danger. Soon I had got him beneath me without having seriously hurt him with my gun. We were at least two kilometers from the front. I thought he intended to land, but there I had made a mistake. Suddenly, when he was only a few yards above the ground, I noticed how he once more went off on a straight course. He tried to escape me. That was too bad.

I attacked him again, and to do so I had to go so low that I was afraid of touching the roofs of the houses in the village beneath me. The Englishman defended himself up to the last moment. At the very end I felt that my engine had been hit. Still I did not let go. He had to fall. He flew at full speed right into a block of houses.

There is little left to be said. This was once more a case of splendid daring. The man had defended himself to the last. However, in my opinion he showed, after all, more stupid foolhardiness than courage. It was again one of the cases where one must differentiate between energy and idiocy. He had to come down in any case, but he paid for his stupidity with his life.

—Captain Manfred Freiherr von Richthofen, trans. by T. Ellis Barker

Not all fighter pilots were as cold-blooded as von Richthofen. Gill Robb Wilson, who flew with both the French and the American air service during World War I recalled the following incident which indicates the occasional startling intimacy of aerial combat:

I was out one day flying as gunner with a friend of mine named Jean Henin. He was kind of a clown, but a very nice boy; he'd been a bank clerk in Paris. We got into a dog fight over the Oise canal. I was firing to the rear. Suddenly I felt Henin beating me on the back, and I could hear him yelling, "Here, over here! Fire, fire!"

I turned around—I had to swing the Lewis guns—and there with his wings locked with ours was a German fighter in a Pfalz. I looked right down this German's throat. He was a man, I would think about forty-five or fifty years old. He had on a black woolen helmet. How in God's name he ever got there I don't know. I could touch the end of his wing.

I swung the guns on him, and the guy just sat there. He had a mustache. I can still see him. There were deep lines on his face, and he looked at me with a kind of resignation. I looked over those machine guns at that guy, and I couldn't kill him. He was too helpless.

Gradually, he drifted off. I said to Henin, "I couldn't kill him." He said, "I'm glad you didn't."

—Oral History Collection of Columbia University

Legends have arisen over the accomplishments of some of the World War I aces with the passage of time, but most historians agree that the role played by air power during this struggle was more romantic than decisive. The most important missions the airplane carried out were reconnaissance and artillery fire control. And when the war bogged down into static frontline trench warfare and increasingly effective camouflage techniques evolved, aerial reconnaissance became less and less significant. No airplane or Zeppelin sank or even seriously disabled any major naval vessel. No war industry was halted by strategic bombing. No major battle's outcome was decided by either control of the air or lack of its control. And so, even though vast technical progress was made in aviation development during World War I, what one celebrates are the men and not so much the machines. Men like Baron Manfred von Richthofen, a brilliant organizer and tactical leader; René Fonck, the leading French ace with 75 victories who would later fail in his attempt to beat Lindbergh across the Atlantic; Britain's Albert Ball, 44 victories, who planted vegetables and raised rabbits around the airfield and who reassured his parents in writing that he was still saying his prayers. Ball died before he reached twenty; one day he simply disappeared. So did Georges Guynemeyer, France's second leading ace. And there was Willy Coppens, the Belgian ace who, incredible as it may seem, actually rolled his wheels on the top of a German observation balloon. Edward Mannock, Britain's top ace with 73 victories, was killed by a German infantryman's rifle bullet. Billy Bishop of Canada survived the war with 72 victories and the Victoria Cross for singlehandedly attacking a German airfield. Eddie Rickenbacker, a former racing car driver and General Pershing's chauffeur, became America's leading ace with 26 victories between the end of April and October, 1918— three months of which he was hospitalized and unable to fly.

The most recent aircraft added to this gallery is the French SPAD XIII, an improved version of the successful SPAD VII. One of the best fighters of World War I, the SPAD XIII was powered by a high-compression 220 hp Hispano-Suiza 8 Be engine geared so that the propeller rotated left instead of right. The bigger engine powered this 1,255-pound fighter to a maximum speed of 138 mph at 6,560 feet and to an altitude of 21,800 feet.

SPAD XIIIs were flown by many Allied aces including Rickenbacker, Luke, Fonck, Nungesser and Baracca each of whom took advantage of the SPAD's superior speed and strength. Not all of the flyers liked them, however. "They were very fast and had a high ceiling," Escadrille Lafayette historians Charles Nordhoff and James Norman Hall wrote of the SPAD XIII in their novel, *Falcons of France*, "but they were far less maneuverable and far less reliable than the one-eighties I was accustomed to. Many of their engines lasted only twelve or fifteen hours."

Given the opportunity to try out this improved-version SPAD, Nordhoff and Hall wrote:

> My new Spad was a formidable little monster, squat and broad-winged, armed to the teeth, and with the power of two hundred and twenty wild horses bellowing out through its exhausts. I felt decided inward trepidations when I took it up for the first time, for the ship was unfamiliar to me . . . It hurtled its way through the air, roaring and snorting and trembling with excess of power, and its speed in a dive took my breath away. But I managed to land without smashing up, and after the first hour or two I got the hang of it, though I never felt for the geared Spad the same affection I had for the one-eighty—my first love.

> —*Falcons of France,* by Charles Nordhoff and
> James Norman Hall

A total of 8,472 SPAD XIIIs were built, including 893 for the American Expeditionary Forces.

Although this gallery honors the brave men and the frail machines in which they flew in World War I, it is a celebration tempered by sorrow, a sadness recognized by a dismayed Orville Wright, himself, who in 1917 wrote, "When my brother and I built the first man-carrying flying machine we thought that we were introducing into the world an invention which would make future wars practically impossible."

Golden Age of Flight

Although the "Golden Age of Flight" is loosely defined as that period in the history of aviation between the two World Wars, the most golden of those years are generally considered to have been those dating from shortly after Charles Lindbergh's 1927 solo crossing of the Atlantic through our entry into World War II.

The Golden Age is deemed "golden" because of the striking technological advances that paved the way for the types of aircraft we know today, the many classic and important aircraft produced during that period, the inumerable record flights, and the mushrooming interest of the public in aviation events. The airlines, the airway system, and air traffic control all began during these years. It was also a period during which an individual with little or no capital could emerge as a leader in aircraft technology. Heroes were made overnight; aircraft manufacturers boomed and busted in the course of a single season; the names of the air race and acrobatic pilots, the explorers and adventurers were as well known as movie stars and presidents; and their achievements were constantly in the headlines and the newsreels.

The most visible and widely publicized aspect of aviation during the Golden Age were the great air races and the air shows that usually accompanied them. The first of the National Air Races was held in 1920; its main event, the Pulitzer Trophy Race, became an annual affair. The trophy, awarded to the winner of an unlimited closed-course race, was offered by publisher Ralph Pulitzer to promote the development of high-speed aircraft. Entries could not have a landing speed of more than 75 mph. The races were held from 1920 through 1925 and all were won by U.S. Army or Navy fighters. In 1926, the Army and Navy decided they could no longer compete and the series ended; but in the course of the six Pulitzer Trophy races the winner's speed rose from 156.5 mph in 1920 to 248.9 mph in 1925.

The Thompson Trophy closed-course unlimited race originated in 1930 and its trophy immediately became the most sought after prize in the National Air Races. The race distance at first was 100 miles; but as aircraft speeds increased, the distance was extended to 300 miles. One of the most flamboyant and talented flyers of the Golden Age was "Colonel" Roscoe Turner, a man who won the Thompson Trophy three times (1934, 1938, and 1939). Turner, whose rank was held in the nonexistent Nevada Air Force, wore flashy military-style uniforms, a carefully waxed mustache, and often flew with "Gilmore," a lion cub.

The Bendix Trophy Race, first held in 1931, was a free-for-all transcontinental race run from Burbank, California, to Cleveland—or from New York to Los Angeles, depending on the location of the National Air Races that year. Landings were permitted during the course of the race, but the winner was declared on the basis of having the shortest elapsed time. Jimmy Doolittle won the first Bendix Trophy in a Laird Super Solution; and Roscoe Turner, who had won the 1933 Bendix Trophy in a Wedell-Williams and set the transcontinental

On September 4, 1922, Jimmy Doolittle set a record flying from Pablo Beach, Florida, to San Diego, California—a 2,163-mile flight—in 22 hours, 35 minutes.

Jimmy Doolittle rounds a pylon in his Gee-Bee R-1 on his way to victory in the 1932 Thompson Trophy air race.

speed record in 1934, was the favorite to win the Bendix again in 1935, the year Benny Howard was flying *Mr. Mulligan*, a high-winged cabin aircraft which looked to the other racers like a weekend pilot's pleasure craft. Howard called it the DGA-6.

When a reporter facetiously asked him whether that was a military designation for a new bomber, "Hell, no," Benny Howard replied. "It stands for my sixth Damn Good Airplane!"

The Bendix was the first of the major races to allow women entrants and women were victorious in the races of 1936 and 1938. In the 1935 race, Amelia Earhart was flying a Lockheed Vega and Jacqueline Cochran a Northrop Gamma; but interest focused on Turner racing a Wedell-Williams, and on Cecil Allen, who was flying the notoriously unstable and dangerous Gee-Bee.

Although Burbank Airport, starting point of the 1935 Bendix race, was blanketed by a thick, swirling fog, Amelia Earhart was able to sneak out through the mists and took off at 12:34 A.M. in the predawn dark. As she headed east, her technical advisor Paul Mantz and engine-maker Al Menasco fixed themselves drinks in the back cabin and settled down to a game of gin rummy.* Then the weather closed back in around Burbank Airport.

> At 2:03 a.m. the race got under way again when the fog lifted sufficiently to see the flares at the end of the runway. Roy Hunt roared off in his *Orion,* one and a half hours behind Amelia Earhart's *Vega;* then one by one the others thundered into the sky—Benny Howard at 2:08; Royal Leonard at 3:42; Roscoe Turner at 3:53; Russell Thaw at 4:01; Earl Ortman at 4:10; Jackie Cochran at 4:22, and finally, one hour later, at 5:18, Cecil Allen.
>
> Then came disaster. The weather killed Allen, minutes after takeoff, when his engine faltered and he lost control [of his Gee-Bee] trying to land. The weather also tricked Ortman who became lost . . . between Albuquerque and Amarillo and finally withdrew from the race, with a damaged cowling, after having landed at Kansas City. Leonard groped through the dirty weather to land . . . at Kansas City with a broken oil line, and Hunt set . . . down in a cow pasture "somewhere in Kansas" with an oil leak after encountering severe icing over the Rockies. . . .

*Amelia Earhart finished fifth, with an elapsed flying time of over five hours more than the winner.

190

All the way from Kansas City to Cleveland Benny [Howard] and his copilot, Gordon Israel, plowed through some of the worst weather ever encountered by Bendix flyers. Rain, sleet and hailstones pounded "Mister Mulligan," and while the plane seemed to wallow, it still bored right through. Approaching Cleveland, Benny began a long, shallow dive.

The air-speed needle seemed to lag. It indicated 250 miles per hour, but no more. Benny frowned; he knew that "Mister Mulligan" could true out at 292 miles per hour at 11,000 feet. Israel looked at Benny and frowned too, then suddenly he froze.

"Benny," he said softly, "the flaps. We left the flaps down!"

They had no way of actually knowing how Roscoe Turner was doing, but they had a good idea. . . .

Rain was pouring over the slick runway when "Mister Mulligan" thundered over the finish line at Cleveland Municipal Airport, at 1:40 p.m. local time. Howard and Israel parked the ship, then climbed out and shook hands with Vincent Bendix, who was waiting dismally under an umbrella.

"Looks like your race, Benny," Bendix sighed.

"Let's wait and see how Roscoe does," Benny replied. Their elapsed time was 8:33:16. Turner had until 3:25 p.m. to equal that time. One hour passed; the small crowd waiting in the grandstands grew restless. The rain was coming down in torrents. Lightning and thunder shook the stands.

At 3:24 p.m. the cluster of aviation writers began drifting off to their phones. It seemed obvious that Howard and Israel had won the $4,500 first prize. But even as the reporters dialed their offices they heard him coming—Roscoe Turner, the grandstand hero, flashing out of the west in his golden low-wing monoplane! Stunned, Benny looked at the official time clock. It was 3:25 p.m.

Turner shot across the finish line, pulled up in a beautiful fighter approach and slipped in for a short-field landing. He leaped out, ran up to the grandstand and grabbed Benny's hand.

"Who won, Benny?" Turner boomed.

"How do I know!" Howard replied. "I think it's a tie!"

Excitedly they bent over the official timers calibrating their watches. The timers scribbled figures on a tally sheet and handed it to Vincent Bendix. The race sponsor looked at the sheet a moment; then turned to the microphone.

"The winner, by twenty-three seconds," he announced, "is Benny Howard!"

Roscoe and Benny hugged each other, but only

Jacqueline Cochran is congratulated by Vincent Bendix in 1938 upon winning that year's Bendix Race in a stock Seversky P-35 at an average speed of over 249 mph.

Benny could see the disappointment in the colonel's eyes. . . .

Turner was not a man to alibi, but he remembered what had happened back at Kansas City. He and Royal Leonard had landed almost simultaneously, and there had been only one gasoline truck to service both ships. Turner had urged Leonard to gas first. . . . For being a gentleman, Roscoe Turner had lost the race. . . .

—*They Flew the Bendix Race,*
by Don Dwiggins

The following year, Louise Thaden entered the Bendix race in a Beechcraft C-17R Staggerwing with Blanche Noyes as her co-pilot. They thought they had been running last and were so discouraged that as they were coming into the finish line at Mines Field, Louise said to her copilot, "Look, we're the cow's tail. Let's not land here!"

"No," Blanche Noyes said, shaking her

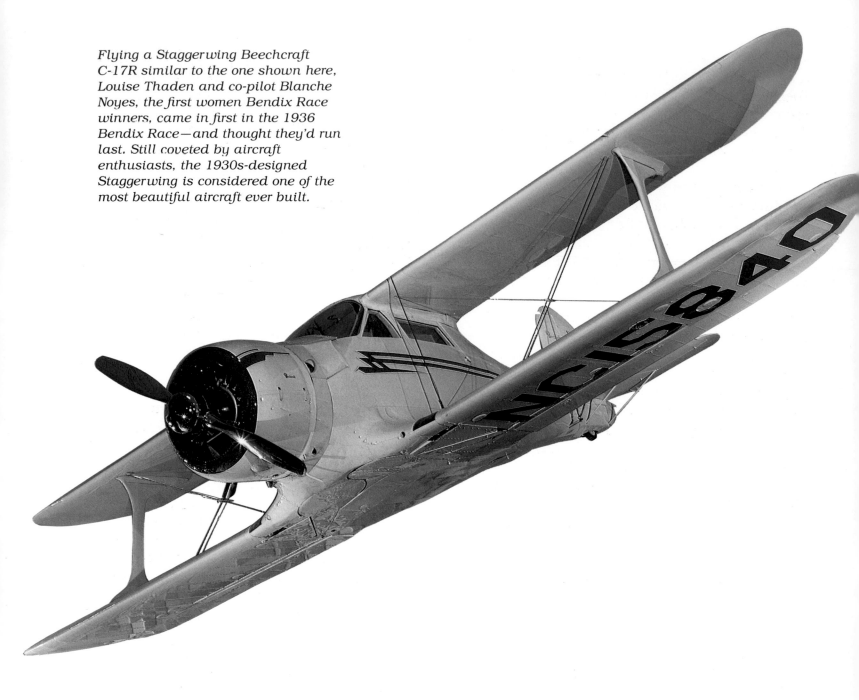

Flying a Staggerwing Beechcraft C-17R similar to the one shown here, Louise Thaden and co-pilot Blanche Noyes, the first women Bendix Race winners, came in first in the 1936 Bendix Race—and thought they'd run last. Still coveted by aircraft enthusiasts, the 1930s-designed Staggerwing is considered one of the most beautiful aircraft ever built.

head, "they don't even expect us to get through! Let's land and fool them."

A moment later they shot across the finish line, between pylon and grandstand—from the wrong direction . . . Louise bit her lip and pulled through a chandelle, dropped her gear and circled back to land.

"Let's sneak over to the parking ramp," she sighed. "I hate to let all those people see us come in last!"

A stream of motorcars swarmed out onto the field. Men were waving their arms excitedly.

"What do you suppose we've done now?" Blanche said nervously. They opened the window and peered out. [Race promoter] Cliff Henderson leaped down and ran to the ship, his face florid.

"Where are you going?" he yelled. "You just won the Bendix Race!" Louise switched off the ignition, looked at Blanche and cried.

Louise Thaden and Blanche Noyes had won the Bendix Trophy, crossed the country in 14 hours, 55 minutes, and 1 second, captured $7,000 in prize money, and averaged 165.32 mph in a second-hand business plane.

The major international air race was the Schneider Trophy closed-course race for seaplanes. The first race was held in 1913; a second in 1914; but the races were then interrupted by World War I. They recommenced in 1920 and continued until 1931. The first aircraft to exceed 400 mph was the 1931 Supermarine S6B Schneider Trophy racer.

All types of flying—except for airline and military aircraft—had emerged slowly after World War I. The lack of demand for private and commercial aircraft, combined with large stocks of war-surplus planes and engines, retarded aeronautical development until the 1920s.

But advances in technology and, in 1927, Lindbergh's stunning non-stop solo crossing of the Atlantic greatly accelerated the development of airframes and engines. (In Milestones of Flight we have noted the impact Lindbergh's flight had on private aviation: that in 1928 the applications for private pilot's licenses in the United States leapt from 1,800 the year before to 5,500!)

The 1930s brought sweeping changes to all areas of aviation. Prior to 1930, ownership of private aircraft in America was limited to the wealthy. Aircraft manufacturers built planes for "sportsman pilots" and advertised them as the ideal way to attend yacht regattas and polo matches. But after 1930, the light airplane movement made private ownership possible for the average individual, and the "sportsman pilot" soon became an anachronism.

High-wing cabin monoplanes superseded open-cockpit biplanes when advances rendered the need for extra wing area obsolete. Without the heavy, inefficient engines of the post-World War I period, there was no need for two wings; and with one wing, there was no aerodynamic drag of struts and bracing wires. Although biplane trainers were built through World War II, by the end of the Golden Age the transition to monoplanes was practically complete.

Adding to the increased diversity of aircraft types during this period was the variety of uses found for airplanes: hauling mail and cargo, carrying passengers,

Howard Hughes stands beside his Hughes H-1 Racer in which, on September 13, 1935, he set a world speed record of over 352 mph.

patrolling pipelines, combating forest fires, transporting company executives, crop dusting, aerial mapping, and skywriting. All were made possible by the technological advances achieved after World War I.

U.S. military aviation, following the end of the war, had begun the transition from an era of combat to another one of peacetime activities. Although air racing, record-breaking flights, and aerial exploration—all areas in which military pilots were involved—were exciting and appealed to the public, the issues that were of overriding importance to the Army during the two turbulent decades that led to World War II were different. These issues were (1) the efforts to establish an air force independent of Army control (this was achieved by the Air Corps Act of 1926, which changed the name of the Air Service to the Air Corps and recognized military aviation as an offensive striking arm rather than an auxiliary service); (2) the development of a strategic bombing doctrine (Brigadier General William "Billy" Mitchell, who had emerged from World War I as the outstanding American air combat commander and the self-proclaimed leader of the fight for greater recognition of airpower, believed that the bomber spelled the end to sea power. And in

The "long-winged" version of the Hughes Racer—one of the Museum's most beautiful aircraft.

195

sea tests held in 1921 and 1923 he demonstrated the vulnerability of major warships to aerial bombing attack); and (3), the search for a heavy bomber.

In 1935, air maneuvers had demonstrated the need for a central striking force of long-range bombardment and observation aircraft to defend the United States and its possessions from seaborne attack. The General Headquarters (GHQ) Airforce, formed in March, 1935, and placed under the command of Brigadier General Frank M. Andrews, removed the combat units of the Air Corps from the control of the local ground-based Corps Area officers and placed these units under one commanding general. Despite Army objections to expensive, long-range, four-engine bombers, growing international tensions forced the beginning of mass American rearmament in 1939. By then Boeing's four-engine bomber, the B-17 Flying Fortress, was already flying. The year before, in fact, the Flying Fortress had demonstrated its long-range capability when, in May, 1938, three B-17s, with then-Lt. Curtiss LeMay as navigator, met and flew over the Italian liner *Rex* at sea 725 miles east of New York.

"I just get my spurs on, go out there, straddle her to let her know I mean business," reported Fred Key on how he performed engine maintenance on Ole Miss *during its record-setting non-stop flight.*

The basic military aircraft of the 1920s had been, still, an open-cockpit, wire-braced biplane. By 1939, as the Golden Age drew to an end, many of the fighters and bombers with which the U.S. would fight the next war had already flown for the first time.

The goals of almost all of the aviators in the Golden Age were to fly higher, faster, and farther than any man or woman had flown before. Exotic geographical locations beckoned: the uncharted, forbidding Polar regions, and the lush, steamy jungles of New Guinea, the Amazon, and the Yucatan. Pilots crossed the deserts of the southwestern United States and northern Africa, and spanned the mighty oceans as well.

Suspended from the ceiling of this gallery is an airplane that took part in one of the strangest, if not most challenging, of events: it is *Ole Miss*, a Curtiss Robin, in which two brothers, Fred and Algene Key, took off from the airport at Meridian, Mississippi, on June 4, 1935, and did not touch ground again until July 1st. They were refueled and received supplies in flight from another aircraft 432 times, braved severe thunderstorms and an electrical fire in the cabin, and remained aloft 653 hours and 34 minutes—a total of 27 days.

A metal catwalk had been built around the front of the fuselage to facilitate lubrication and emergency repairs to the engine. Fred Key, the smaller of the two brothers, was given the job of climbing out. On June 17, they finished their second full week in the air and it was determined the Key brothers had flown 21,120 miles, a straight-line distance equivalent to a flight around the world. That was also the day Fred nearly fell from the catwalk:

. . . The weather was very turbulent. At 3,000 feet just as he stepped outside the plane, and before he attached the safety strap to the catwalk railing, the "Ole Miss" hit a tremendous airpocket. Luckily, Fred had his feet planted firmly enough on the "runningboard" to hold him on until he caught the rail. He never wore a parachute at anytime when he ventured outside the plane on the catwalk because wind resistance prevented his wearing one.

Maybe the near-death jolt served as a friendly reminder for Fred to take his work more seriously. He became so astute at climbing out on the catwalk and then straddling the engine to inspect gas lines (the only vantage point from which he could see the lines leading to the No. 3 cylinder) that he began to grow perhaps too lax with the procedure. In a radio interview he told a reporter, "I just get my spurs on, go out there, straddle her to let her know I mean business." As the flight progressed, the *Meridian Star* stated he "took to playing about the plane like children on a sandpile." To support this notion, Ben Woodruff observed Fred from another plane, "sitting astride the motor mount on the exposed point of the plane, cleaning the windshield with one hand and holding a cold drink in the other." When he saw Woodruff, he "waved . . . and proceeded to jump about like a monkey on the exposed nose of the ship as it traveled 80 miles an hour."

—*The Flying Key Brothers and Their Flight to Remember*, by Stephen Owen

By June 30 it had become evident that the Key brothers would not be able to stay aloft, as they had hoped, until July 4th:

The constant vibration was causing two types of fatigue: metal fatigue in which wires and braces were threatening to weaken and mental fatigue which created disorientation in the pilots. Nerve shock and sheer weariness caused the Key who was resting to have to take as many as five minutes before he could wake and get his bearings. Under such circumstances, emergencies could not be dealt with easily. They agreed to nurse the plane along until the next evening and land at 6:30 p.m. after they had safely passed the [unofficial] Jackson-O'Brine record at 12:02.

News spread throughout the community that the Keys would land . . . throngs of A-Models and pedestrians flocked toward the airport, causing dust to thicken in the summer sky. All totaled, 35,000 to 40,000 went to pay tribute to the exhausted aviation heroes the entire nation and the world were reading about. . . .

Inclement weather threatened to the extent that Dr. Key talked his sons into landing twenty-four minutes early. At 6:06 p.m., July 1, 1935, the bedraggled flyers circled the airport three times at 100 feet, and then touched down on the soggy runway on the newly named Key Field. . . .

As the victorious pilots taxied towards the hangar, local police and national guardsmen could not hold the enthusiastic crowd back. Reminiscent of Lindbergh's landing at LeBourget Field in Paris, the mob swept past news camera crews and lifted their heroes triumphantly from the plane. Before them stood two scraggly specimens of humanity. Their eyes were severely bloodshot . . . their hair "had grown 'wild' . . . Their clothing hung limp on them from weight loss. . . . Yet . . . there was no appearance of despair on their proud faces. Their boyish grins proclaimed a typically American air of hope and triumph."

The Wright J-6-5 Whirlwind engine that had powered the Key brothers' airplane *Ole Miss* to the world's endurance record of 27 days aloft had consumed 6,500 gallons of fuel and 300 gallons of oil during its estimated 52,320 miles of flying. Although the propeller had made 52,860,000 revolutions during the flight, the engine needed only new exhaust valve guides and a new set of rings to put it in top running condition afterward. *Ole Miss* burned an average of 10 gallons of fuel an hour, necessitating four in-flight refueling contacts each day.

The endurance record of the Key brothers' *Ole Miss* was just one of dozens of aeronautical records set during this Golden Age, the first of which might be considered Jimmy Doolittle's 22-hour, 35-minute, 2,163-mile flight from Pablo Beach, Florida, to Rockwell Field, San Diego, on September 4, 1922.

Doolittle's transcontinental crossing in less than one day was a far cry from the 49 days it had taken Calbraith Perry Rogers to cross the country in the *Vin Fiz* in 1911.

Another indication of the enormous progress being made during this golden period can be seen by comparing the achievement in May, 1923, of Army Lieutenants Kelly and Macready, who, piloting a giant Fokker T-2, were able to fly across the U.S. without stopping, whereas Rogers, who had made 69 stops during his transcontinental flight, had had to land 23 times in Texas alone. And the year after Kelly and Macready's flight, four Douglas World Cruisers left Seattle, Washington, and

Powered by a Wright J-6-5 Whirlwind engine, the Key brothers' Curtiss Robin airplane Ole Miss *set a world's endurance record by remaining aloft for 653 hours and 34 minutes—a total of 27 days spent flying in circles over Meridian, Mississippi. The Whirlwind engine burned an average of ten gallons of fuel an hour, necessitating four in-flight refueling contacts each day. By the end of its estimated 52,320 miles of flying* Ole Miss *had consumed 6,500 gallons of fuel, 300 gallons of oil, and its propeller made over 52,860,000 revolutions.*

five months and 27,553 miles later, two of them safely returned, having completed the first successful round-the-world flight.*

On May 9, 1926, Lt. Comdr. Richard E. Byrd and Floyd Bennett reportedly made the first airplane flight over the North Pole; one year later Lindbergh in *Spirit of St. Louis* crossed the Atlantic. Crossings of the South Atlantic and the Pacific followed. In 1929, the first airship flight around the world took

---

*Many of the famous aircraft from the Golden Age of Flight appear in galleries throughout the Museum: The Fokker T-2 and the Douglas World Cruiser *Chicago* are in the Museum's Pioneers of Flight gallery, as is the Curtiss R3C-2 fitted with pontoons in which Jimmy Doolittle won the 1925 Schneider Trophy Race. Earlier that year, Cyrus Bettis won the Pulitzer Race in that same aircraft with a fixed landing gear (designated the R3C-1). Also in Pioneers of Flight is Amelia Earhart's Lockheed Vega in which, in 1932, she made the first solo flight across the Atlantic by a woman; and the Lockheed Sirius *Tingmissartoq* in which Charles and Anne Lindbergh made two major flights in 1931 and 1933 to survey possible overseas airline routes in the early days of airline travel.

In Milestones of Flight hangs Lindbergh's Ryan NYP *Spirit of St. Louis;* and in Air Transportation are the Douglas DC-3, the Boeing 247D, the Ford Trimotor, the Northrop Alpha, the Fairchild FC-2, and the Pitcairn Mailwing.

A Boeing P-26A and the Grumman *Gulfhawk II* hang in Special Exhibits; Wiley Post's Lockheed Vega *Winnie Mae* is in Flight Testing; a Boeing F4B-4 is in Sea-Air Operations.

A Curtiss F9C-2 is in Balloons and Airships.

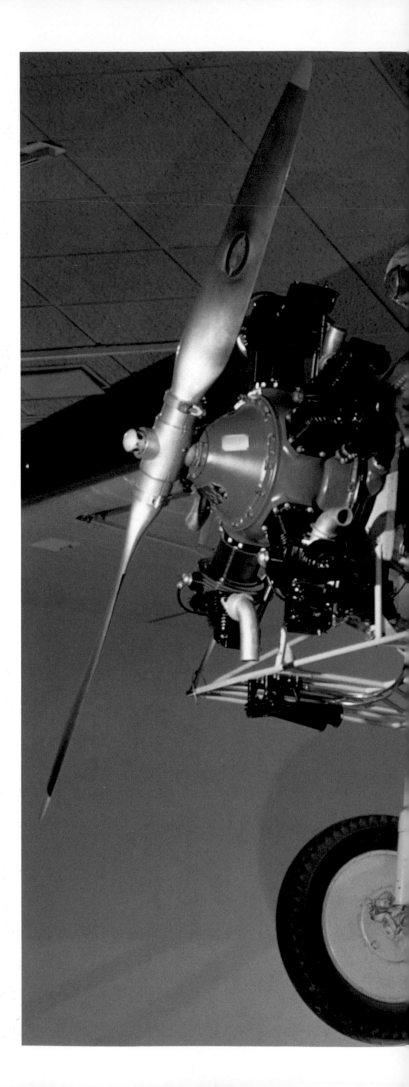

Lincoln Ellsworth's Antarctic record-setting
Northrop Gamma Polar Star.

place as did Commander Byrd's flight over the South Pole.

Exactly five years after Lindbergh's solo flight across the Atlantic, Amelia Earhart achieved the first solo transatlantic flight by a woman and on January 11–12, 1935, she made the first solo flight from Hawaii to the U.S. mainland. Later that year, Lincoln Ellsworth and Herbert Hollick-Kenyon in this gallery's Northrop Gamma *Polar Star* flew 2,400 miles from Dundee Island in the Weddell Sea across the Antarctic continent, but they ran out of fuel and were forced to land 25 miles short of their goal.

Their successful flight across the Antarctic continent came after Ellsworth's two previous failed attempts. His first attempt, in January, 1934, with Bernt Balchen as pilot, had to be cancelled when the *Polar Star* was damaged by the breaking-up of the ice upon which it rested; and the aircraft had to be returned to its factory in the United States for repairs. Ellsworth's second attempt, again with Balchen, in January, 1935, was cancelled due to bad weather and the *Polar Star* was placed back on its ship, the *Wyatt Earp*, and taken home again.

"The *Wyatt Earp* had carried me and my plane in the last three years 48,000 miles in search of a suitable taking-off ground for our 20-hour flight across Antarctica," Ellsworth later wrote. "A good-natured friend had remarked to me when he bade me farewell in New York, 'Your *Polar Star* has traveled farther and flown less than any other plane!'"

On November 21, 1935, on Ellsworth's third attempt, he took off in perfect weather but was forced to land when they had trouble with one of the *Polar Star's* instruments—but not before Ellsworth had a taste of what lay a little farther on:

There, just ahead, lay a great unknown mountain range, with peaks rising majestically to 12,000 feet. These had never before been seen by the eyes of man. I had lived for this moment. Only one who has known intense anticipation of some great event can imagine the depth of my despair at being forced back after coming so far along the path of victory.

—"My Flight Across Antarctica," by Lincoln Ellsworth in *The National Geographic Magazine*

Two days later, at 8:05 on the morning of November 23rd, the aircraft was repaired, the weather was clear, and the Northrop Gamma *Polar Star* took off again. Ellsworth's account of the early portion of the trip captures the romance all such endeavors held:

At 12:22 we crossed Stefansson Strait and took compass bearings of the continental coast. The low, black, conical peaks of Cape Eielson rose conspicuously out of a mantle of white on our left.

We climbed to 13,000 feet, where the temperature was 10 degrees, Fahrenheit. We were now over the unknown.

In November, 1935, Lincoln Ellsworth and Herbert Hollick-Kenyon attempted to fly this Northrop Gamma Polar Star *2,400 miles across the Antarctic continent and ran out of fuel just 25 miles short of their goal. It took them six days on foot to reach Little America, Richard Byrd's abandoned Polar headquarters. "The love of great adventure is not an acquired taste," Ellsworth later wrote, "it is in the blood."*

This midget racer with its 15-foot, 1-inch wingspan enjoyed one of the most successful careers in air-racing history. From 1931 until its retirement in 1954 the Wittman Buster set racing records, won numerous class races, and twice won the Goodyear Trophy, in 1947 and again in 1949.

It falls to the lot of few men to view land not previously seen by human eyes. It was with a feeling of keen curiosity and awe that we gazed ahead at the great mountain range which we were to cross. Bold and rugged peaks, bare of snow, rose sheer to some 12,000 feet above sea level.

Again I felt a supreme happiness for my share in the opportunity to unveil the last continent in human history.

We were indeed the first intruding mortals in this age-old land, and looking down on the rugged peaks, I thought of eternity and man's insignificance. So these first new mountains we saw I named Eternity Range. The three most prominent peaks on our right I named Faith, Hope, and Charity, because we had to have faith, and we hoped for charity in the midst of cold hospitality. . . .

We fully realized that this was the most dangerous area of our flight, for on one side lay the frozen Weddell Sea, which no ship could penetrate, and on the other an unknown continent larger than the United States and Mexico.

After flying nearly 14 hours during which they crossed approximately 1,500 miles Ellsworth and Hollick-Kenyon made the first of their four landings.

We climbed out of the plane rather stiffly and stood looking around in the heart of the Antarctic.

There we were—two lone human beings in the midst of an ice-capped continent two-thirds the size of North America. Perhaps this thought brought us closer together.

Suddenly I noted the fuselage was crumpled. Kenyon thought it must have been done on the take-off, but I had been writing my notes and had felt no jar then. Now I recalled that when we came down here I thought my teeth would go through the top of my head.

Visitors to this gallery can clearly see the dent in the *Polar Star*'s fuselage just behind the engine caused by that hard landing on polar ice. During the three-day blizzard that raged during their stay at their third camp, the inside of the plane was packed solid with drifting snow. The temperature was −5° and a 45-mph gale threatened to blow away their tent. For eight days the storm confined them to their tent and when it abated Ellsworth and Hollick-Kenyon spent an entire day with a teacup scooping out the powdery snow that had filled the *Polar Star*.

For some unexplained reason all three radios Ellsworth carried failed and there was no way for the crew of the *Polar Star* to contact their ship. And so, with the passing of the days when no word came the party on board the *Wyatt Earp* began to fear Ellsworth and Hollick-Kenyon had perished.

The final adventure resulted when the *Polar Star* exhausted its fuel short of its goal. Ellsworth and Hollick-Kenyon, dragging a sledge filled with supplies, were forced to march 25 miles to Byrd's Little America camp where they holed up until men from the Royal Research Ship *Discovery II* found them alive and, with the exception of Ellsworth's frozen foot, well.

"The love of great adventure is not an acquired taste," Ellsworth later wrote, "it is in the blood. . . . Who has known heights and depths shall not again know peace, for he who has trodden stars seeks peace no more."

During the two decades of the Golden Age of Flight the airplane was transformed from a rudimentary, wood-and-fabric machine with an uncertain future into an established and developed commodity with an all-metal semi-monocoque skin.

Visitors to the Golden Age of Flight gallery see not just how aviation expanded and flourished during the two decades of the 1920s and 1930s, but they will see and hear—through film clips, newspaper headlines, artifacts and radio broadcasts—that changes were taking place in all aspects of American life. And if the upbeat mood of America at the start of the 1920s was followed by the Depression at its close, and the euphoria that had marked the 1920s gave way to the despair, massive unemployment, and breadlines of the 1930s, by the end of that decade President Franklin Delano Roosevelt's New Deal had given this country hope—but it was an optimism tempered by the realization that Hitler's armies were marching across Europe and that the U.S. might once again be at war.

FLIGHT TESTING

VERTICAL FLIGHT

Flight Testing

If the visitor to the Flight Testing gallery comes away with nothing more than an awareness that a test pilot is a highly skilled aerospace professional and *not* the reckless, devil-may-care, high guts to low brains ratio "show-me-where-the-stick-is-and-I'll-fly-it" character depicted by the Hollywood movies of the 1930s, then one of the primary purposes of this gallery has been realized.

More than fifty years ago Edward P. Warner and F.H. Norton of the National Advisory Committee for Aeronautics (NACA, the National Aeronautics and Space Administration's forerunner) wrote: "Test flying is a highly specialized branch of work, the difficulties of which are not generally appreciated and there is no type of flying in which a difference between the abilities of pilots thoroughly competent in ordinary flying becomes more quickly apparent." The main aim of the Flight Testing gallery—also known as the Hall of X-Airplanes, "X" for Experimental—is to demonstrate the importance of the test pilot's work to the development of new aircraft and to the derivation of new knowledge in the aeronautical sciences.

Suspended above the entrance to the gallery is an 1894 Lilienthal glider, one of eight machines of this type built by that towering figure of early flight testing, Otto Lilienthal himself, whose studies of the lift created by a curved surface on a moving stream of air, *Der Vogelflug als Grundlage der Fliegekunst* (Bird Flight as the Basis of Aviation), had such an influence on the Wright brothers' work. The glider's frame was constructed of willow and bamboo, its wings and surfaces covered with plain cotton cloth. Its wings are designed to fold backward for ease in transportation and storage. Lilienthal considered the glider on display here to be one of his safest and most successful designs.

Lilienthal would suspend himself between the wings by placing his arms through padded cuffs attached to bars which, in turn, were attached to the glider's frame. Two additional padded braces in line with his ribs were also secured to the frame. Lilienthal was able to provide some degree of control by swinging his legs and torso to alter the center of gravity and made flights of up to 1,150 feet in length from the artificial hill he had constructed near his home in the Berlin suburb of Gross-Lichterfelde.

In 1896, Lilienthal wrote, "One can get a proper insight into the practice of flying only by actual flying experiments." His experimentation ended tragically on August 8th, that same year, when the glider he was testing, one similar to the 1894 glider on display, stalled at an altitude of fifty feet and crashed to the ground, mortally injuring its

Otto Lilienthal built this hill near his home in a Berlin suburb to test his gliders.

inventor. The German aviation pioneer died the following day, August 9, 1896, in a Berlin hospital. Otto Lilienthal was fully aware of the risks involved testing flying machines, but, as he said shortly before his fatal accident, "Sacrifices must be made."

Each of the three full-size aircraft on exhibition in this gallery represents a separate aspect of flight research. The first, Wiley Post's modified Lockheed Model 5-C "Vega," the *Winnie Mae,* is displayed with its special jettisonable landing gear and Post's pressure suit used during his high-altitude research flights. He also completed two round-the-world flights in this aircraft.

The second aircraft is the Bell XP-59A Airacomet, America's first experimental turbojet-propelled aircraft. The Airacomet, a direct ancestor of all American jet aircraft, made its maiden flight on October 1, 1942, piloted by Bell Aircraft Corporation's test pilot Robert M. Stanley at California's Muroc Dry Lake. Because of wartime security, the XP-59A was tested in complete secrecy and, at one point, officials disguised the aircraft when it was on the ground by fitting a dummy four-bladed 7-foot-diameter propeller to its nose to make onlookers think the Airacomet was a conventional propeller-driven airplane.

The other aircraft is the Hawker-Siddeley Kestrel, a vertical and short take-off and landing (V/STOL) research aircraft which is here as a representative of a highly successful approach to vertical flight and jet-vectored aircraft. The first Kestrel flew on March 7, 1964, and the Museum's specimen wears the NASA markings it bore during its 210 test flights made under the auspices of the National Aeronautics and Space Administration at the Langley Research Center. Directly beneath the Kestrel a continuous-loop movie permits the visitor to see this aircraft's amazing ability to rise vertically off the ground like a helicopter, hover, and dart away.

Another aim of the Flight Testing gallery is to explain what flight research is and what it involves. Flight research is undertaken to determine whether an actual machine's performance is what has been predicted or hoped-for through its design, to provide basic information and experience that will benefit aeronautical science in general, to furnish additional data that will be useful in the continuing development of the particular series of aircraft being tested, and to provide "proof of concept" validation through the in-flight testing of new concepts or developments. From the very earliest days of flight the aims of flight research have been the same.

Orville and Wilbur Wright, who combined extensive ground research in their wind-tunnel experiments on the behavior of various types of wings and propellers with flight research aloft, recognized the need for acquiring reliable, verified data. Along with Otto Lilienthal they acknowledged the elemental partnership between ground research and flight testing. Wilbur, whom we recall compared the testing of a flying machine to riding "a fractious horse," stated: "If you are looking for perfect safety you will do well to sit on a fence and watch the birds, but if you really wish to learn you must mount a machine and become acquainted with its tricks by actual trial." The Wrights' determination resulted, on December 17, 1903, in man's first successful powered flight. Orville, who was the test pilot on that flight, wrote:

America's first flying turbojet, the Bell XP-59A Airacomet at Muroc Dry Lake, California, in 1942.

Wilbur ran at the side, holding the machine to balance it on the track. The machine, facing a 27-mile wind, started very slowly. Wilbur was able to stay with it until it lifted from the track after a forty-foot run. The course of the flight up and down was exceedingly erratic. The control of the front rudder was difficult. As a result, the machine would suddenly rise to about ten feet and then as suddenly dart for the ground. A sudden dart when a little over 120 feet from the point at which it rose into the air, ended the flight. The flight lasted only twelve seconds, but it was nevertheless the first in the history of the world in which a machine carrying a man had raised itself by its own power into the air in full flight, had sailed forward without reduction of speed, and had finally landed at a point as high as that from which it started.

As Dr. Richard P. Hallion of NASM's Department of Science and Technology has written, Orville Wright's post-maiden flight analysis "is a model test flight report. It presents the test conditions, a critical examination and analysis of the airplane's stability and control, and finally, a summation of the flight's significance."

Another purpose of this gallery is to introduce the visitor to some of the major accomplishments of flight research, such as the advent of blind flying, stratospheric flight research, the introduction of turbojet aircraft, the supersonic breakthrough, and the development of practical V/STOL aircraft such as the Kestrel exhibited in this hall.

Also on display here is an exact 1:16 scale replica of the Consolidated NY-2 in which Jimmy Doolittle made the world's first blind flight on September 24, 1929. On that day, Doolittle climbed into the NY-2's back cockpit and closed the hood; Army Lt. Benjamin Kelsey rode in the front cockpit as safety pilot, but never needed to touch the controls. Then, guided only by his gauges, including three new aviation instruments— a Kollsman precision altimeter, a Sperry Gyrocompass, and Sperry artificial horizon—and using special radio receivers, Doolittle completed the world's first blind flight from take-off to landing at Mitchell Field, Long Island.

If the average Museum visitor knows anything at all about Wiley Post, it is probably only that he was the pilot of the plane that crashed near Point Barrow, Alaska, killing the humorist Will Rogers, on August 15, 1935. And yet Wiley Post, who died with Rogers in that crash, was the first person to make a solo round-the-world flight (July, 1933) and was the winner of aviation's top awards. Post was honored by two New York ticker-tape parades, was received by two Presidents at the White House; he created the world's first practical high-altitude pressure suit, discovered and was the first to take advantage of the jet stream, and at the time of his death had flown more hours at ground speeds above 300 mph and had more flight hours in the stratosphere than any other man. Lauren D. Lyman, then aviation editor of the *New York Times*, observed at Post's death that it was Wiley Post who moved stratospheric flight into the realm of reality. Post did it in the *Winnie Mae*.

Following his round-the-world flights, Wiley Post made several modifications on the *Winnie Mae* to better enable it to make long-distance high-altitude flights. Because he could not pressurize the airplane's cabin, Post asked the B.F. Goodrich Company to help him develop a full-pressure suit he could wear while flying the plane. In spite of the suit being created in response to the limitations of his aircraft, Post's suit, consisting of three layers (long underwear, an inner black rubber air pressure bladder, and an outer cloth contoured suit), and helmet (containing a special oxygen breathing system and outlets for earphones and a throat microphone) must be considered the ancestor of the sort of full-pressure suits used in the X-15 research aircraft and those worn by the astronauts. The *Winnie Mae* was equipped with a supercharger, a special jettisonable landing gear, and a metal-covered spruce landing skid glued to the bottom of the fuselage. In late July, 1934, Post had declared his intention to fly across the country at an altitude of more than 30,000 feet, thereby to take advantage of the high-altitude winds he knew existed.

After months of preparation Post was ready, and early in the morning on February 22, 1935, he put on his pressure suit and helmet, entered the *Winnie Mae*, and took off into the darkness and dense fog surrounding Burbank, California, for New York. At an altitude of about 200 feet, just before climbing through the clouds, Post jettisoned his landing gear to lighten and streamline his plane and, shortly thereafter, just 31 minutes into the flight, his engine began throwing oil. Post prepared for an emergency landing on Muroc Dry Lake in the Mojave Desert. The *Winnie Mae* was carrying more than 300 gallons of fuel and Post had no way to dump it prior to the

forced landing. In spite of the danger, Post, superb pilot that he was, skillfully glided down to so smooth a landing that H.E. Mertz, who was but 400 yards away tinkering on a wind-powered "sail car," did not even hear the *Winnie Mae* land. When Wiley Post, still wearing his high-altitude full-pressure suit and helmet, walked up to Mertz to ask his help in removing the helmet's rear wing nuts, Mertz nearly fainted in terror. It was subsequently learned that the *Winnie Mae* had been sabotaged; a quart of emery dust had been placed in the supercharger's air intake the night before the flight. The moment the supercharger was cut in, the emery dust was sucked into the engine where it ground down the piston rings and wreaked such

Famed aviator Wiley Post completed two round-the-world record flights in 1931 and 1933 and a series of special high-altitude substratospheric flights in the Winnie Mae, *a modified Lockheed Vega.*

The direct ancestor of all American jet-propelled airplanes, this historic Bell XP-59A Airacomet was a closely guarded wartime secret when it flew in 1942.

havoc that the engine had to be completely rebuilt. The sabotage had been performed at the instigation of a jealous pilot who thought Post's successes were jeopardizing his own chances for sponsorship.

Three weeks later, on March 15, 1935, Wiley Post took off on a second transcontinental record attempt. One hundred miles east of Cleveland, Post ran out of oxygen and had to turn back. He landed the *Winnie Mae* again on her belly, and after he climbed out and his pressure suit was removed Post learned that he had covered the 2,035 miles in 7 hours and 9 minutes. That meant that not only had the *Winnie Mae* averaged 279 miles per hour,

which was over a hundred miles faster than her normal maximum speed, but that at times the *Winnie Mae* had been traveling as fast as 340 miles per hour! There was no question about it: Wiley Post and his *Winnie Mae* had been in the jet stream. As Stanley R. Mohler, M.D., and Bobby H. Johnson, Ph.D., wrote for a NASM Smithsonian Annals of Flight monograph, "Within a quarter of a century of Post's high altitude flights, men, women, and children would be hurtling through the stratosphere at almost the speed of sound in the comfortable pressurized cabins of jetliners, wholly ignorant of the frustrating labors of 1934 and 1935, unmindful of the man who met

Thomas Edison's complaint about the Wright brothers' flight was that no aircraft could be considered truly practical until it could rise from the ground and settle back again, vertically. Using a special "vectored thrust" turbo-fan engine, this 1964 Hawker XV-6A Kestrel could rise straight up, hover, dart away in horizontal flight, slow and settle vertically back down on the ground. The success of the experimental Kestrel led to the development of the world's first Vertical Take-Off and Landing (VTOL) jet fighter, the Hawker Siddeley Harrier now in service with the British Royal Air Force and the U.S. Marine Corps.

the difficulties in their rudest shapes. Yet every time a contrail runs its white chalkline across the blue, it deserves recollection that it was Wiley Post who pointed the way to putting it there."

Although flight research is a business oriented around science, fact, and actual observation, and the test pilot is a highly skilled professional, there are moments when even the most disciplined test pilot cannot resist showing off a little. Test pilot Scott Crossfield, who smashed numerous speed barriers in such supersonic research aircraft as the Douglas D-558 Skyrocket and North American's X-15, tells of a different sort of barrier he broke through while testing an F-100 Super Sabre of the sort that had already cost the life of George Welch, North America's top test pilot:

When I reached 35,000, I leveled the ship. At that very instant a blaze of red flashed on my instrument panel. Fire in the compressor section.... There was an old and tired axiom about it at Edwards [Air Force Base]: "If you see a compressor fire-warning light and you haven't blown up, well, you're going to in just a second." I got busy fast. I throttled back on the engine. As I did, the fire-warning light flickered and dimmed. Then it flashed back on again full strength. However, I saw no other signs of real fire, so I concluded that it was a false warning. I would bring the ship down dead stick.... There was a big debate raging among pilots at Edwards about whether or not a F-100 could be landed dead stick. I was not concerned. Every test pilot develops a strong point. I was certain that my talent lay in dead stick landings.

I called Edwards and declared an emergency. All airborne planes in the vicinity were warned away. I held the ailing F-100 on course, dropping swiftly, lining up for a dead stick landing. I flared out and touched down smoothly. It was in fact one of the best landings I ever made. I then proceeded to violate a cardinal rule of aviation: never try tricks with a compromised airplane. I had already achieved the exceptional, now I would end it with a flourish. I would snake the stricken F-100 right up the ramp and bring it to a stop immediately in front of the NACA hangar. It [would be] a fine touch. After the first successful dead stick landing in a F-100 it would be fitting.

According to the F-100 handbook, the hydraulic brake system was good for three

"cycles" (pumps on the brake) engine out. The F-100 was moving at about 15 mph when I turned up the ramp. I hit the brakes once, twice, three times, the plane slowed but not enough. I hit the brakes a fourth time—and my foot went clear to the floorboards. The hydraulic fluid was exhausted. The F-100 rolled on, straight between the yawning hangar doors!

The NACA hangar was then crowded with expensive research tools—the Skyrocket, the X-3, X-4 and X-5. Yet somehow, my plane, refusing to halt, squeezed by them all and bored steadily on toward the side wall of the hangar.

The nose of the F-100 crunched through the corrugated aluminum, punching out an eight-inch steel I-beam. I was lucky. Had the nose bopped three feet to the left or right, the results could have been catastrophic. Hitting to the right I would have set off the hangar fire-deluge system, flooding the hangar with 50,000 barrels of water and ruining all the expensive airplanes. Hitting to the left I would have dislodged a 25-ton hangar door counterweight, bringing it down on the F-100 cockpit, and doubtless ruining Crossfield.

Chuck Yeager never let me forget that incident. He drew many laughs at congregations of pilots by opening his talk, "Well, the sonic wall was mine. The hangar wall was Crossfield's." That's the way it was at Edwards. Hero one minute, bum the next.

—Always Another Dawn, by Scott Crossfield

Wind-tunnel displays and models are exhibited to demonstrate the importance of ground research and its connection with flight research. Any object that moves through the air creates pressure disturbances; the disturbance waves expand much like the bow waves of a boat as it cuts through the water. An aircraft's disturbance wave moves ahead of it at the speed of sound, and as the vehicle's velocity approaches that speed, it will catch up with its own forward-moving pressure wavefront. At transonic speeds—those transitional speeds between subsonic and supersonic (velocities between Mach 0.7 and Mach 1.3)—shock waves form on the airplane, stream from its fuselage, wings, and tail. The center of lift shifts, which results in changes in trim, stability, and control forces. During World War II fighter pilots in high-speed dives would suddenly and terrifyingly

The milestone X-15 rocket-powered aircraft was drop-launched from a B-52 Stratofortress.

encounter transonic phenomena: severe buffeting would shake the plane, controls would "freeze," sometimes planes broke up in flight.

The problem was how could scientists and engineers acquire the sort of factual data they needed on transonic and supersonic flight when conventional wind tunnels "choked," that is, made it impossible to gain accurate measurements because shock waves from the model being tested and its supports bounced like echoes off the wind tunnel's walls? It was known that the airspeed over the upper surface of a moving wing is greater than in the general air stream. And models were attached on the upper wing of actual airplanes to take advantage of this phenomenon. An aircraft flying at from Mach 0.5 to Mach 0.75 would have an airflow speed around the model of from Mach 0.8 to Mach 1.1. Models, too, were dropped like bombs. The bright yellow 2/5 scale model of a Grumman XF8F Bearcat displayed was carried up to 35,000 feet in a B-17, then released. Telemetry equipment carried in the model's forward compartment transmitted information on airspeed, aerodynamic loads, control positions, and control forces as the model dove, reaching speeds of 600 mph, on its way down to 20,000 feet. The model would then be pulled out of its dive, its parachute would be popped, and it would float gently down into the ocean off Wallops Island, Virginia, where it would be retrieved and used again. With the development of the porous and slotted-throat wind tunnel, which eliminated the problems conventional tunnels had with "choking," aerodynamics researchers were again able to gain much more useful transonic and supersonic data from ground testing.

One of the most dramatic examples of how ground research can result, in some cases, in basic changes in an aircraft's design is the exhibit related to the Convair F-102 interceptor. When this aircraft completed its initial flight tests in 1954, because of unforeseen high-drag characteristics it could not exceed the speed of sound. Richard

215

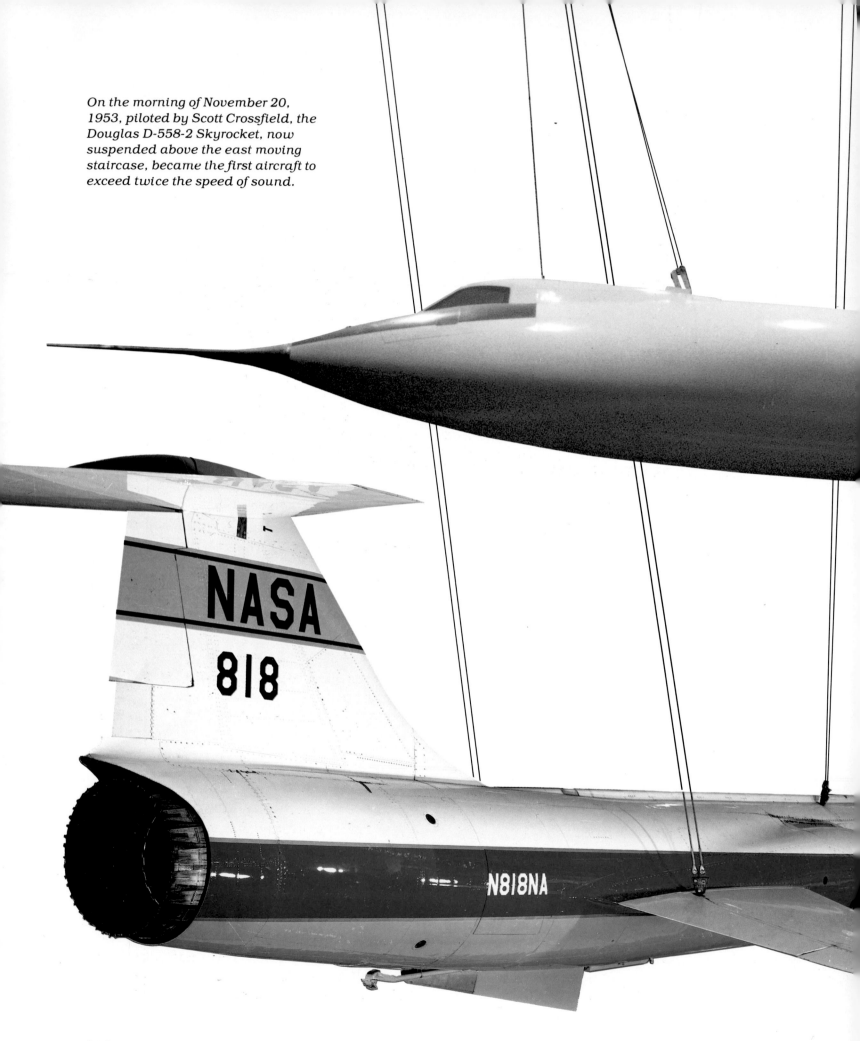

On the morning of November 20, 1953, piloted by Scott Crossfield, the Douglas D-558-2 Skyrocket, now suspended above the east moving staircase, became the first aircraft to exceed twice the speed of sound.

Whitcomb's ground research in wind tunnels revealed the value of "area ruling." The NASA scientist's discoveries led to the F-102's having a new "coke bottle" fuselage design, and when this new form was tested, the F-102 handily broke the sound barrier. Flight research thereby validated an important new principle of aircraft design that ground research had discovered. Displayed nearby is a wind-tunnel model of the Bell X-2 tested in 1946 by NACA at the Langley Aeronautical Laboratory's 300 mph wind tunnel. Ten years after wind tunnel tests were begun on this model, after ten years of extended, laborious ground testing resulted in major modifications to its design, this now nearly forgotten research aircraft became on September 27, 1956, the first airplane to fly three times the speed of sound. The Bell X-2 was also the first aircraft to reach an altitude of 126,200 feet.

The Flight Testing gallery blends serious scholarship with lighter, popularly oriented exhibit units to demonstrate to the Museum visitor the importance of flight testing to the

The stubby-winged Lockheed F-104 Starfighter, hanging above the west moving staircase, was the United States' first interceptor capable of flying at sustained speeds above Mach 2. When the F-104 joined operational fighter squadrons in 1958, it set so many world speed and altitude records, the aircraft was nicknamed "the missile with a man in it."

The Tomahawk cruise missile was designed to be launched from aircraft, submerged submarines, surface ships, and land platforms. It has the capability of navigating over land and performing low-altitude terrain-following and evasive maneuvers while delivering either a conventional or nuclear warhead on its target.

A wind-tunnel model of the now nearly forgotten research aircraft, the Bell X-2, which on September 27, 1956, became the first aircraft to fly at three times the speed of sound.

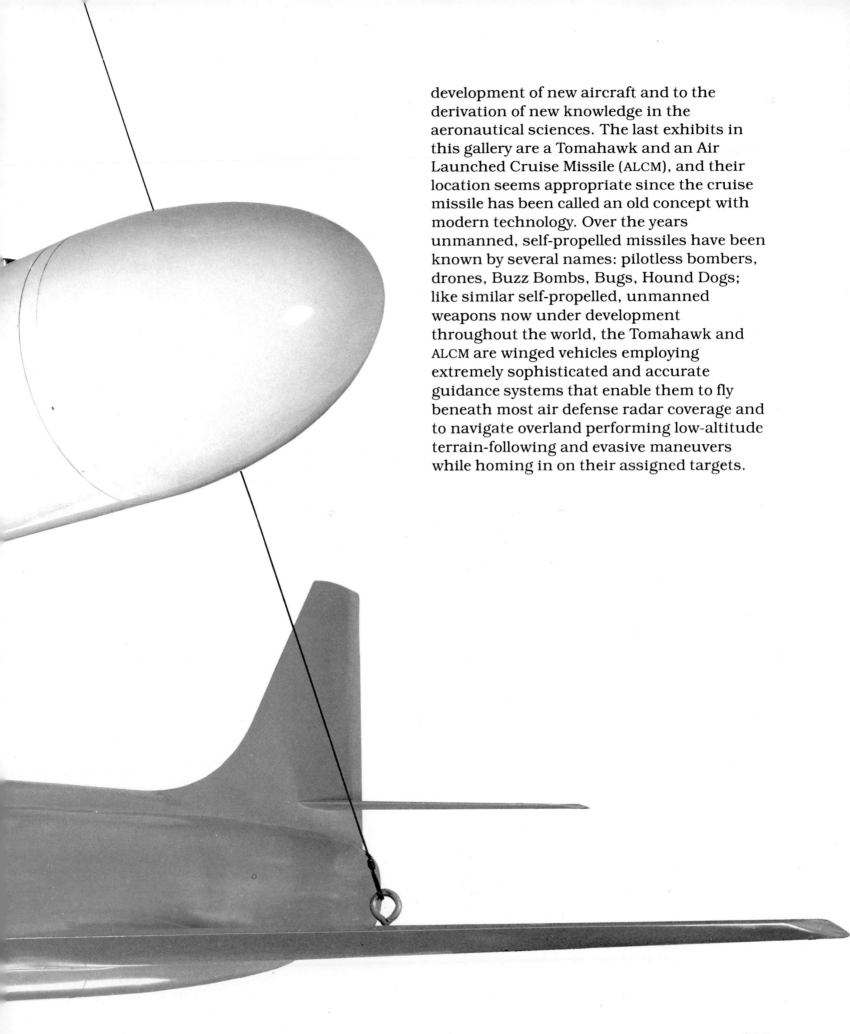

development of new aircraft and to the derivation of new knowledge in the aeronautical sciences. The last exhibits in this gallery are a Tomahawk and an Air Launched Cruise Missile (ALCM), and their location seems appropriate since the cruise missile has been called an old concept with modern technology. Over the years unmanned, self-propelled missiles have been known by several names: pilotless bombers, drones, Buzz Bombs, Bugs, Hound Dogs; like similar self-propelled, unmanned weapons now under development throughout the world, the Tomahawk and ALCM are winged vehicles employing extremely sophisticated and accurate guidance systems that enable them to fly beneath most air defense radar coverage and to navigate overland performing low-altitude terrain-following and evasive maneuvers while homing in on their assigned targets.

Chet Engle. Chuck Yeager. 1964. Oil on panel, 31½ x 39½″.
Gift of Lockheed Aircraft Corporation

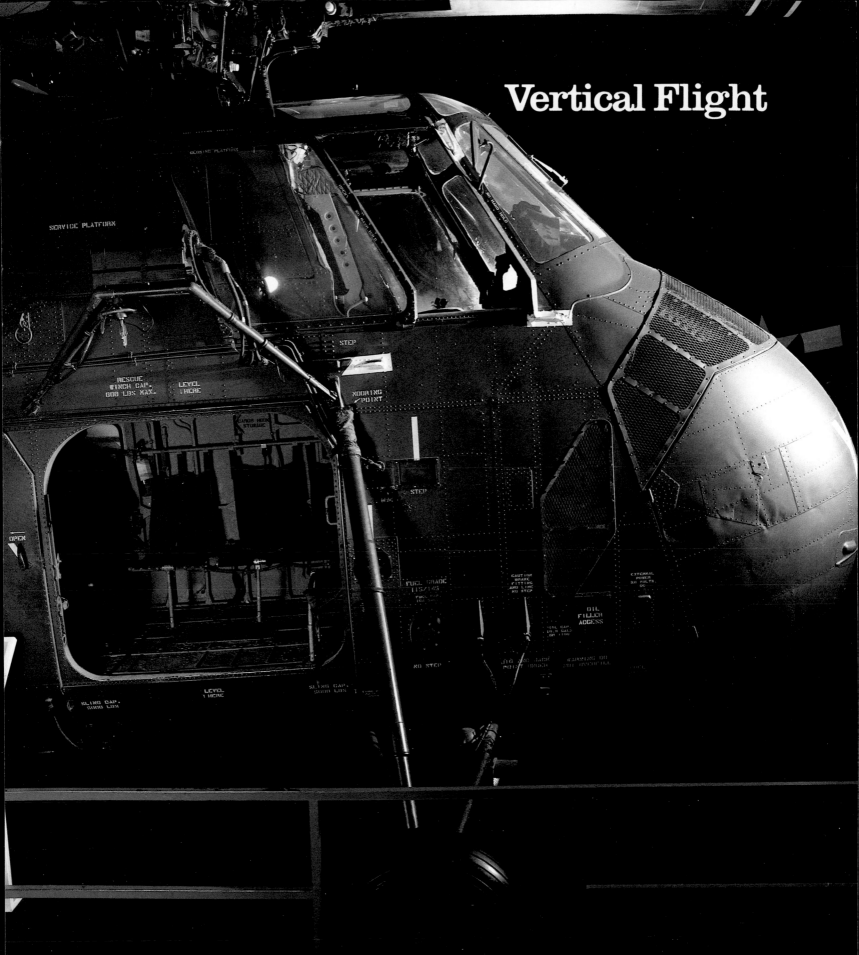

Vertical Flight

Thomas Edison's comment about the Wright brothers' flight was that no aircraft could be considered truly practical until it could rise from the ground and settle back again, *vertically*. It must then have seemed an impossible task since, as far back as the twelfth century, when the Chinese inserted wooden wings into the top end of a stick, then spun them into the air, men had been trying unsuccessfully to achieve vertical flight. And yet on November 13, 1907, not quite four years after the Wrights' twelve-second flight near Kitty Hawk, a twin rotor-bladed, rigidly trapeze-braced, bicycle-wheeled, and seemingly perfectly symmetrical machine built by Paul Cornu of France developed enough lift to raise itself and its pilot about six feet off the ground for almost twenty seconds. Still, another thirty years would pass before the first practical helicopter, the Focke-Achgelis FW-61, would fly. This machine was built by Germany's Dr. Heinrich Focke, whose previous experiments with autogiros were evident in the FW-61's design. The Focke-Achgelis FW-61 could achieve a speed of 76 mph and remain aloft for nearly an hour and a half.

The National Air and Space Museum has such an extensive collection of helicopters, autogiros, and other types of vertical-flight aircraft that it was difficult to choose which vehicles to display in this gallery. The decision was made to exhibit only rotor-wing aircraft here although some of the other forms are pictured or displayed as models. The layout of the gallery is not oriented to the progressive development of rotor-wing aircraft since the majority of the machines on exhibit were developed between the mid-1940s and mid-1950s. Emphasis instead has been placed on displaying the various design configurations rotor-wing aircraft have taken and to demonstrate the principles of flight and control associated with those configurations.

Immediately upon entering the gallery the visitor is confronted by a flock of somewhat bizarre-looking machines suspended from wires or atop platforms set at various levels upon the gallery floor so that one's first image is of an air space crowded with rotor-wing flying machines. This impression reflects the momentary post-World War II euphoric belief that this newly perfected machine was going to become as abundant and commonplace as the automobile and, like the automobile, every family would want to have one. Unfortunately the machines cost too much for mass acceptance.

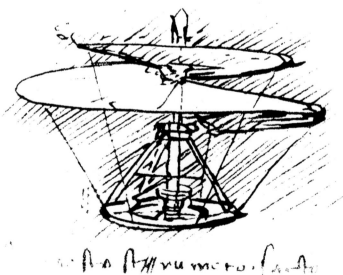

Leonardo da Vinci's fifteenth-century sketch of a Helix, a precursor of modern helicopter design.

Near the entrance to the gallery the visitor comes upon illustrations and models of early rotor-wing attempts: a replica of the Chinese "flying top," which was quite possibly the first man-made object designed to fly under its own power. There, too, is a reproduction of one of Leonardo da Vinci's fifteenth-century notebook sketches of a Helix, his proposed helicopter design. Four men would, theoretically, push turn bars attached to a shaft atop which a large "airscrew" was fixed. The men, racing around the circular platform at the Helix's base, would literally screw their machine into the air. Da Vinci was much more fascinated with designing ornithopters (devices with flapping wings) than helicopters, however, and dozens of drawings of ornithopters fill his notebooks. It has been suggested that if Leonardo da Vinci had devoted one-tenth of the attention and time he spent designing ornithopters on inventing fixed-wing gliders instead, he might have developed a working, man-carrying glider four hundred years before Lilienthal. But then, like Lilienthal, he would have been frustrated by the lack of a suitably powerful but lightweight means of propulsion. That same problem haunted all the early aviation pioneers whether they were attempting to design fixed-wing or rotor-wing craft or, like the Englishman Sir George Cayley, both. It was Cayley who wrote in 1809, "The whole problem is confined within these limits, viz.—to make a surface support a given weight by the application of power to the resistance of air." A full-size reproduction of Cayley's 1796 feather-and-cork "helicopter" is in this gallery. It could rise about fifteen feet into the air. Twelve years earlier, in 1784, two Frenchmen, Launoy and Bienvenu, had demonstrated the lifting capability of two pairs of contrarotating blades which spun about a vertical shaft. Cayley's and the Frenchmen's models recognized and took advantage of the basic principles of vertical flight and were fundamental steps in the development of helicopters.

Encouraged by Cayley, David Mayer, another Englishman, built a full-size man-powered helicopter in about 1825. Mayer anticipated the sort of explanation forwarded by government contract-seeking experimental aircraft designers by about 150 years when he described his machine's failure to leave the ground as "very flattering, if not perfectly successful."

And then in 1842 yet another Englishman, W.H. Phillips, flew what may have been the world's first engine-driven model aircraft that did not depend upon the use of springs, elastics, or other power devices of short duration. Steam from a small boiler in Phillips' machine pushed through a hollow rotor shaft, then out through rear-facing vents at the tips of the propeller blades. This primitive form of jet propulsion caused the blades to spin and the model craft to rise. Unlike Langley's Aerodome #5, however, Phillips' model was incapable of any sustained flight.

As the twentieth century commenced, full-scale helicopters began to meet with some success as designers from all over the world focused their attention upon the problems of vertical flight. In 1907 two Frenchmen, Louis Breguet and Charles Richet, built the first manned helicopter to leave the ground. Their machine, the Breguet Gyroplane #1, rose about three and a half feet into the air, but assistants placed at each of the machine's four corners were required to provide control and stability. Paul Cornu's helicopter, which rose six feet into the air that same year and did not require tie-down ropes, must be considered the first "successful" helicopter although, like the Breguet Gyroplane, Cornu's helicopter lacked adequate control. In Russia, in 1909, Igor Sikorsky, then twenty years old, built his first helicopter but it lacked power to fly. The following year he built a helicopter that could lift its own weight—but not, regrettably, the additional weight of its pilot. Disillusioned, Sikorsky abandoned helicopters for airplanes and, in 1913, built the world's first four-engined airplane.

In 1912, Jacob Ellehamer, a Danish

Igor Sikorsky's original XR-4 (1942), the first helicopter in the world to enter production.

aviation pioneer, constructed a helicopter with two coaxial rotors, but in the four years he devoted to it, his machine never rose more than a few inches off the ground and he gave up. Ten years would pass as helicopter inventors continued to be frustrated by their inability to control their machines, and then in 1923, rotor-wing aircraft took a new turn with the development and testing of autogiros.

An autogiro requires the forward pull of a conventionally mounted aircraft engine and propeller to create the moving airstream through which its "lifting wings"—a set of windmilling rotor blades—glide. The important distinction between helicopter blade action and autogiro blade action is that air flows *down* through helicopter blades and *up* through autogiro blades. Juan de la Cierva, a Spaniard, created a successful autogiro in 1923 when he attached flapped hinges to each of his autogiro's four rotor blades, thus providing his machine with the balance and stability essential to helicopter control.

Oemichen's 1924 helicopter must have been one of the most astonishing machines. Powered by a single 120 hp engine, the Frenchman's craft had thirteen separate transmissions going to four main rotors for lift, two propellers for forward propulsion, a vertical propeller for steering, and five auxiliary horizontal propellers for control. Incredibly enough this twelve-propellered machine made over 1,000 flights, many of which lasted several minutes in duration. One is tempted to say it was later converted into a sawmill, but there is no evidence to support that.

In 1930 von Baumhauer, a Dutchman, built a machine that had a single main rotor powered by a 200 hp engine and a small tail rotor powered by an 80 hp engine to provide control. This small tail rotor would counteract the torque of the main rotor and although Baumhauer's helicopter was damaged before his design could prove itself, Sikorsky's subsequent success utilizing Baumhauer's principle as a prototype shows the importance of the Dutchman's vision. Seven years later, in 1937, Germany's Dr. Heinrich Focke, as mentioned, built his

Igor Sikorsky proudly sits in the rescue sling of one of his R-4s.

The Kellett YO-60 autogiro was developed in 1942 as an observation aircraft for the United States Army Air Force.

Focke-Achgelis FW-61, the world's first practical helicopter.

In 1940, Igor Sikorsky returned to the study of helicopters after an absence of thirty years. Now an American citizen and already famous for his multi-engine aircraft and flying boat designs, Sikorsky, in 1941, established a helicopter world endurance record with the VS-300, his first successful helicopter design. Based on this success, the U. S. Army awarded a development contract to Sikorsky for a machine produced in sufficient quantity to fill military needs. The VS-300 led to the XR-4, which completed the first extended cross-country flight by covering 761 miles from the Sikorsky plant in Connecticut to the Wright Field Army Base in Ohio on May 18, 1942. With a few modifications and minor design refinements the XR-4 became the R-4, the first helicopter in the world to enter production. One hundred and thirty-one R-4s were built during World War

II and the original XR-4 is now on display in the Vertical Flight gallery.

Among the other major artifacts on display in this gallery is the Kellett XO-60 autogiro, which crouches like some giant insect near the entrance door. An autogiro's ability to provide a slow, stable platform, to fly close enough to the ground for observation, and to make near-vertical take-offs and landings was quickly recognized as having military potential. In the Kellett autogiro the pilot sat forward of the observer, whose seat could be swiveled to face the front or rear. The cockpit canopy bulged over the fuselage so that the occupants could look down over the sides of the aircraft without sticking their heads into the wind. Windows were also provided in the fuselage floor. After several modifications to the rotor pylon, landing gear, and tail fins, the aircraft was accepted by the Army Air Forces in 1943. Unlike earlier autogiro designs, the Kellett's rotor wing was linked to the engine for "jump" starts.

The Kellett XO-60 autogiro crouches like a giant insect near the Vertical Flight gallery's entrance door.

The rotor would be turned to near-takeoff speed while the autogiro was still at rest. This, combined with normal propeller thrust, made a short takeoff possible. On the ground the blades could be folded back for easy storage and transport.

Also on exhibit is the Pitcairn AC-35 roadable autogiro, a good idea whose time never came. Designed in response to the dream of having an airplane in everyone's garage, the autogiro could fly at about 75 mph, land in a confined space, and then the pilot would emerge, fold back the rotor blades, disengage the propeller, climb back in and drive off down the road at a modest 25 mph, turn into his driveway and park his Pitcairn in his garage. The cabin seated two persons side by side with space for hand luggage behind them. The engine was connected by a shaft to the tailwheel, which provided the motor power on the road; the front wheels were used for steering.

James G. Ray, vice-president and chief test pilot of the Autogiro Company of America, flew the Pitcairn autogiro to the capital on October 2, 1936, landed in a small downtown Washington park near the Department of Commerce building, and then converted the autogiro to its roadable configuration. With a slightly unhappy looking police escort, Ray drove the autogiro to the main entrance of the Commerce building, where John H. Geisse, chief of the department's Aeronautics Branch, accepted the craft for the Department of Commerce and it was taken to the department hangar at Bolling Field for testing. As recently as the mid-1960s a company attempted to produce autogiros based on the Pitcairn design, but it went out of business.

A second Sikorsky helicopter in this gallery is his larger, later UH-34D medium-assault helicopter—a machine familiar to post-Korean War and Vietnam-era servicemen. Designed in 1954 initially for antisubmarine warfare (ASW) and designated the HSS, this design's performance was so outstanding it achieved world speed records: 100 kilometers at 141.9 mph; 1,000 kilometers at 132.6 mph. Navy pilots called it the "pushbutton" helicopter since when placed on automatic pilot it would maintain an 80-knot airspeed two hundred feet above the ocean and then, at a preselected spot, automatically dip down to fifty feet, hover and lower its sophisticated sonar gear into the water. The Marines and Army used this type of helicopter for assault missions, wire laying, artillery spotting,

October 2, 1936: James G. Ray of the Autogiro Company of America driving down Washington, D.C.'s Constitution Avenue on his way to the Commerce Building in a Pitcairn AC-35 roadable autogiro.

A Marine Corps UH-34D was used in the recovery of America's first astronaut, Alan B. Shepard.

medical evacuation, and as a troop and supply transport. A UH-34D in 1961 fished Alan Shepard and his *Freedom 7* spacecraft out of the water following Shepard's first American in space suborbital testing of the Mercury spacecraft. This event marked the beginning of helicopter use for the recovery of space vehicles. The Museum's specimen wears the markings of Marine Medium Helicopter Squadron 163, a combat unit which served in the Da Nang area of Vietnam in 1965 and was one of the most highly decorated Marine helicopter squadrons of the Vietnam War.

That little bubble-canopied Bell VH-13J in the gallery is the first helicopter to carry a President of the United States. Although then-President Dwight D. Eisenhower made only one trip in it—a quick hop from the south lawn of the White House to a military command post not far from Washington as part of a military exercise on July 13, 1957— the significance of this flight is that it marked the coming of age of helicopters. If officials were willing to entrust the life of the President of the United States to a helicopter then those machines must be safe.

That bright yellow small twin coaxial counter-rotating helicopter-bladed aircraft was built by a teenager, Stanley Hiller, Jr. It is an XH-44, more popularly known as a "Hiller-copter," and he was seventeen in 1942 when he started work on it. By the time he was nineteen he was using it to make numerous demonstration flights. The Hiller-copter was the first successful twin coaxial counter-rotating blade helicopter in the United States. Hiller wanted to design a helicopter without a tail rotor because he believed tail rotors robbed almost 40 percent of an engine's power from the lifting rotor. Counter-rotating blades, Hiller felt, would eliminate much of the vibration existing in then-conventional helicopters which made them so difficult to fly. Hiller's counter-rotating blades canceled out each other's torque, and the craft was so stable that Hiller liked to demonstrate the success of his design by sticking both hands out the craft's windows while in flight. Only one XH-44 was built since subsequent models were so improved.

Another Hiller creation in the gallery is his "Flying Platform," which was designed to be an "airborne motorcycle." It operated on the ducted-fan principle; two counter-rotating

Young Stanley Hiller, Jr., lands his float-equipped XH-44 in a backyard swimming pool.

Hiller's Flying Platform was designed as an airborne motorcycle for both military and civilian use.

blades in the duct provided lift and propulsion and the craft would move in the direction the pilot leaned. Its inherent tendency to right itself made sustained flight in any one direction difficult to achieve. Another seemingly good idea was the Pentecost "Hoppicopter," which was a pair of engine-driven counter-rotating blades a man wore over his shoulders like a parachute harness. In theory a man wearing this engine-and-blade assembly could go 80 mph and reach 12,000 feet. The problem lay in the landing gear: the wearer's legs. If the pilot stumbled or tripped on landing or take-off the whirling blades would strike the ground and shatter into a mass of jagged splinters, thus ending the flight and quite possibly the wearer's life.

The Bensen Gyro-Glider (1954) and Focke-Achgelis FA-330 (1942) were rotor-winged gliders designed to be towed. The Focke-Achgelis was pulled behind a submarine to provide the German U-boat with long-range visibility. The pilot-observer, being towed along at an altitude of 200 to 500 feet, could telephone the submarine commander information from his higher vantage point. The FA-330's use was restricted by the discovery that they, too, could be seen from greater distances: they showed up on radar.

Although the dream of a flying machine in

The Pentecost Hoppicopter was supposed to propel a man up to 80 mph at a maximum altitude of 12,000 feet.

*The bubble-nosed Bell VH-13J Ranger became, on July 13, 1957,
the first helicopter to carry a president of the United States.*

every man's garage has never been realized, visitors to the Vertical Flight gallery emerge with a greater understanding of the variety of uses to which vertical-flight aircraft have been put. Helicopter airliners shuttle between this nation's major airports and downtown terminals, eliminating the hassle of crowded rush-hour streets. Helicopters save businessmen time, moving them swiftly from point to point. The helicopters' suitability for rescue work is well known; their ability to hover over otherwise inaccessible areas makes them indispensable for this task. The downwash of a helicopter's rotor blades is useful for fighting fires. A helicopter can reach a crash fire-site quickly and once there the downwash helps disperse the heat and flame so that rescuers can approach a crash fire in greater safety. This same downwash evenly distributes agricultural chemicals over foliage and hard-to-reach ground areas. Traffic and police helicopters crisscross over our cities. "Flying cranes" assist in an endless variety of aerial lifting tasks. The military application of helicopters as gunships, air assault weapons, and medical evacuation and supply ships has changed the face of war.

Although the Vertical Flight gallery naturally concentrates on rotor-wing aircraft, other forms of vertical flight are mentioned and shown. The increase in engine thrust-to-weight ratio has made possible vertical flight that is not dependent upon the aerodynamic lift of rotating wings. Several concepts including aircraft with tilt-wings, tilt-props, and directed thrust lift reflect designers' attempts to replace the helicopter—which is limited in forward speed—with high-performance vertical take-off-and-landing (VTOL) machines. Photographs depict the 1953 Bell ATV, a gimbal-mounted jet engine aircraft that was a feasibility test model only, and the propeller-driven Convair XFV-1 vertiplane of 1954 known as the "Pogo Stick," which was built as a test vehicle. A Hawker-Siddeley P. 1127 Kestrel is on display

The German Focke-Achgelis FA 330, here being ground-tested, was designed in 1942 to be towed behind U-boats to provide an observation platform.

234

The Museum's specimen Pitcairn AC-35 is an example of a good idea whose time never came. Designed in 1935 to be the "aircraft in everyone's garage," the Pitcairn roadable aircraft could reach a modest 25 mph maximum speed on the ground and about 75 mph in the air.

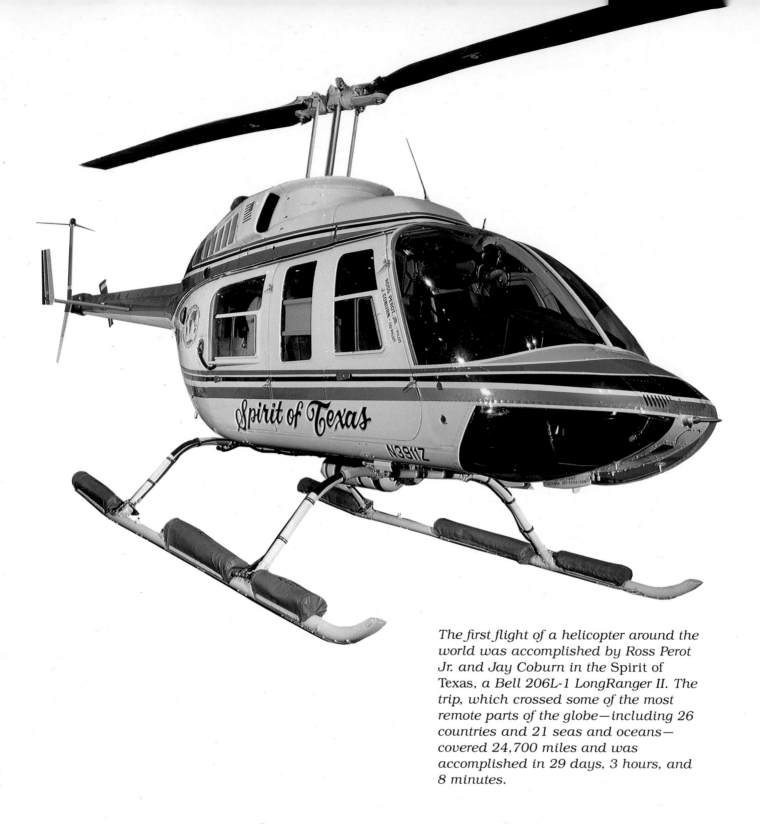

The first flight of a helicopter around the world was accomplished by Ross Perot Jr. and Jay Coburn in the Spirit of Texas, *a Bell 206L-1 LongRanger II. The trip, which crossed some of the most remote parts of the globe—including 26 countries and 21 seas and oceans— covered 24,700 miles and was accomplished in 29 days, 3 hours, and 8 minutes.*

in the Flight Testing gallery on this same floor. The Kestrel, a 1965 aircraft which in its advanced version, the Harrier, serves with U.S. Marine Corps units, is a British design employing the modified jet-engine vectored thrust principle, which permits the pilot to direct his thrust vector upward or forward by deflecting the jet exhaust downward or rearward.

The Vertical Flight gallery touches upon most, if not all, of the many facets by which vertical flight can be achieved. Today all shapes and sizes of helicopters dart like dragonflies through the air. Perhaps, in the not so distant future, a Volkscopter will be developed for everyone's garage; but somehow, the resulting chaos might be one aspect of progress we could do without.

Sea-Air Operations

Model builder Stephen Henninger worked about 1,000 hours a year for 12 years between November, 1970, and August, 1982, to construct this 1:100-scale 11' 1½"-long and 2'6¾"-wide model of the USS Enterprise (CVN-65). Henninger spent about 4,000 hours just building the ship's 83 aircraft; the four E-2 Hawkeyes were scratch-built and required about 200 hours labor each, and the last aircraft was completed 11 years after the first was started.

R.G. Smith. Curtiss SOC-1. *1975. Oil on canvas, 33½ x 23½". Gift of the M.P.B. Corporation*

April 5th. At sea.

I've never counted how many times a day the bos'n's mates pass the word, but I'd bet we never have 15 consecutive minutes without their "Peeeeep! Now, hear this—" Between sunrise and the end of the working day, we must hear 50 or 60 calls. They are so familiar by now that all we need is the opening phrase; the rest we can supply from memory. As soon as the pipe peeps and the bos'n's mate says, "Turn to!", we know that this will follow: "All sweepers, man your brooms! Clean sweep-down, fore and aft! Empty all trash cans and spit-kits"— delivered in a singsong cadence, with the last syllable of each phrase drawn out and falling.

Here are some of the calls we hear most often:

"All extra-duty men lay down to the master-at-arms' shack."

"General Quarters! General Quarters! All hands man your battle stations on the double!"

"Now, the smoking lamp is out throughout the ship while taking aboard aviation gasoline and fuel oil."

"Relieve the watch!"

"Now five hands from the K division and ten hands from the fifth division report to the First Lieutenant at Number 1 crane."

Pat Garvan told me he knew an officer of the deck who got so fed up with the whole business, he had this word passed: "Now all those who have not done so, do so immediately." My own favorite is, "Now the man with the key to the garbage-grinder lay below and grind same."

—*Aircraft Carrier,*
by Lt. Cmdr. J. Bryan III

The entrance to the USS Smithsonian's Hangar Deck is through the traditional Quarter Deck.

A Douglas SBD-6 Dauntless is suspended above a Grumman FM-1 (F4F-4) Wildcat in the gallery's realistic Hangar Deck.

GALLERY
DECK
115 V. OUTLET

CVN-76 USS SMITHSONIAN

The Sea-Air Operations gallery is unquestionably one of the most ambitious exhibits in the National Air and Space Museum since it attempts to recreate the environment of a United States Navy aircraft carrier at sea.

As the Museum visitor approaches the USS *Smithsonian* (CVM-76), a "carrier for all times," he can hear the bos'n's pipes before he even reaches the quarter-deck, the traditional entrance to Navy ships. There he passes between brass stanchions hung with Navy macramé, a highly polished ship's bell glints overhead, the pipes *peep* him aboard, and the visitor finds himself thrust into an aircraft carrier's hangar deck crowded with U.S. Navy and Marine Corps aircraft. All about, one sees actual carrier hangar deck artifacts, wing tanks, munitions, hoses, controls, watertight doors. Stencils direct one's attention to fire extinguishers, lifejackets, safety equipment. No attempt has been made to model the hangar deck after any specific aircraft carrier or period since the airplanes on exhibit span some forty years of Navy flight, but the atmosphere seems to fit more the World War II period than any other.

An aircraft carrier's hangar deck serves primarily as a protected space for working on aircraft, and on an actual ship it is a vast area covering almost two acres that spans the ship's width and nearly two-thirds of its length. Space on a hangar deck is at a premium; parked and tied-down aircraft, wings folded, are wedged inches apart. Deck space not filled with aircraft is occupied by the sort of equipment needed for the maintenance of highly complex aircraft and weapons systems: electronic test equipment, service parts, spare engines, air compressors, et cetera; and the surrounding sections or bays contain workshops, storage rooms, engine test facilities, and fire mains. These bays are separated from the hangar deck by huge retractable steel doors which can be shut in case of fire or emergency. Access doors in the steel deck open when munitions from the storage rooms many decks below are brought up on the bomb elevators.

During flight operations and peak maintenance hours, a hangar deck is a scene of frenzied activity with aircraft being transferred from the hangar deck to the flight deck by the elevators which extend out from the sides of the ship. The atmosphere in the *Smithsonian*'s hangar deck is enhanced by the open hatch against the gallery wall, through which the visitor sees the sea rushing past. A continuous-film loop of about five minutes' duration projects scenes of escort destroyers taking up stations, rescue helicopters returning to the ship. Standing there, one can almost feel the hangar deck heave.

The first aircraft the visitor sees in the hangar deck is a stubby Grumman F4F Wildcat.* The Wildcat was the Navy's and the Marine's basic fighter and the only carrier-based fighter operating with the Navy at the

---

* Actually, since the Museum's specimen was manufactured under license by the Eastern Aircraft Division of General Motors, this version was designated FM-1.

outbreak of the Second World War. F4Fs first saw action at Wake Island when that tiny Pacific outpost was attacked on December 8, 1941, and Marine Fighter Squadron VMF-211 lost eight of its twelve F4Fs that first day. For two more weeks the remaining Wildcats fought heroically against the Japanese Zeros, which could outmaneuver and outrun them, and before December 22nd, the day the Japanese landed on Wake Island and the last two Wildcats were destroyed, the Marine's F4Fs even managed to sink a Japanese cruiser and a submarine while continuously breaking up air attacks on their island.

Saburo Sakai, Japan's greatest fighter pilot to have survived the war, recalls the day he saw a single Wildcat attacking three Zeros 1,500 feet below him:

The Zeros should have been able to take the lone Grumman without any trouble, but every time a Zero caught the Wildcat before its guns the enemy plane flipped away wildly and came out again on the tail of a Zero. I had never seen such flying before.

I banked my wings to signal [my wingman] Sasai and dove. The Wildcat was clinging grimly to the tail of a Zero, its tracers chewing up the wings and tail. In desperation I snapped out a burst. At once the Grumman snapped away in a roll to the right, clawed around in a tight turn, and ended up in a climb straight at my own plane. Never had I seen an enemy plane move so quickly or so gracefully before; and every second his guns were moving closer to the belly of my fighter. I snap-rolled in an effort to throw him off. He would not be shaken. He was using my own favorite tactics, coming up from under.

I chopped the throttle back and the Zero shuddered as its speed fell. It worked; his timing off, the enemy pilot pulled back in a turn. I slammed the throttle forward again, rolling to the left. Three times I rolled the Zero, then dropped in a spin, and came out in a left vertical spiral. The Wildcat matched me turn for turn. Our left wings both pointed at a right angle to the sea below us, the right wings to the sky.... On the fifth spiral, the Wildcat skidded slightly. I had him, I thought. But the Grumman dropped its nose, gained speed, and the pilot again had his plane in full control. There was a terrific man behind that stick.

He made his error, however, in the next moment. Instead of swinging back to go into a

sixth spiral, he fed power to his engine, broke away at an angle, and looped. That was the decisive split second. I went right after him, cutting inside the Grumman's arc, and came out on his tail. I had him. He kept flying loops, trying to narrow down the distance of each arc. Every time he went up and around I cut inside his arc and lessened the distance between our two planes. The Zero could outfly any fighter in the world in this kind of maneuver.

When I was only 50 yards away, the Wildcat broke out of his loop and astonished me by flying straight and level. I pumped 200 rounds into the Grumman's cockpit, watching the bullets chewing up the thin metal skin and shattering the glass...the Wildcat continued flying as if nothing had happened. A Zero which had taken that many bullets into its vital cockpit would have been a ball of fire by now. I could not understand it. I slammed the throttle forward and closed in...until our planes were flying wing-to-wing formation. I opened my cockpit window and stared out. The Wildcat's cockpit canopy was already back, and I could see the pilot clearly. He was a big man, with a round face. He wore a light khaki uniform. He appeared to be middle-aged, not as young as I had expected.

For several seconds we flew along in our bizarre formation, our eyes meeting across the narrow space between the two planes. The Wildcat was a shambles. Bullet holes had cut the fuselage and wings up from one end to the other. The skin of the rudder was gone, and the metal ribs stuck out like a skeleton. Now I could understand his horizontal flight. Blood stained [the pilot's] right shoulder, and I saw the dark patch moving downward over his chest. It was incredible that his plane was still in the air.

But this was no way to kill a man! Not with him flying helplessly, wounded, his plane a wreck. I raised my left hand and shook my fist at him, shouting, uselessly, I knew, for him to fight instead of just flying along like a clay pigeon. The American looked startled; he raised his right hand weakly and waved.

I had never felt so strange before...I honestly didn't know whether I should try to finish him off. Such thoughts were stupid, of course. Wounded or not, he was an enemy, and he had almost taken three of my men a few minutes before. However, there was no reason to aim for the pilot again. I wanted the airplane, not the man.

I dropped back and came in again on his tail...

—*Samurai!*
by Saburo Sakai with Martin Caidin

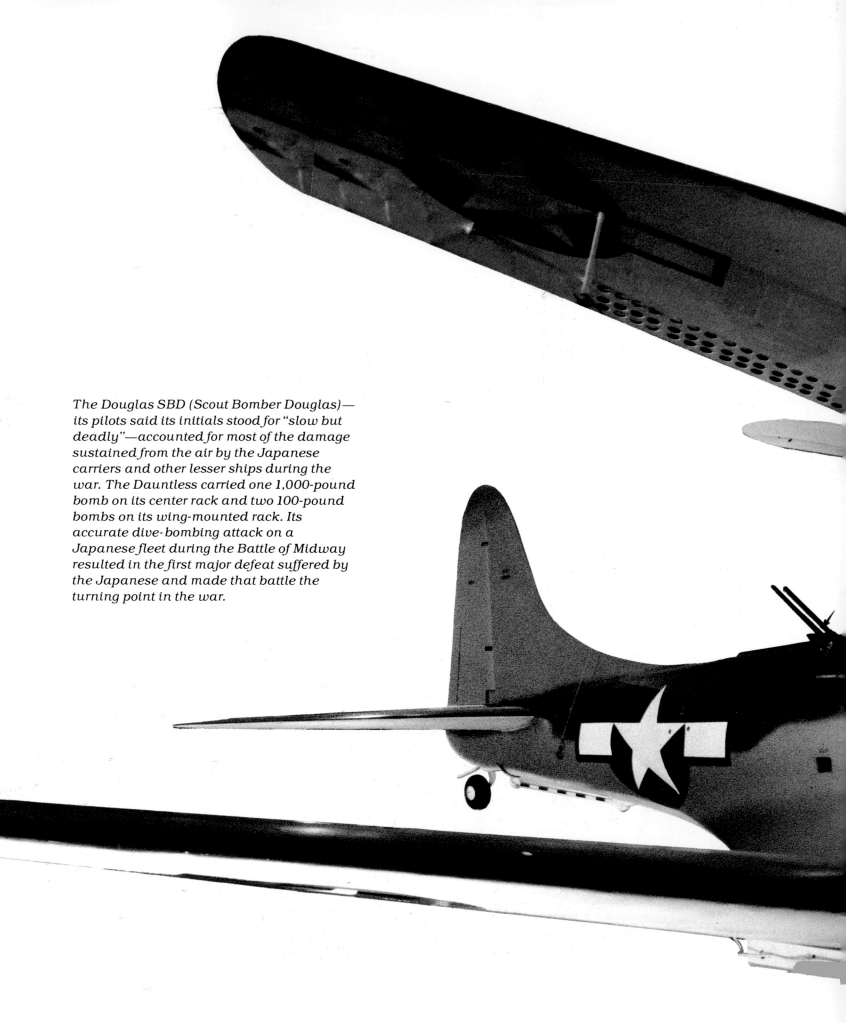

The Douglas SBD (Scout Bomber Douglas)—
its pilots said its initials stood for "slow but
deadly"—accounted for most of the damage
sustained from the air by the Japanese
carriers and other lesser ships during the
war. The Dauntless carried one 1,000-pound
bomb on its center rack and two 100-pound
bombs on its wing-mounted rack. Its
accurate dive-bombing attack on a
Japanese fleet during the Battle of Midway
resulted in the first major defeat suffered by
the Japanese and made that battle the
turning point in the war.

244

The F4F, called the Martlet *by the British, during its duty with the Fleet Air Arm was the first United States aircraft in British service to shoot down a German plane (a JU.88) during World War II.*

Saburo Sakai set fire to the Wildcat's engine and he saw the pilot bail out and drift down toward the beach at Guadalcanal. Despite the Zero's higher performance, which was made possible by the sacrifice of armor plating, self-sealing fuel tanks, and lighter armament in order to gain long range, good maneuverability, and high speed, in the hands of a skilled pilot the Wildcat could hold its own. It was a tough little fighter and its ratio of victories to losses during World War II was a surprising 6.9 to 1. And even when it was superseded by higher-performance carrier aircraft such as the Grumman F6F Hellcat and the Chance-Vought F4U Corsair, the little Wildcat

continued in operation off short-decked escort carriers throughout the war. The cowling on the Museum's F4F was from one of Wake Island's defenders and until the mid-1960s had served as part of the Wake Island Memorial "dedicated to the gallant Marine, Naval, Army and Civilian personnel who defended Wake against overwhelming Japanese invasion armadas, 8 thru 23 December 1941."

Hanging above the Wildcat, as though suspended in flight, is a Douglas SBD-6 Dauntless, the standard Navy dive bomber at the outbreak of the war, whose accurate dive-bombing attacks upon the Japanese fleet during the Battle of Midway resulted in the

An SBD-5 photographed making a twilight patrol over Wake Island.

first major naval defeat suffered by the Japanese.

A special section at the left of the Sea-Air Operations gallery is given over to an exhibit on the Battle of Midway, the battle considered by naval historians to have been the turning point in the war against Japan, and to have been a battle typical of World War II naval battles in the Pacific in that it was fought by carrier forces and submarines without direct contact between surface forces. During that battle the Japanese lost four of their fleet carriers, two-thirds of their fleet carrier total, thereby suffering a blow from which their naval power in the Pacific would never recover. The

SBD—its initials stood for Scout Bomber Douglas although its pilots said it was for "Slow But Deadly"—accounted for most of the damage from the air sustained by Japanese aircraft carriers and other surface ships during the war. It was a compact, rugged, and easily serviced aircraft that gave an excellent account of itself in every engagement in which it took part—and it took part in more than was intended since it took longer than expected for its replacement, the Curtiss SB2C, to become acceptable for aircraft carrier service. There was a saying about the SB2C, too: "If there's a harder way to build a plane, Curtiss will find it." As for the gull-winged Corsair, one

The Boeing F4B-4, one of the best-looking biplane fighters, was the last fixed-landing-gear fighter in the Navy. F4B-4s saw active carrier service until 1937 when they were replaced by Grumman's faster biplane fighters. The Museum's specimen wears the marking of aircraft number 21 Marine Corps Squadron VF-9M. The Army version was designated the P-12.

The Douglas A-4C Skyhawk was sometimes known as "Heinemann's Hot-Rod" after its chief designer, Ed Heinemann, whose philosophy was "simplicate and add lightness." From the late 1950s through the 1960s, the A-4 was the backbone of Navy and Marine Corps light jet attack forces and saw wide use in Vietnam. Unlike many carrier aircraft, the A-4, with its relatively small wingspan (27 1/2 feet), does not have folding wings.

This view of the underside of the Douglas A-4C Skyhawk shows part of the large bomb load carried on combat missions. Also visible are an external fuel tank and a nose wheel tow bar.

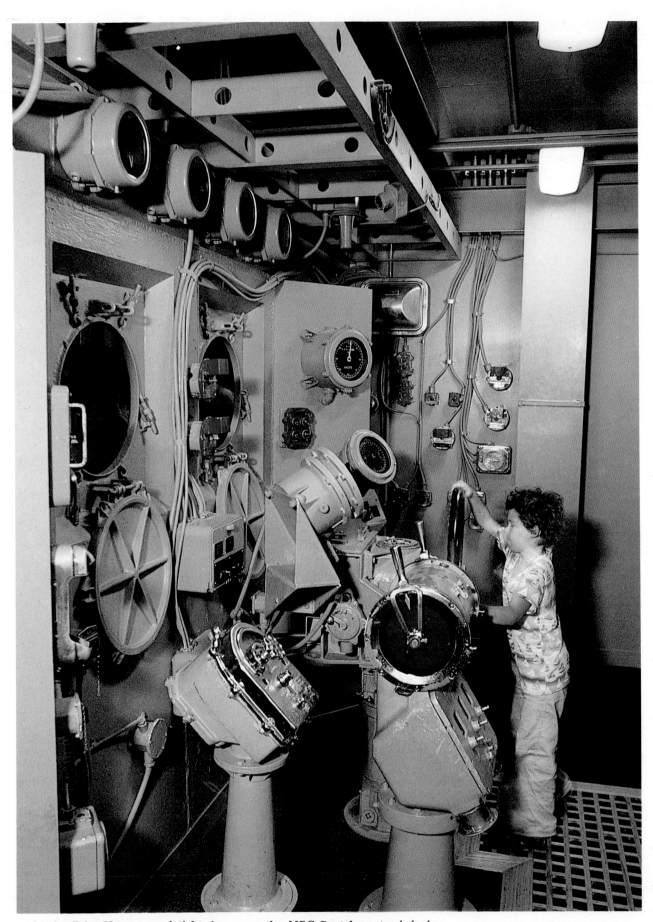

In the Pilot House, a child takes over the USS Smithsonian's helm.

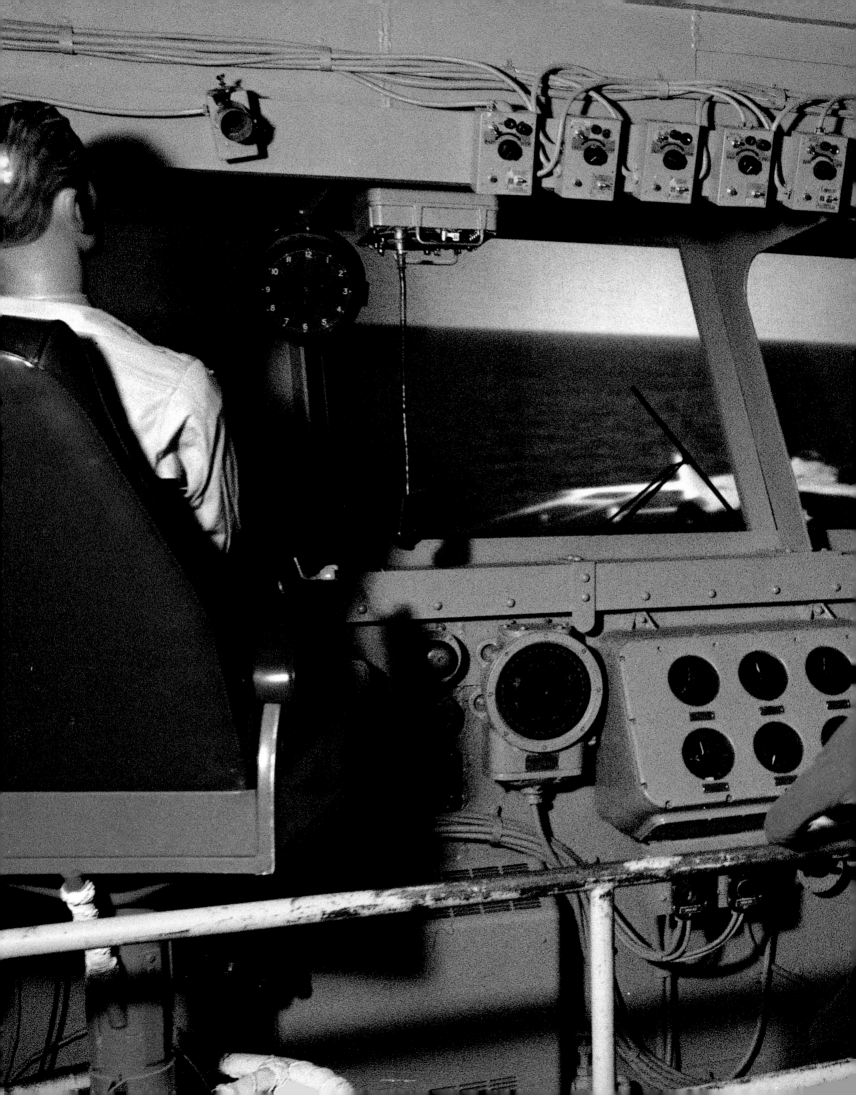

NASM's A-4C is shown on the deck of the USS Bon Homme Richard *off the coast of Vietnam in 1967.*

fighter pilot described it as climbing "like a homesick angel." The SBD Dauntless could carry a 1,000-pound bomb on its center rack and two 100-pound bombs on wing-mounted racks, all externally mounted. Two .50-caliber machine guns were in the nose and the gunner/observer/radioman had a flexible mount carrying twin .30-caliber machine guns in the rear. Unlike the majority of World War II carrier aircraft, the Dauntless' wings did not fold. NASM's specimen wears the markings of aircraft 109, which served in combat with VS-51 on the USS *San Jacinto*.

The sporty-looking little biplane suspended from the ceiling behind the Dauntless is a Boeing F4B-4, the last fixed-gear shipboard fighter to see service in the Navy (the Army version of the F4B was the P-12). The F4B first flew in 1928 and went through various design modifications until production of the model displayed in the gallery began in 1932. F4B-4s remained in active carrier service until 1937 when they were replaced by faster biplane fighters built by Grumman. The F4B-4 on exhibit wears the markings of U.S. Marine Squadron VF-9M.

The one jet on the hangar deck floor is a Douglas A-4C whose wingspan, surprisingly, is two and a half feet shorter than the biplane over its head. And yet, despite its relatively small size, the Skyhawk displayed was able to carry some 5,000 pounds of bombs, missiles, fuel tanks, and gun pods in three stations. Later models, beginning with the A-4E, could carry 8,200 pounds in five stations. The A-4 was sometimes known as "Heinemann's Hotrod" after Ed Heinemann of Douglas Aircraft whose design philosophy was "simplicate and add lightness." This was in response to his concern at the trend toward increasing weight and complexity in combat aircraft. When he and his design team proposed a new attack plane with a gross weight of about one-half the official specification weight of 30,000 pounds, their design was accepted by the Navy and the initial contract was let on the A-4 in June

1952. Because of its small wingspan the A-4 does not have folding wings; elimination of this feature made it possible to build a much simpler, lighter wing and, therefore, a much lighter plane. The A-4 was the primary Navy and Marine light jet attack plane from the late 1950s through the 60s and its versatility, simplicity, and performance made it a valuable asset to both services. Throughout the Vietnam War the A-4 was noted for its accuracy in attacking selected ground targets. The Museum's specimen wears the markings it wore when it was actually assigned to Navy Attack Squadron VA-76 on board the USS *Bon Homme Richard* while operating off the coast of Vietnam from March to June, 1967. The A-4 is displayed with extra fuel tanks.

After strolling around the aircraft in the hangar deck area, the visitor might ascend to the upper level "balcony" where he can get a different perspective on the planes on display. He would then continue along the balcony platform through a simulated watertight door and continue on into the broad passageway between the PRIFLY (Primary Flight Control) area and the Navigation Bridge.

An aircraft carrier's navigation bridge is located high on the forward edge of the carrier's island superstructure and it is the normal duty station of the ship's commanding officer during sea operations. From his large swivel chair on the port side of the bridge the Captain has an unobstructed view of the flight deck and the surrounding sea and air space. The navigation bridge and the pilot house are open to NASM visitors, and one can stand looking over the Captain's shoulder as jets are catapulted into the air and spotted about the flight deck prior to launch. The rear projection movie screens provide an extraordinarily realistic idea of what the bow of an aircraft carrier looks like during air operations.

The PRIFLY, or Primary Flight Control, section of an aircraft carrier serves essentially the same role as an airport's control tower. Here on the Sea-Air Operations gallery's second level, a visitor can watch aircraft approach the flight deck and make their landings. Like the Navigation Bridge, the PRIFLY area is also equipped with the same radio gear, electronic equipment, and telephones which were taken from actual carriers. Although he is protected from the deafening roar of the jets on the flight deck and those making their approaches by walls of armor plate and glass, the visitor can hear what is going on. PRIFLY is the nerve center of a carrier during day operations; it is where the Air Boss can see the entire flight deck, from the two steam catapults at the bow all the way back to the stern where the Landing Signal Officer monitors the approaches of the returning aircraft. And while the visitor watches all the landing activity, he can hear the clatter of teletype, radio chatter, the telephones ringing, the Air Boss' orders to the deck crews. From the instant "Launch Aircraft" is ordered by the Air Boss at dawn until the last aircraft has been landed and secured for the night, constant, precise coordination, both physical and mental, is demanded between PRIFLY, the navigation bridge, the deck crews, and the pilots as they go about their work in a world where there is little margin for error.

The "Ship's Museum" and the Ready Room are located below PRIFLY and the Navigation Bridge on the same level of the Sea-Air Operations gallery as the hangar deck. The Ready Room is a theater in which the NASM visitor receives a pilot's eye view of carrier activity. He is "strapped" into his seat behind the pilot while the jet is hooked onto the catapult and, as the camera watches over the pilot's shoulder, the jet with the visitor as a passenger is shot aloft. The visitor experiences a mock aerial combat, and then, after an inflight refueling, the film produced by the Navy has as its most sensational sequence a carrier landing. The jet banks and turns, the aircraft carrier appears ahead and far below, like a small wooden chip afloat in the ocean, and

although the carrier grows larger and larger as the jet enters the traffic pattern and landing approach, its flight deck never seems large enough for the jet to land safely. One finds oneself inhaling sharply as the jet screams in far too fast for comfort until suddenly the tail hook catches the cable and the jet slams to a stop. Unfailingly, along with the sound effects that accompany this film, there is a gasp from those watching. The Ready Room in the gallery does not attempt to duplicate in detail one found on an actual carrier. Each squadron on a carrier is assigned a Ready Room; it is the pilot's living room, his office, classroom, movie theater. The walls are covered with boards, charts, maps, briefing materials. In one corner a large teletype machine clatters away, providing a constant stream of meteorological data, vital navigation information, and flight procedures which are projected onto a screen for flight crews to make notes on during briefings.

A visitor can browse among photographs, memorabilia, models, and other exhibits tracing the development of sea-air operations in the "Ship's Museum." Here one can find photographs of Henri Fabre's "hydroplane" which, in 1910, became the first airplane to take off from the water. Here, too, are photographs of Eugene Ely's take-off from the deck of the cruiser USS *Birmingham* on November 14, 1910, and his considerably more difficult achievement two months later of landing on a wooden platform built above the afterdeck of the USS *Pennsylvania,* two events of historic importance in the development of aircraft carriers. One sees the history of catapults and the development of Glenn Curtiss' flying boats. The outline of a Nimitz-class aircraft carrier is imposed upon an outline of the National Air and Space Museum, both drawn to the same scale so that the visitor has an idea of how huge modern carriers now are.

So all-pervading is the sense of being on board an actual carrier in the Sea-Air Operations gallery that one feels somewhat disoriented and stunned upon emerging

Visitors examine the cockpit area of the A-4C Skyhawk.

from the hangar deck and reentering the Museum to find oneself on land.

March 16th. At sea.
Babe Herman comes on the squawk-box: "Now, Sullivan, bos'n's mate thoid class, foist division, dial zero—belay that woid!" A long pause, then another whistle, and "Sullivan, bos'n's mate *foist* class, *thoid* division, dial 760!"

They say that smell is the strongest stimulus to memory, but if I ever wanted to recreate shipboard life, I'd be hard put to find a scent that would summon it. Except for coffee and burnt powder, I don't believe a warship has any characteristic smell.

I could do it by sounds, though. There are a dozen to choose from, any one of which would make the *Yorktown* or the *Lexington* take shape before my mind's eye: the irregular rattle of the shutter on a blinker; the ticking of a 40mm director, and the clamor of the 40s themselves, like a regiment of recruits trying to keep step up an iron stairway; the muffled roar of the blowers; the clank of a tool dropped on a metal deck; the riveting hammer of the water taps; the grinding SLAM! of the catapult; the soft iambic *pop* of the line gun; and the ripple of the barriers going down, exactly like the ripple of reef points as a mainsail comes about.

As for the gong that calls us to General Quarters, if I heard that same tone ten years after the end of the war, I'd automatically grab for a helmet.

—*Aircraft Carrier*,
by Lt. Cmdr. J. Bryan III

It is easy to spot among the visitors to the Sea-Air Operations gallery the men who have served on aircraft carriers: they are the ones with their backs to the airplanes and soft smiles on their faces as they read white paint stencilled signs on the abandon ship lockers overhead.

World War II Aviation

Probably no six-year span in the history of aviation saw more astounding advances than the period between 1939 and 1945 during World War II. Biplanes such as the Gloster Gladiator and Fairey Swordfish were operational when the war began; by the war's end, aircraft propelled by jet and rocket engines swept the skies and guided missiles of various types had been employed.

Since the Museum could not possibly show in one gallery more than a very small fraction of the myriad types of aircraft used during World War II, and since none of the 100-foot-plus wingspan multi-engine bombers would fit, the Museum has concentrated on a selection of the best-known fighter aircraft of the major powers—with the exception of Russia. (NASM has no representative Soviet fighter in its collection.) Visitors to the World War II Aviation gallery will find the only Italian Macchi C.202 Folgore (Lightning) known to be on exhibit, an almost equally rare Japanese Mitsubishi A6M5 Zero, Germany's Messerschmitt Bf.109G, Britain's Supermarine Spitfire Mark VII, and America's most outstanding fighter, the North American P-51D Mustang. Representing bombers is the nose section of a twin-engine bomber with one of the most impressive combat records ever achieved by an individual aircraft; it is from the B-26B Marauder *Flak Bait*, which flew 202 combat missions over Germany and German-occupied Holland and France.

Dominating the gallery, however, is Keith Ferris' 25' × 75' mural of the Boeing B-17G *Thunder Bird*, which is so incredibly realistic that the unwary NASM visitor might get the dismaying impression he has entered the World War II Aviation gallery at the precise moment as has a formation of heavy bombers that is under attack by German fighters and flak at 25,000 feet.

In "Cowboys and Indians," his fascinating and impressively researched account of the mission depicted by the Ferris mural, Jeff Ethell tells how in March, 1975, NASM's Curator of Art, Jim Dean, telephoned Keith Ferris with the request that he paint a mural showing a formation of B-17 Flying Fortress bombers "representative of the war they fought in." Ferris accepted and asked artist John Clark to be his painting assistant and Ethell to be his researcher. Ferris, Clark, and Ethell agreed to attempt to document a precise moment in history accurate down to the last detail. They first drew up a list of requirements for the aircraft to be depicted: (1) a veteran camouflaged B-17G of the Eighth Air Force with much wear evident; (2) photos of the aircraft available for reference; (3) a known battle record; (4) name and nose art in reasonable taste; (5) a good combination of mission tally symbols and markings—air division, wing, group, squadron and tail markings; and (6) the specific mission to be depicted had to meet the following requirements: a) it had to take place between July and December, 1944, the height of Eighth Air Force activity; b) during good weather, when there were c) contrails, d) flak, and e) enemy fighters.

The field of eligible B-17s was initially narrowed down to nine aircraft and eventually to the 303rd Bomb Group's veteran B-17G *Thunder Bird* because of her long mission history, her colorful markings, the availability of photographs, and her condition. "Was she ever beat up!" Ethell recalled. "Patches all over her skin, paint that was bleached badly, entire replacement parts that were natural aluminum standing out against the olive drab. And she was

strictly 'GI' having been flown by so many crews." Once the aircraft was selected, Ethell had the chore of going through the records of each of the 116 missions flown by *Thunder Bird* between July and December, 1944. Only one mission fit their requirements: Mission 72, flown August 15, 1944, against the Luftwaffe fighter airfield at Wiesbaden, Germany.

At 11:22 that morning more than thirty-five years ago, the B-17s dropped their bombs and turned back toward their base at Molesworth, England. Although the raid on Wiesbaden is *Thunder Bird*'s seventy-second bombing mission, it is one of her crew's first exposures to combat. They are flying now at 25,000 feet, visibility is between 10 and 15 miles. By 11:45 German anti-aircraft fire is intense, but the flak is inaccurate, bursting behind the bombers and to their left. Three Luftwaffe fighter wings have already scrambled. JG 300, composed of heavily armed Focke-Wulf Fw 190A-8s and Messerschmitt Bf.109Gs, is the primary unit responsible for attacking the bomber formations. JG 300 is led by Major Walter Dahl. His Sturmgruppe—literally, a Storm Group, a large wedge formation used to penetrate the tight bomber formations—has become famous throughout Germany for its pilots' tactic of ramming bombers if necessary to bring them down, then parachuting to safety. Dahl sees that the bombers are without fighter escort and notifies his division, which gives permission to attack.

With Dahl at the point, the German fighters scream in. They sweep through the formation, the heavy cannon in Dahl's Fw 190 rips the wing off a B-17, and as the Fortress tumbles, Dahl sees three chutes open.

Thunder Bird is in the lead section of the 303rd Bomb Group's thirty-nine B-17s when Dahl's fighters strike the lower section.

Pulling back on his stick after his firing run, Unteroffizier Leopold Bigalke in his Bf.109G *White 12* climbs for an attack on the lead section, then banks to sweep across *Thunder Bird*'s right flank toward a B-17 ahead. Lieutenant Klaus Bretschneider in

his Fw 190A-8 *Red One* swings in behind Bigalke's Bf.109 and to its left to close in on the B-17 *Bonnie B* just below *Thunder Bird*'s inboard starboard engine. Both fighters are taking advantage of the bomber formation's contrails as they press in for their attacks. This is the precise moment frozen in Keith Ferris' mural.

A touching footnote to the extraordinary accuracy of the mural lies in an incident that occurred shortly after the Museum opened. A lady from Doylestown, Pennsylvania, visited the World War II gallery because her husband Bill had flown B-17s during the war until his plane was shot down and he was killed. As she stood before Ferris' mural she was stunned by the *Thunder Bird* co-pilot's resemblance to her late husband. Although the man's face was nearly covered by an oxygen mask, she was ready to swear that he was Bill. When she examined the crew photo in front of the mural she had the shock of seeing her late husband again. Only two of the crew that had flown *Thunder Bird* on the mission depicted survived the war; the others, Bill included, were killed just ten days after the Wiesbaden mission when a German 88mm flak round exploded within the fully loaded bomb bay of *Myasis Dragon*, the B-17 they were assigned to on that mission.

The B-17 was one of the best-known and most widely used heavy bombers of World War II. As General Carl "Tooey" Spaatz, the Air Force's first Chief of Staff, told wartime Chief of the United States Army Air Forces General "Hap" Arnold, "The B-17 was the single weapon most responsible for the defeat of Germany." More than 12,700 B-17s were produced by Boeing. Armed with thirteen .50-caliber machine guns including those in its chin, nose, top, and belly (ball) turrets, the turbo-supercharged four-engine bomber could carry 6,000 pounds of bombs some 2,000 miles at 30,000 feet. By the end of 1943, B-17Gs such as those in the mural were being used by many units in the European Theater.

Just as the Sea-Air Operations gallery across the way concentrates on the Pacific

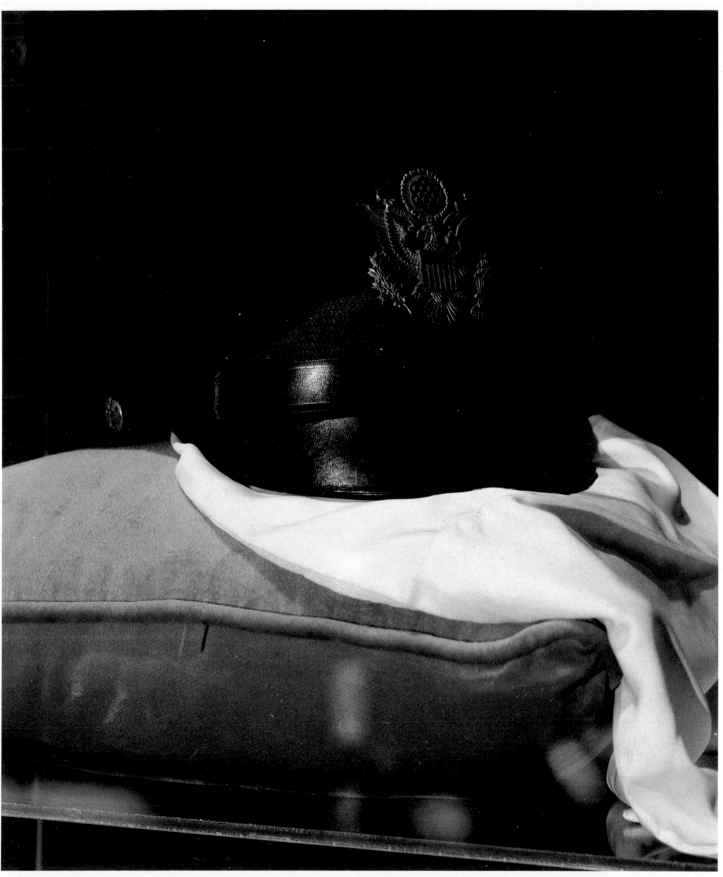

The "50 Mission Cap" with its pronounced crushed appearance became the World War II mark of a veteran Army Air Force flight crew member with many combat missions behind him.

Theater of operations, the emphasis in the World War II gallery—despite the presence of the Zero suspended from the ceiling—is on the European Theater. No chronological history is attempted, but significant actions involving the various aircraft displayed are highlighted. Near the Spitfire, for example, are Battle of Britain artifacts and photographs.

An exhibit related to the 332nd Fighter Group, the first all-black fighter group, is displayed near the P-51, an aircraft they flew. At the entrance to the gallery, just before one passes beneath the Zero's wing, there is a glass display case containing memorabilia honoring the exploits of the Flying Tigers, the group of American volunteers led by General Claire Lee Chennault. Their outstanding skill and courage in the early stages of the war gave American morale a much-needed boost by proving that the Japanese could be beaten in the air. There is also a display relating to the WASPS (Women Airforce Service Pilots), who did ferrying for the Air Transport Command and logged some 60 million miles in the air, and one relating to the famous "Tokyo Raid" led by Jimmy Doolittle in which sixteen B-25, twin-engine Mitchell bombers were carrier-launched on April 18, 1942, against the Japanese homeland in retaliation for the Pearl Harbor attack.

The World War II gallery is a large, open room designed to provide the visitor with the best possible vantage points from which to view the aircraft. Since so many visitors to the Museum come with cameras, every effort has been made to provide a variety of angles from which good photographs can be taken. But perhaps the most extraordinary view in this gallery is that from the B-26 bomber *Flak Bait*'s forward section.

There may not be a more fitting name for an airplane than the one given this particular bomber; during its 21 months of combat *Flak Bait* collected over 1,000 enemy hits.

The Martin B-26B Marauder started its career with the worst possible reputation. It was a very hot airplane, and its comparatively short wings necessitated high speed, power-on landings, and dangerously high stalling speeds. Due to pilot inexperience, engine failures, runaway propellers, structural weakness in its tail, and so on, there were so many training accidents that Marauders were nicknamed "One-a-Day-in-Tampa Bay," "widow-makers," and, since the wings provided no visible means of support, "The Baltimore Whore." Because of the large number of accidents, on at least four occasions production nearly ceased while Congressional committees—including one headed by then-Missouri Senator Harry S. Truman—investigated its design. And yet it was a beautiful machine with symmetrical curves and extremely clean lines. General Jimmy Doolittle was such an outspoken supporter of the Marauder that he flew the aircraft on one engine at training bases around the country to prove to young pilots that any emergency could be safely handled in that plane. When the training command ensured that fledgling pilots would be given extensive experience in transitional twin-engine aircraft before flying the Marauder, the accident rate dropped dramatically. But its reputation took a long time to improve, and no amount of training or time could prevent the apparent affinity *Flak Bait* had for attracting bits of enemy metal.

Jim Farrell, *Flak Bait*'s pilot, recalled, "It was hit plenty of times—*all* the time. I guess it was hit more than any other plane in the Group." Other aircraft in the 322nd Bomb Group could return from a mission unscathed; *Flak Bait* would come back peppered. The first time B-26s were used in the European Theater was on May 14, 1943, against a power plant in IJmuiden, Holland. Twelve B-26s from the 322nd Bomb Group took part; one plane was lost, every plane was hit—one returned with more than 300 holes. Their target was little damaged. Another mission against IJmuiden was scheduled for three days later. On May 17th, eleven B-26s took off; one was forced to turn back before reaching the Dutch coast. Not one of the remaining ten returned. The raid was a disaster. Every operational B-26, sixty men, and the unit's commanding officer were lost.

"Right after we lost the ten planes at IJmuiden," Farrell recalled, "morale was devastated because we had no idea that we weren't going to have every raid just like that—all out, none back. So fellows would take every bit of money they had, go on leave and spend every nickle of it because they didn't think they would be around tomorrow. Others tried to get transferred to the [heavy bombers] and used every bit of influence they had."*

The heavy losses were due to an almost criminal misunderstanding of how best to tactically use this superb medium bomber. Marauders could carry 3,000 pounds of bombs, half the load carried by a B-17, but because they lacked superchargers they could not operate at the high altitudes the heavy bombers regularly used above the worst of the flak. Eighth Air Force therefore chose the opposite extreme: B-26s were sent in hedge-hopping at 200+ mph, and German defenses were simply too good. Light flak, machine gun, and even small-arms fire came at them from all sides. After the terrible losses at IJmuiden, tactics were changed. Crews were trained in flying tight, high formations where, even though there was more flak, there was less smaller arms fire. "All the airplanes came back pretty well shot-up," Farrell said, "a few wounded, but no losses."

After Farrell and his crew had flown their un-named B-26 for several missions they decided they'd have to christen it somehow. Farrell's brother's dog was nicknamed "Flea Bait" and from there to "Flak Bait" was but a short, imaginative hop. As soon as the name was painted on her nose, *Flak Bait* began to receive hits. On September 6, 1943, a Bf. 109 took *Flak Bait* on with a head-on pass. A 20mm shell smashed through the B-26's Plexiglas nose and exploded behind the left

---

*Jay P. Spenser, "Flak Bait!", *Airpower Magazine*, vol. 8, no. 5, Sept., 1978.

instrument panel. Miraculously, neither Farrell, nor his co-pilot, Thomas Owen Moore, nor his bombardier/navigator/nose gunner, O.J. "Red" Redmond, was severely wounded. On another mission, a week later, *Flak Bait* returned with a huge chunk taken out of her horizontal stabilizer. Soon after, Don Tyler's tail gun position was hit and the sheet of fuselage metal with his name on it was blown away. Tyler evened the score.

"We were in the 'tail-end-Charlie' position in the low flight," Tyler remembered. "The attack came from about eleven o'clock high and in a shallow dive. The boss only had time to say 'Here they come.' A head-on attack only lasts seconds because the relative speed is so great. But from that angle of attack the Bf. 109 was right in my field of fire. My last glimpse of him was he went into a steeper dive and veered left. There were several enemy fighters in the air and you didn't have time to watch anything that was out of your field of fire." A Spitfire pilot observed the enemy crash; confirmation subsequently came through channels.

By October, 1943, six months after the disastrous IJmuiden raids, the morale of the B-26 pilots was high. The Marauders were achieving a bombing accuracy unequaled by any other aircraft and, although the planes were still taking heavy punishment, the losses were astonishingly low. Farrell explains,

Our strong points were our speed and our heavy armament....We had the advantage over the heavies because we were moving much faster than they were in relation to the German fighters. We had many encounters with them early in the war and my feeling is we became too tough for them. For the most part the fighters left us alone and the Germans came at us with flak. Because of our altitude it was more effective with us than it would ever be with the heavies. They were at twenty-seven to thirty thousand feet and we were down at twelve and thirteen thousand feet, but they had to fly pretty much straight and level while we had the advantage of being able to maneuver as a group....

We figured it took the flak guns seventeen seconds to target us and another thirteen before the shells started bursting. For that reason, we

271

The forward fuselage of the famed B-26B
Marauder *Flak Bait* proudly wears its
symbols for 202 bombing missions, 5 decoy
missions, and tail-gunner Don Tyler's
confirmed kill. This American medium
bomber flew more combat missions than any
other American bomber and while earning
her name, endured 21 months of combat,
and suffered more than one
thousand enemy hits.

Flak Bait *crew was Lt. Jim Farrell, pilot;*
Lt. Thomas F. Moore, co-pilot; Lt. Owen
J. Redmond, bombardier/navigator/nose-
gunner; Sgt. J. D. Thielan, top turret gunner/
armorer; Sgt. Donald Tyler, flight engineer/
tail-gunner; and Sgt. Joseph B. Manuel,
radio operator/waist gunner, whose position
is depicted on the following page.

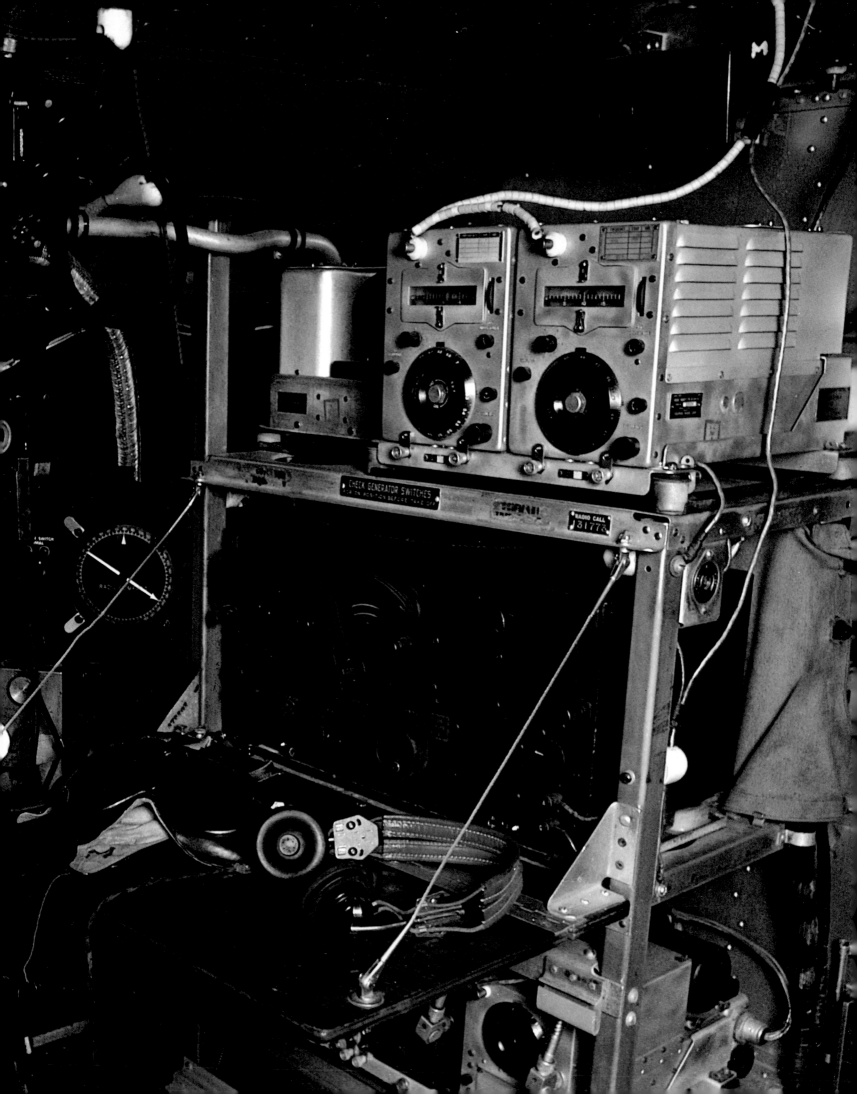

Alfred Owles. Northrop Black Widow Night Fighter. *c. 1943. Watercolor on paper, 20½ x 15½".*
Gift of Captain Fred Slightam

wouldn't go thirty seconds without doing something—climbing and turning to the right, descending and turning to the left, etc.... There was armor plating bolted to the fuselage on the pilot's side and he had an armor shell behind his seat. I wore a flak vest—it was like an apron—and had the instrument panel in front of me. It was thick enough to stop that 20mm [cannon shell] which is good enough for me.

A B-26 co-pilot, however, was not so well protected. The instrument panel was open on his side and the rudder pedals folded back so that the navigator/bombardier could crawl

through to the nose. On March 26, 1944, more than 350 B-26 Marauders returned to IJmuiden, Holland, and had their revenge: they dropped over 700 tons of bombs on a Nazi torpedo boat and submarine pens, and only one B-26 was lost.

The B-26B Marauder *Flak Bait* flew its 202nd and final mission in May, 1945. Among its missions were 31 against Pas de Calais V-1 rocket-launching sites, 5 against shipyards and E-boat pens, 45 against railway marshaling yards and communications centers, 35 against

Flak Bait's *cockpit frames an apparent
confrontation between the Museum's
Macchi C.202 Folgore and the B-17G
Thunder Bird.*

The only known surviving aircraft of its type, the Italian Macchi C.202 Folgore (Lightning) was one of the most effective fighters in the early stages of World War II; it could out-maneuver any of its opponents and out-perform all but the late model Mustangs and Spitfires.

A Rolls-Royce Merlin Mark 64 (1,705 horsepower) V-12 engine, used in Spitfires, Hurricanes, Mosquitoes, Lancasters, and Mustangs.

German airfields, 31 against Seine River and other bridges, 15 against fuel dumps, 30 against defended towns and cities, and 8 against gun emplacements. She logged 725 hours combat time, dropped over 375,000 pounds of bombs, and flew approximately 178,000 miles. The remainder of *Flak Bait* is in the National Aeronautical Collection and will be restored in the future.

That fighter plane one can see cutting across *Flak Bait*'s nose on its way to Keith Ferris' bomber formation is the Macchi C.202 Folgore, one of the best fighter planes made in the early stages of the war and certainly the best Italian fighter to have been produced in quantity. But despite its high reputation in Italy, it never became as famous as other nations' fighter aircraft even

though when the Folgore (Lightning) entered battle in North Africa in the summer of 1941 against the British, in the hands of a skilled pilot, this new machine was more than a match for its adversaries. The Folgore was superbly maneuverable and light-fingered to handle. It was substantially superior to the Curtiss P-40 and Hawker Hurricanes it encountered, and by war's end it could still outperform all but the late-model Spitfires and Mustangs. The Folgore's major drawback was its light armament compared to other fighters. A Spitfire, for example, carried two 20mm and four .303-caliber machine guns; the Folgore mounted only two 12.7mm and two 7.7mm machine guns. After the Italian Armistice in September, 1943, a number of Folgores were flown

Ron Kleemann. Mustang Sally Forth. 1973. Acrylic on canvas, 50 x 72''. From the Stuart M. Speiser Collection

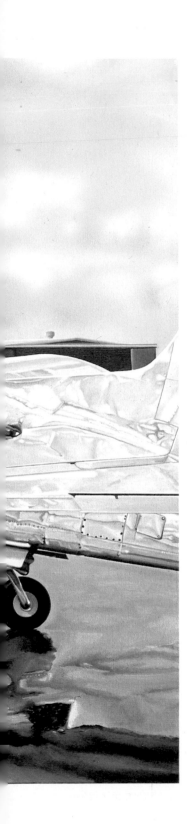

The B-17G was armed with 13 .50-caliber machine guns and included chin, nose, top, and belly- or ball-turrets like the B-32 turret shown above.

against the Germans by the small Italian Co-belligerent Air Force.

The Macchi C.202 Folgore suspended from this gallery's ceiling is perhaps the sole remaining aircraft of this type anywhere in the world. Although its early history is obscure, it was known to have been one of the many captured enemy aircraft to have been brought to this country's Army Air Technical Service Command at Wright and Freeman Fields for evaluation, then placed in storage. In 1975 NASM technicians restored the Folgore to exhibit condition and painted it in the markings of the 40° Stormo, 10° Gruppo, 90° Squadriglia, which operated in Libya during the summer of 1942. The *F. Baracca* painted on its nose was in honor of Major Francesco Baracca, Italy's leading World War I ace with 35 victories. A curious, but hardly noticeable, facet is that the Folgore's left wing is 8⅜ inches longer than its right, making it one of the rare aircraft to have used this asymmetrical method of counteracting its engine's rotational torque to assist pilot control.

Not many people know that the P-51 Mustang, which many consider the best fighter plane of World War II, was originally built for the British, who had approached North American early in 1940 requesting a

The Museum's P-51 D Mustang, considered by many to be the finest fighter plane of World War II, wears the yellow and black checkerboard colors of the 351st Fighter Squadron, 353rd Fighter Group, Eighth Air Force. This unit, stationed at Raydon, Suffolk, England, was assigned to escort bombers such as Thunder Bird *on missions deep into Germany.*

The Mustang's cockpit.

The two major British fighters of the Battle of Britain were the Hawker Hurricane (left) and the Supermarine Spitfire (below).

284

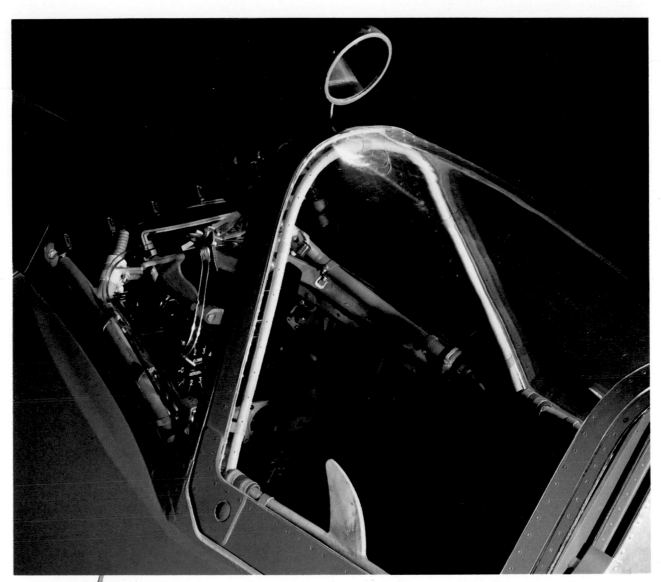

The Spitfire's cockpit.

A head-on view of the German Messerschmitt Bf.109G reveals its two MG 131 machine guns and single 30mm MK 108 cannon which fired through the spinner. Although slightly slower and less maneuverable than its first major opponent, the Spitfire, the Bf.109's performance was superior at high altitudes. The Museum's specimen wears the markings and camouflage of ship number 2, 7th Squadron, 3rd Group, 27th Wing that operated in the Eastern Mediterranean during late 1943.

Expert craftsmen restored the Bf.109's cockpit to near-operational condition.

The shark-mouthed Curtiss P-40 E Warhawk achieved its greatest fame in China and Burma with Lt. Gen. Claire Chennault's Flying Tigers.

quantity of P-40s to alleviate the RAF fighter shortage. North American proposed instead to build an entirely new and superior plane. The British agreed, and were astonished at the speed with which the plane was built at the West Coast manufacturer's plant. By the summer of 1942 the first Mustangs were in combat and the British pilots' enthusiasm was soon obvious. The United States Army Air Forces, however, were still cool to the P-51, regarding it a "foreign" design, and were concentrating on perfecting their Lockheed P-38 Lightnings and Republic P-47 Thunderbolts. But as the Mustang began to prove itself in combat, the USAAF took a closer look. When the P-51's Allison engines were replaced by Rolls-Royce Merlins with two-speed blowers, and the three-bladed propellers were exchanged for four-bladed props, there was no better fighter performing in the air.

That Japanese Mitsubishi A6M5 Zero, which appears to be clawing for air after strafing the Sea-Air Operations gallery's USS *Smithsonian* across the hall, was, to the dismay of Allied pilots who were astounded by its exceptional maneuverability and speed, the primary Japanese Naval fighter throughout World War II. When aircraft carriers were no longer available to the Japanese, the Zero (properly

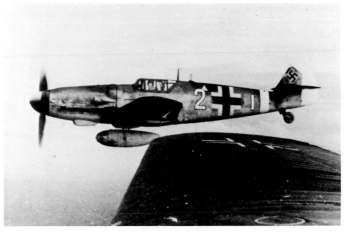

The actual aircraft after which the National Air and Space Museum patterned the markings and camouflage for its Messerschmitt Bf.109G.

code-named Zeke) was shifted to land bases. Only when the Grumman F6F Hellcat and the North American Mustang reached the Pacific in quantity was the Zero's relative superiority diminished.

Divided by floor space in this gallery, as the English Channel separated them during World War II, are two of the Battle of Britain's legendary adversaries: Britain's Supermarine Spitfire Mark VII and Germany's Messerschmitt Bf.109G-6. The Spitfire, though slightly faster and considerably more maneuverable than the thinner and smaller-winged Bf.109, could not match the German plane at higher altitudes, and until the Spit's engine was modified, British pilots often lost their opponent through his dives. But the British had the advantage of fighting over their home island, whereas the German fighter's critical range limited his time over Britain to less than twenty minutes. The Messerschmitt Bf.109E was already in mass production when Germany went to war in 1939, and over a thousand 109s were with operational fighter units. Only 400 Spitfires were in service. Both aircraft went through various modifications to improve performance. NASM's Spitfire has the extended wingtips of the high-altitude version and its Bf.109 is of the "G" (for "Gustav") series, which had two MG-131 and

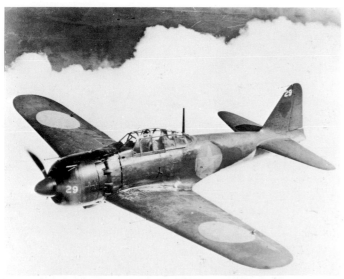

A captured Zero undergoing evaluation tests. Note the American star within the rising-sun fuselage insignia.

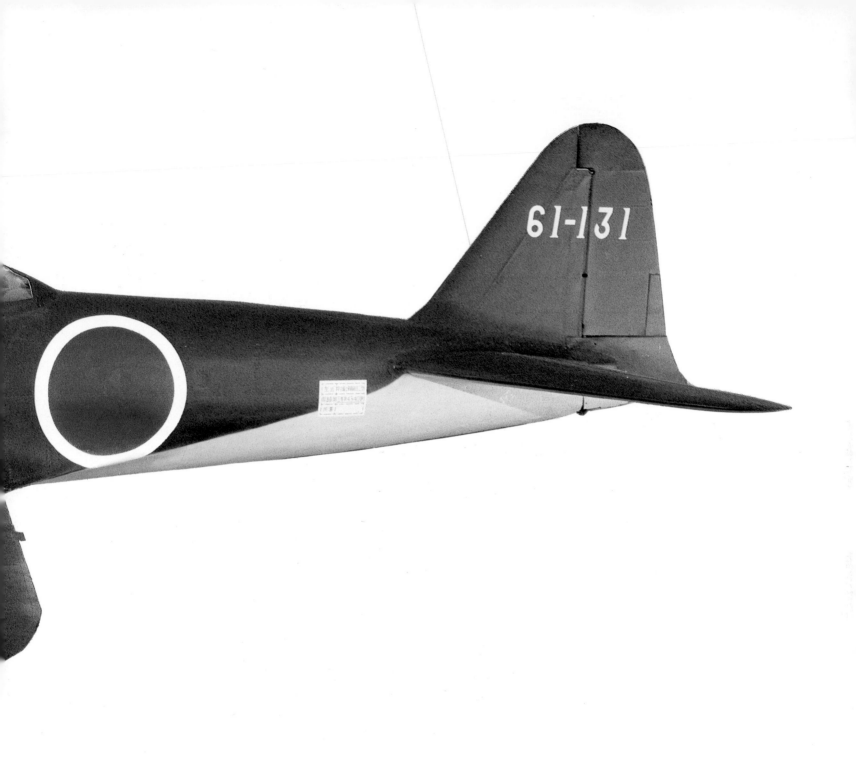

Early reports of the Japanese Zero's speed, maneuverability, firepower, and range were so incredible that they were rejected as incorrect and unbelievable by American aeronautical experts. Pearl Harbor and the appearance of the Zero in seemingly countless numbers throughout the Pacific in the opening days of the war proved the validity of those reports. The Museum's Mitsubishi Zero Fighter, Model 52 (code named Zeke by the Allies), is presumed to have been one of 12 late-model Zeroes captured on Saipan Island in April, 1944, and sent to the United States for evaluation.

two MG-151 wing-mounted 20-mm machine guns and one 30mm Mk-108 cannon firing through its spinner. This armament was ideally suited for bomber interception, but the added weight and drag reduced the 109's efficiency against other fighters.

The Norden Bombsight, one of America's most closely guarded secrets during World War II, is also on exhibit in this gallery. The Norden was a key to the USAAF's daylight high-precision bombing strategy, since only with an extremely accurate bombsight could small military targets be hit from high altitudes. The bombardier set altitude, airspeed, and other factors into his Norden Bombsight (which was basically an automatic speed and distance calculator) and then kept his telescopic viewer's crosshairs centered on his target. As the aircraft moved, ground speed and wind drift were fed into the computer. A bombadier then connected the bombsight to the autopilot, which kept the aircraft on course and released the bombs at the proper time.

Perhaps the one exhibit that will bring a smile to the gallery visitor is illustrated near the beginning of this chapter; it is the "50-Mission Cap" resting on a silk scarf on the pillow in a glass case. The 50-mission cap was the symbol of an Army Air Force flight crew member with many combat-mission hours behind him. The look of a veteran was achieved by removing the stiffening grommet from the standard service cap so that a radio headset could be worn. With each successive mission the cap would then achieve a more and more pronounced crushed appearance.

That white parachute-silk scarf, typical of those wrapped about fighter pilots' necks, was not just worn as an affectation, or even for warmth. The silk kept the pilot's neck from chafing as he swiveled his head back and forth in search of enemy aircraft.

Jet Aviation

Enter the Jet Aviation gallery through the nacelle covering for a giant Boeing 747 jet engine and you enter the age of jet aviation—an era that commenced on August 27, 1939, just four days before the outbreak of World War II. On that day, test pilot Erich Warsitz climbed into the cockpit of a stubby metal-fuselaged, wooden-winged Heinkel He 178 experimental aircraft, taxied onto the runway at Marienehe Airfield in Germany, and several seconds and several thousands of feet of takeoff roll later, took to the air to become the first man in the world to fly in a turbine-powered aircraft. Ironically, as Warsitz carried out his simple around-the-pattern flight, he also became the first pilot to suffer a jet engine failure due to striking a bird. Warsitz landed safely, however, and the flight of the He 178 marked the first practical application of turbojet technology to manned flight and led to Germany's dominance in the development of jet combat aircraft during World War II—a dominance that need not have been.

Ten years earlier British Pilot Officer and aeronautical engineer Frank Whittle had envisioned a turbojet "of a type," he later wrote, "which produced a propelling jet instead of driving a propeller." In other words, instead of using a gas turbine engine to power a propeller, Whittle foresaw using a gas turbine engine in combination with a compressor to force the exploding exhaust gasses out of a nozzle with enormous thrust—like air escaping from the neck of a balloon—thereby creating enough power to drive an aircraft forward. Britain's Air Ministry, however, was not impressed by his idea; so Whittle continued to work alone.

On January 16, 1930, Whittle applied for a patent on his turbojet engine. The English continued to remain so disinterested that approximately eighteen months later, when Whittle's patent was granted, the government did not even bother to classify its details on its "secret list." And by mid-1932 the particulars of Whittle's published patent were available worldwide. Even Whittle, at that time, seems to have become distracted for he simply permitted his patent to expire by failing to pay the renewal fee. However, in March, 1936, encouraged by friends to reconsider his jet-turbine engine, Whittle, along with some others, formed Power Jets Ltd., and a contract was given to the British Thomas-Houston Company to build a test engine.

On April 12, 1937, this engine, known as the W.U. (Whittle Unit), was run under its own power for the first time.

Jet Aviation gallery's former curator E. T. Wooldridge later reported:

> "The experience was frightening," said Whittle. With a scream like a banshee, the engine began to accelerate uncontrollably, threatening to bring the entire proceedings to a hasty and unpleasant conclusion. "Everyone around took to their heels except me," Whittle recalled. "I was paralyzed with fright and remained rooted to the spot." The engine began to slow down after reaching about half design speed. Investigation revealed that fuel had pooled in the main part of the combustion chamber; ignition of this had caused the

"runaway." The first successful run was the precursor of many to come, although there were times when Whittle was convinced he had truly created a monster beyond his control.

—Jet Aviation: Threshold to a New Era

By 1938, Whittle's centrifugal-flow turbojet engine was capable of producing up to 16,500 hp for as long as half an hour; success, however, remained uncertain. Finally, after a stunning demonstration of his engine before a Senior Air Ministry Official on June 30, 1939, Power Jets Ltd. was awarded a contract to manufacture a flight engine; and the Gloster Aircraft Company was chosen to build the aircraft to carry it: the E.28/39. Within two months, of course, Germany's He 178 would already be flying. Britain's Gloster E.28/39 would not make its maiden flight until May 15, 1941—and by then it would be not the second but the third jet-propelled aircraft to fly.

How did the Germans achieve such a headstart? In the mid-1930s, while Frank Whittle was an undergraduate at Cambridge University, Hans von Ohain was studying for his doctoral degree at Göttingen University. Like Whittle, von Ohain was convinced that higher airspeeds could be achieved by replacing the conventional piston-driven engine and propeller by some form of compressor-driven gas turbine engine. It was not so much any difference between their engines that was critical as it was the difference between the two designers: von Ohain had contacts; Whittle did not.

In 1936, one year after von Ohain had, with his mechanic and machinist Max Hahn, built a promising test engine, von Ohain's Göttingen professor, a Dr. Pohl, introduced von Ohain to the German aircraft industrialist Ernst Heinkel. As former National Air and Space Museum director Walter J. Boyne wrote of Heinkel:

. . . Heinkel was a great buccaneer in the industry—obsessed with speed and money, and capable of defying any government on any issue. He wanted to enter the lucrative field of engine manufacture but had not been allowed to do so under the Luftwaffe's "normalization" plans, which allocated tasks to manufacturers. Von Ohain's invention promised not only superior speed, but acceptance in an industry on the best possible basis: an exclusive product and a relatively small investment.

Heinkel provided von Ohain with direct access to the boss (Heinkel), a well-equipped shop, and unlimited funds. The result was that the HeS 3b engine was ready for flight in the specially designed Heinkel He 178 airframe on August 27, 1939. . . .

—The Leading Edge, by Walter J. Boyne

The second jet-propelled aircraft to fly was the Caproni Campini N.1 which had an unusual combination of a piston engine and ducted fan compressor. This Italian aircraft's first flight took place on August 28, 1940, and although the N.1 was jet-propelled, its lack of a true turbojet engine hampered its performance.

Adolph Galland, who was later to become commander of all the Luftwaffe fighters, wrote, "I shall never forget May 22, 1943, the day I flew a jet aircraft for the first time in my life." Early that morning Galland had met Willi Messerschmitt at Lechfeld, the testing field near the Messerschmitt main factory at Augsburg. After listening to speeches by the specialists who had taken part in the construction of the Messerschmitt Me 262, the world's first operational jet fighter, Galland wrote:

We drove out to the runway. There stood the two Me 262 jet fighters, the reason for our meeting and all our great hopes. An unusual sight, these planes without propellers. Covered by a streamlined cowling, two nacelles under the wings housed the jet engines. . . . If the characteristics of performance which the firms had calculated, and partially flown were only approximately correct, then undreamed possibilities lay ahead. And this was all that counted! . . .

The chief pilot of the works made a trial demonstration with one of the "birds," and after refueling I climbed in. The mechanics started the turbines with many manipulations, which I followed with the greatest interest. The first one started quite easily. The second caught fire, and in no time the whole engine was burning. Luckily, as

Although British Pilot Officer Frank Whittle's centrifugal-flow turbojet engine (right) known as the W.U. (Whittle Unit) became on April 12, 1932, the first turbojet engine to run under its own power, German designer Dr. Hans von Ohain's centrifugal-flow Heinkel HeS 3b turbojet (below) became, on August 27, 1939—just four days prior to the outbreak of World War II—the first to actually power an aircraft into the air.

a fighter pilot I was used to getting in and out of a cockpit quickly, but in any case the fire was soon put out again. The second plane caused no trouble. I took off along a runway . . . at a steadily increasing speed, but without being able to see ahead—this was on account of the conventional tail wheel of the mass-produced Me 262. Also I could not use the rudder for keeping my direction: that had to be done for the time being with the brakes. A runway is never long enough! I was doing 80 m.p.h. when at last the tail rose, I could see, and the feeling of running your head against a wall in the dark was over. Now, with reduced air resistance, the speed increased quickly, soon passing the 120-m.p.h. mark and long before the end of the runway the plane rose gently off the ground.

For the first time I was flying by jet propulsion! No engine vibration, no torque, and no lashing noise from the propeller. Accompanied by a whistling sound, my jet shot through the air. Later when asked what it felt like, I said: "It was as though angels were pushing."

—"The First and the Last," by Adolph Galland
in *The Saga of Flight*

Galland recognized that the Me 262, nicknamed "Schwalbe" (Swallow), if used properly, would be a weapon with extraordinary potential and immediately called for its quantity production. From the moment in July, 1944, that the first German Messerschmitt Me 262s appeared in combat in the skies over western Europe, the days of the piston-engined fighter were numbered.

Fortunately for the Allies, relatively few Me 262s appeared in the closing year of the war and never really threatened Allied dominance of the air. Although 1,443 Me 262s were produced, it is estimated that only about 300 saw combat, the others having been damaged or destroyed in Allied

With its appearance on the Western Front in late 1944, Germany's Messerschmitt Me 262, the world's first operational jet fighter, signalled the end of the conventional piston-engine fighter. With its jet engines, swept wings, and extremely high speed, the Me 262 set the pattern for aircraft design of the future.

bombing raids or training accidents. Nevertheless, in a remarkable effort, the Messerschmitt engineers continued to build the "Schwalbe" long after their factories had been bombed out. Me 262s were built in forest clearings, fueled, then taxied out onto the German autobahn for delivery to combat units.

The first American jet was the Bell XP-59A Airacomet (it is on display in the Flight Testing Gallery). Designed around two Whittle-type turbojets built by General Electric, the XP-59A first flew on October 1, 1942; and although that aircraft made it possible for America to become the fourth nation to produce a jet-propelled airplane, its performance was disappointing. Because of the Airacomet's slow speeds (389 mph), it was inferior from the beginning to faster, already operational piston-engined aircraft of its day, and few were produced. As a result, the Army Air Forces turned to Lockheed which, since 1940, had been trying to get an Air Corps contract to build a pure jet-powered airplane.

Under the leadership of Clarence L. "Kelly"

Although the Allies dominated the skies over Europe in 1944, there was much concern over the appearance of Germany's jet-powered Me 262, an aircraft superior to any fighter the Allies possessed. America's response was the P-80 whose prototype, here, was designed and constructed in just 143 days by Lockheed Chief Research Engineer Clarence L. "Kelly" Johnson's "Skunk Works" group of engineers and technicians. America's first operational jet fighters, the P-80s, did not see action in World War II, but were used extensively in the ground support role during the Korean conflict.

Johnson, Chief Research Engineer for Lockheed, a group of engineers and technicians designed and constructed the XP-80 prototype of America's first operational jet fighter in just 143 days! The workshop in which this remarkable feat was accomplished and its members were to become collectively known as "The Skunk Works."

How did it get that name? "Kelly" Johnson would later write that he wasn't sure, but that:

> . . . in the strict secrecy of wartime, and simply for efficiency and to avoid distractions, we allowed no one who wasn't working on the project to wander in and out. The legend goes that one of our engineers . . . was asked, "What the heck is Kelly doing in there?"
>
> "Oh, he's stirring up some kind of brew," was the answer.
>
> This brought to mind Al Capp's popular comic strip of that day, "Lil Abner," and the hairy Indian who regularly stirred up a big brew, throwing in skunks, old shoes, and other likely material to make up his "kickapoo joy juice." Thus the Skunk Works was born and named.
>
> —*Kelly: More Than My Share of It All,*
> by Clarence L. "Kelly" Johnson
> with Maggie Smith

Over the years out of "The Skunk Works" would roll out such stunning aircraft as the F-104 Starfighter, the first operational jet aircraft to fly twice the speed of sound in level flight; the U-2 high-altitude reconnaissance aircraft; and its successor, the SR-71A which, capable of achieving three times the speed of sound, is the fastest operational turbojet aircraft in existence. These aircraft, however, were still in the future.

On January 8, 1944, a cold, damp morning on Muroc Lake's dry bed, Lockheed test pilot Miro Burcham pushed forward the throttles of the XP-80 nicknamed "Lulu-belle" and the aircraft took to the air.

Five months later, an improved XP-80A with an airframe that was heavier so as to contain a 4,000-pound thrust General Electric I-40 turbojet, was finished.

Somewhat later, four YP-80As were sent overseas in early 1945; two went to England and two to Italy to demonstrate their performance to aircrews stationed there. The eagerly anticipated duel between the awesome Me 262 "Schwalbe" and the acclaimed P-80 Shooting Star—those two unique aircraft now parked wingtip to wingtip in this Jet Aviation gallery—was never to take place.

During the Korean War, the F-80 (as it was known after 1948) flew a wide range of combat missions, including close air support, bomber escort and reconnaissance. Although the Shooting Star was phased out of the active forces shortly after the Korean War, variations of the F-80 remained active in the United States and Allied nations for at least another two decades.

The first wholesale use of jet aircraft in combat occurred during the Korean War; and the first aerial combat between jet fighters occurred on November 8, 1950, when an American F-80 shot down a swept-wing MiG-15 despite that Soviet-designed aircraft being considerably faster. Soon American fighters like the North American F-86 Sabre and Republic F-84 Thunderjet guaranteed Allied domination of the skies over Korea; and those Soviet-built MiGs that ventured out of their Chinese sanctuaries and crossed the Yalu River into North Korea fell at the rate of 10 to 1 to the American jets. And on May 20, 1951, Major James Jabara, USAF, became the world's first jet ace when he destroyed his fifth MiG-15.

If the stunning technological advances, striking personalities, classic-designed aircraft, and remarkable record-breaking flights occurring between 1927 and 1939 resulted in that period becoming known as the Golden Age of Aviation, then, based on these same measurements, the decade of the 1950s might be considered the Golden Age of *Jet* Aviation.

On May 2, 1952, the British Overseas Airways Corporation (BOAC) inaugurated the first passenger service between London and Johannesburg, South Africa, with its brand new de Havilland DH 106 Comet. In building the world's first jet-powered airliner de Havilland had, as R.E.G. Davies wrote in *A History of the World's Airlines*, "staked its reputation on a daring attempt to leap-frog every other competitor in the civil markets and take a major share of a lucrative trade which had by 1945 become almost exclusively American."

Orders for the aircraft poured in from airline companies all over the globe quick to realize the extraordinary breakthrough de Havilland's sleek, new aircraft had achieved. The breakthrough was not only that the Comet was faster (with its 500 mph cruising speed, the Comet cut nearly in half the time the 270+ mph cruising speed piston-powered Super-Constellations, DC-7Cs and Argonauts required to make the London-Johannesburg flight), but that it was more economical, too. Although fuel consumption was still high, the compensating factor was that jet fuel was considerably less expensive than the high-octane gasoline required by piston engines, and jet fuel consumption at the higher altitude flown was less than expected. Maintenance costs on engines plummeted because jet engines were so much simpler and more efficient; instead of metal parts knocking each other about as they do in reciprocating engines, jet engine parts simply go round and round. Furthermore speed, which the jet engine's increased power provided, also gave aircraft designers the opportunity to create "cleaner," more aerodynamic shapes. Within a year, de Havilland had received orders for 50 Comets and another 100 were being negotiated.

During their first year in service, Comets, flying at an average of 88 percent capacity, carried nearly 28,000 passengers 104.6 million air miles—but not without loss. Two Comets had crashed taking off; a third—on the anniversary of the London-Johannesburg inaugural flight—had flown into a violent thunderstorm six minutes after its takeoff from Calcutta and disintegrated in mid-air. Both the accidents

taking place during takeoff had occurred when the airliner had simply failed to become airborne. Procedures were changed to require higher takeoff speeds, and a modification was made to the Comet's wing to provide increased lift at lower speeds. The third accident was ascribed to the risks of flying: no plane could be expected to withstand the turbulence encountered in a freak thunderstorm.

Comets continued to fly until January 10, 1954. On that day BOAC Comet G-ALYP— call-sign "Yoke Peter"—bound for London from Rome's Ciampino Airport, was climbing through the clouds. A second, piston-driven BOAC airliner, seeking less turbulent air, had asked the Comet pilot Alan Gibson to report at what height the clouds ceased. Gibson had radioed the Rome airport that he was coming out of the overcast at 26,000 feet on his way to his cruising altitude of 36,000 and had just begun his radio transmission to the second BOAC, registration G-ALHJ, below.

"George How Jig from George Yoke Peter," Gibson radioed. "Did you get my . . ." And then there was silence.

Not quite one minute later, before the eyes of horrified Italian fishermen at sea off the coast of Elba, the burning remains of "Yoke Peter" plunged from the clouds into 500 feet of water.

Less than 40 hours later, BOAC's chairman announced, "As a measure of prudence, the normal Comet passenger services are being temporarily suspended to enable minute and unhurried technical examination to be carried out at London Airport." Still no one seriously believed that there might be a problem with the plane, and on March 23rd service was resumed.

And then, two weeks later, a second Comet, one belonging to South African Airways, disappeared at night in similar circumstances off the coast of Sicily.

"In the whole history of air transport," Davies would write, "no event stands out more vividly than the second crash of a de Havilland Comet 1 in the Mediterranean off

On July 21, 1946, the XFD-1 Phantom, designed by McDonnell Aircraft Corporation, became the first U.S. pure jet aircraft to take off from and land on an aircraft carrier. In 1947, fleet squadrons began receiving the production model Phantom, designated the FH-1, and the Phantom subsequently became the first jet fighter to enter operational service with both the U.S. Navy and U.S. Marine Corps.

Stromboli on 8 April 1954. . . . Some eighteen years before," Davies would continue:

> an air disaster of similar magnitude provided the world's press with banner headlines and dramatic pictures when the German airship "Hindenburg" exploded as a result of a static discharge whilst mooring at Lakehurst, the United States terminal of the regular Atlantic service. That was on May 6,

1936, and since then there has never been any serious suggestion of an airship revival. On 8 April 1954 there were no pictures but the headlines told of a disaster as far-reaching as it was unexpected.

> —*A History of the World's Airlines,*
> by R.E.G. Davies

Because the two Comet crashes had occurred so close together in time and under such similar conditions, it was hoped an explanation for one might solve the mystery of both. By that time, enough of the Elba crash had been recovered for a careful, painstaking investigation to be possible. The results of these findings were published in the autumn of 1954. The Comet's cabin had ruptured as a result of metal fatigue brought upon by the stress of repeated pressurization and depressurization. The Court of Inquiry concluded no one was to blame. "The Comet had been made in accordance with all the principles of aircraft engineering and with the knowledge of metal fatigue then accepted," Davies wrote. ". . . This was an indictment of the lack of knowledge in certain aspects of metallurgy rather than of aerodynamics or aircraft design."

On July 15, 1954, fourteen weeks after the crash of the second Comet, the United States regained its leadership of commercial jet aviation when model 367-80, the prototype of the Boeing 707—known as the "Dash Eighty" after the last two digits in the Boeing engineer model number—flew for the first time. After an hour and 24 minutes aloft, Boeing test pilot A. M. "Tex" Johnston exulted, "It wanted to climb like a rocket!" (It is said that Johnston was so excited by the 707's maiden flight performance that when he returned to the pilot's ready room to clean up, he stepped into the shower still wearing his cowboy boots.)

The "Dash Eighty" had grown out of Boeing's experience in the 1940s developing the B-47 and B-52 bombers—the only large jet aircraft built by any of the five major American aircraft manufacturers. Both the B-47 and the B-52 were military planes, of course, but many of the problems encountered in producing large military jet aircraft would necessarily be encountered in creating a large civilian jet-powered transport. And so, with good reason, Boeing felt it was ahead of its American competitors. And in 1950, as Curator E. T. Wooldridge noted, "Encouraged by the success of the bomber program, Boeing

President William M. Allen decided to gamble a considerable portion of the company's funds on the premise that the 'jet was right' for commercial aviation."

The "Dash Eighty" prototype was huge for an aircraft of its time. Its nose gear, alone, weighed as much as the Wright Flyer, and its 128-foot fuselage was eight feet longer than the distance covered by Wilbur Wright during man's first powered flight. But even though it weighed 80 tons, the "Dash Eighty" handled like a dream. Once "Tex" Johnston was asked by Boeing president William M. Allen to make a low flying pass over the Gold Cup Hydroplane Boat Races on Seattle's Lake Washington. Allen wanted to impress members of the Aircraft Industries Association who were present at the races aboard a chartered yacht. He got more than he bargained for:

> As the 707 screamed into view, Allen's composure dissolved. He watched aghast as Johnston put the only 707 in existence through two complete barrel rolls at high altitude, then dived and rolled the huge plane again, directly over the yacht. It was an impressive demonstration of the plane's structural strength, but for a heart-stopping moment Allen had visions of a $16 million investment splitting at every rivet and seam. Later he is supposed to have chastised Johnston. But the industry officials who had witnessed the stunts gave the 707 valuable word-of-mouth advertising.
>
> —*The Jet Age*, by Robert J. Serling

On October 26, 1958, Pan American World Airways inaugurated trans-atlantic service with the Boeing 707-120, and leadership of the commercial jet age by U.S. manufacturers had begun. True, three weeks earlier, an improved Comet 4 had begun regularly scheduled trans-oceanic service; but the de Havilland aircraft was overshadowed by the superior performance and adaptability of the Boeing 707. And over the next two decades the 707 came to be used by virtually every one of the world's major airlines. Boeing's gamble had paid off.

A family of Model 707s emerged with various combinations of fuselage length,

Photo by MacGillivray Freeman Films,
courtesy of Continental Oil Company

Frank Wootton. Night Reconnaissance. *1978. Oil on linen, 29½ × 39½". Gift of Phyllis S. Corbitt in Memory of Colonel Gilland W. Corbitt, USAF*

wingspan, wing area, and powerplant. The growing need for jet performance on regional and short-haul lines resulted in new transport designs, among them, in 1964, the rear-mounted 3-engine configuration of the Boeing 727 used for intermediate- and short-range service, and in 1968 the twin-engine short-range 737 that, under favorable conditions, can take off and land at any airport that could accommodate the old piston-engined DC-3 transport seen hanging in the Hall of Transportation.

The development of smaller, short-range commercial jet transports was paralleled by the development of the "wide-body" jet-liners such as the Boeing 747, Douglas DC-10, and the European Airbus A300 "jumbo jets" and supersonic transports (SSTs). The Soviet Union became the first nation to build and fly an SST, the Tupolev TU-144, on December 31, 1968.

Eager to show off their new jet, they displayed it at the Paris Air Show in 1971 and again, in 1973 when, after making a low pass over the airfield, the TU-144's pilot pulled the SST jet up sharply, climbed to 4,500 feet and somehow lost control. The pride of the Soviet airfleet suddenly plummeted to the earth, its tail breaking off as it fell, and it crashed in a ball of fire in a nearby village. Seven villagers, mostly children, were killed as was the SST's crew of six.

The TU-144 continued to have problems and although in 1977 it was put in service between Moscow and Alma-Ata in Soviet Central Asia, it was grounded after just ten months, seemingly for good. By then the Anglo-French Concorde SST was in service.

In a 1974 demonstration of its speed, a French Concorde took off from Boston's Logan Airport at the same moment that an Air France 747 left Paris. The Concorde was able to fly to Paris, spend an hour and eight minutes refueling and discharging its passengers, take off for Boston, and reach that city eleven minutes before the 747 had arrived. Despite such savings in time, only sixteen Concordes were built. Other airlines were reluctant to commit themselves to purchasing so expensive a plane (the Concorde cost $65 million compared to $21 million for a 747 which carried three times as many passengers), environmentalists objected vigorously to sonic booms and engine noise, and the cost of passenger tickets limited Concorde service largely to those with expense accounts.

During the 1960s the third generation of jet fighters—the first generation to include supersonic jets such as the F-100, F-101, and F4D—was replaced by aircraft capable of flying at twice the speed of sound such as Lockheed's F-104 and the versatile McDonnell Douglas F-4 Phantom II which set numerous performance records and became the standard fighter for many air

forces around the world. This fourth generation of Mach 2 fighters armed with missiles in addition to guns also included such pivotal aircraft as the General Dynamics F-111, the first supersonic United States bomber with variable sweep wings, and the subsonic Hawker Siddley Harrier currently in service with the Royal Air Force, the Royal Navy, and the United States Marine Corps. The Harrier is the only operational vertical takeoff and landing jet fighter.

Although the performance of today's jet fighters and bombers has not significantly improved over that attained by those of the 1960s, the 1970s and 1980s generation aircraft, such as the F-14, F-15, F-18, and the B-1 bomber have achieved major improvements in electronics, computers, and weaponry.

Fifty years have now passed since Heinkel test pilot Erich Warsitz lifted that first stubby, wooden-winged, jet-turbine-powered He 178 off the runway in Germany. Not only has jet aviation revolutionized military warfare, but it has also altered immeasurably the world in which we live by virtually shrinking the size of the globe.

Looking at Earth

A certain Captain [John Randolph] Bryan, a
young aide-de-camp of [Confederate] General J. B.
Magruder's, was borrowed by Gen. Joseph E.
Johnston in the spring of '62 and sent up in a
balloon to map [General George B.] McClellan's
dispositions around Yorktown. Bryan had never
seen a balloon before. This one was made of tarred
cotton; hot air from a pineknot fire gave it lift, and
it was controlled by a cable from a drum and a
windlass. Well, the balloon bounced and whirled
and rocked, and the Yankees shot at it with
everything they had, but Bryan managed to do his
job and get down again. He gave the maps to
General Johnston and asked (I still remember the
curiously stilted language), "Will you not now, sir,
reassign me to General Magruder?"

"Sir," Johnston said—and his smile was lined
with sharp teeth—"you forget that you are my only
experienced aeronaut. Pray hold yourself in
readiness for another ascension at any moment."

—"Brave Deeds of Confederate Soldiers"
as reported in *The Sword over the Mantle*,
by J. Bryan, III

From the moment the first man stood and
looked around him, then walked to the
top of the highest nearby mound for an even
better look, mankind has sought to expand
his horizons by viewing his world from
above.

In the beginning he used hills, trees, a
companion's shoulders, later a fortress's
towers; and as technology progressed, man
continued to climb higher and higher using
ever more complex platforms. Today aircraft
and spacecraft look down on the Earth to
help us predict the weather, survey the
terrain, monitor crops and forests, locate
resources, plan cities and gather military
intelligence. In this gallery, the Museum
explores the various means we have used

from balloons and birds to supersonic
aircraft and globe-circling satellites in our
unceasing effort to provide ourselves with
new perspectives and increasingly
sophisticated means for "Looking at Earth."

In July, 1861, President Abraham
Lincoln, impressed by the potential of
balloons in determining the strength and
movements of military forces, urged Lt. Gen.
Winfield Scott, one of his senior military
commanders, to meet with Thaddeus Lowe,
a pioneer in balloon reconnaissance. Scott
was skeptical. Joseph Henry, however, was
not. Henry, the first Secretary of the
Smithsonian Institution, was a firm believer
in Lowe; and it was through Joseph Henry's
aid that Lowe was able to launch his balloon
at the site now occupied by the National Air
and Space Museum. On June 18, 1861,
from his vantage point 500 feet above the
ground, Thaddeus Lowe could see nearly 25
miles in every direction. In a further
demonstration of the value of his balloons,
Lowe telegraphed the following message to
President Lincoln: "The city with its girdle
of encampments presents a superb scene. I
have pleasure in sending you this first
dispatch ever telegraphed from an aerial
station."

The gathering of military intelligence
using manned observation balloons during
the Civil War was not without its risks. In
1862, George Armstrong Custer, then a
junior lieutenant with the Union forces, was
assigned duty as an observer on the
Constitution, one of Thaddeus Lowe's gas-
filled balloons. Despite his grave doubts
about the safety of such balloons, Custer

Thaddeus Lowe, a pioneer in balloon reconnaissance, flew high above the battlefield to observe troop movement during the Civil War. In this Mathew Brady photo, Union soldiers are seen inflating Lowe's balloon at Fair Oaks, Virginia.

performed his duty and made notes and sketches of Confederate troop positions near Yorktown visible from the balloon. At the same time and in the same area but on the opposing side, General Johnston's "only experienced aeronaut," Capt. John Randolph Bryan of the Confederate Army of the Peninsula held himself, as ordered, "ready for another ascension." And on May 5, 1862, Bryan was sent aloft again for what became an inadvertent free-flight. As his tarred cotton hot air balloon was being winched back down, an awestruck young soldier got his sleeve caught between the observation balloon's cable and the drum. The soldier's arm was being drawn in and crushed when a companion snatched up an ax and chopped the cable. Freed of its tether the hot air balloon:

> . . . bounded two miles into the air. First it drifted over the Union lines, then was blown back toward the Confederate lines near Yorktown. The Confederates, seeing it coming from that direction, promptly opened fire. Finally it skimmed the surface of the York River, its guide-rope splashing

in the water, and landed in an orchard. On this trip the balloon made a half-moon circuit of about fifteen miles, about four miles of which was over the York River.

—*Photographic History of the Civil War,* edited by Francis Trevelyan Miller

After the Civil War and throughout the latter part of the 19th century, so keen was the interest in how cities might look from above that towns throughout the United States commissioned idealized "bird's-eye-view" portraits of their streets and buildings. But idealized sketches were not enough. People wanted to see how such places *really* appeared.

Before the development of practical aircraft photography, aerial views of the Earth were obtained by kites, rockets, balloons, and even pigeons. In 1895, Lt. Hugh D. Wise of the 9th Infantry Division built an 18-foot-high kite, attached a box camera with a shutter triggered by a timing device to its string, and took photographs of New York's Madison Barracks from an altitude of 600 feet.

In 1897, Alfred Nobel, whose dynamite fortune led to his establishment of the Nobel Prize, designed a photo rocket; he was also experimenting with balloon-borne cameras at that time.

In 1903, Dr. Julius Neubronner patented a miniature camera activated by a timing mechanism that he attached to pigeons. In 1908 camera-equipped pigeons photographed a castle in Kronberg, Germany. One such bird's-eye-view photograph is on display in this gallery. Displayed here as well are examples of early airplane photography including Boston *Journal* cameraman George T. Murray's June 26, 1914, photograph of the disastrous fire that swept Salem, Massachusetts. Murray's photograph, which was enlarged to cover the newspaper's entire front page, is thought to be the first known use of aerial photography in journalism.

The de Havilland DH-4 suspended in the gallery is not just *any* de Havilland DH-4; it

In 1895, Lt. Hugh D. Wise of the 9th Infantry Division built an
18-foot-high kite, attached a box camera (with a shutter triggered
by a timing device) to its string and took this photograph of New
York's Madison Barracks from an altitude of 600 feet.

The Museum's de Havilland DH-4 is the first of the more than 4,000 DH-4s equipped with Liberty engines that were built in this country between 1917 and 1919.

The DH-4 played many roles in both military and civilian capacities. During World War I, it served as a bomber, photo-reconnaissance and observation aircraft; and after the War these "Liberty" planes, as the DH-4s were known, were utilized for forest patrols, geologic reconnaissance, aerial photography, and mapping.

Aerial cameras in the DH-4 could be hand held or mounted either inside or outside the rear cockpit.

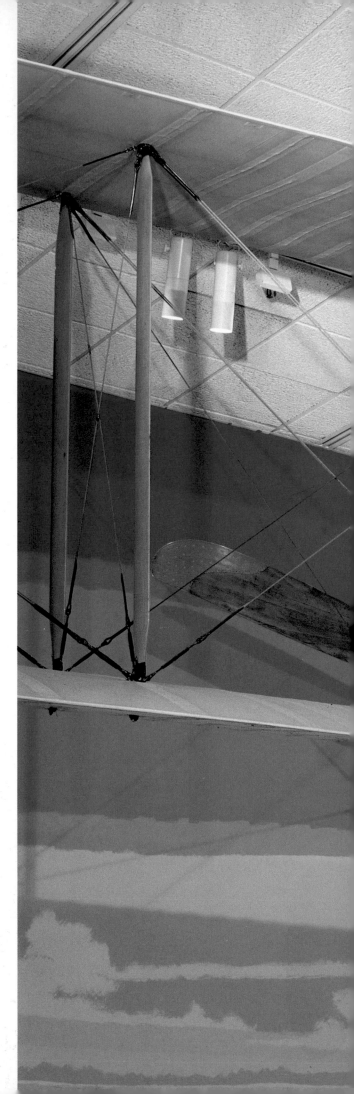

With aerial photography the desire for a "bird's-eye-view" of city landmarks became a practical reality.

is the first of the more than 4,000 DH-4s equipped with a Liberty engine that were built in this country between 1917 and 1919.

The DH-4 played many roles in both military and civilian capacities. During the war it served as a bomber, photo-reconnaissance and observation aircraft; and after the war these "Liberty Planes," as the DH-4s were known, were utilized for forest patrols, geological reconnaissance, and aerial photography. When the post-war Army Air Service was called upon to supply the Army Corps of Engineers and the Geological Survey with aerial photographs

for maps and stereo viewing, the DH-4 was employed because it was both the most available and the most suitable for the job. The DH-4 became, therefore, the standard airplane used by the Army Air Service for aerial mapping for the next ten years.

When the Earth is viewed from the air, patterns, boundaries and landmarks appear that are often not visible at close range; as a result, aerial photography has practical applications in many fields. In geology, for example, rugged and otherwise inaccessible terrain can be studied through aerial photographs. In 1933, nine years after George Leigh Mallory and his partner

Andrew Irvine vanished into the mists on Mt. Everest's northeast ridge during Mallory's third expedition to reach that mountain's 29,028-foot summit, two especially redesigned British camera-equipped biplanes took off from a landing field near Purnea in Bihar, India, for the 154-mile flight north to Mt. Everest in Nepal. In his *National Geographic Magazine* account of the journey, Lt. Col. L.V.S.

Blacker, O.B.E., wrote of clearing Everest's amazing, immaculate crest:

> I looked down through the open floor and saw what no man since time began had ever seen before. No words can tell the awfulness of that vision.

Blacker's photographic record of that flight provided the first film images of the rocks and ridges of Everest's peak viewed

Aerial photography has practical applications in many fields. For example, in geology, rugged and otherwise inaccessible terrain can be studied.

When cities are viewed from the air, patterns, boundaries, and landmarks appear that are often not discernible at close range. This view of Washington, D.C., shows the city at the time Memorial Bridge was under construction.

from above. In addition to providing the means to study otherwise inaccessible terrain, other geological applications of aerial photography are the monitoring of shoreline erosion, documentation of recent volcanic activity, charting of a river's shifting path, and so on.

Just as from aerial photographs geologists can easily recognize rock structures and predict what resources may lie underneath, this same sort of photography enables archeologists to identify ancient natural and man-made features such as roads, canals,

dams, lakes and cities whose boundaries are no longer visible from the ground. Aerial photos are helpful in pinpointing devastation and organizing relief efforts after major disasters such as earthquakes, volcanic explosions, tidal waves, landslides, floods, and storms. Environmentalists study historic aerial photos to determine old hazardous waste sites beneath more recent construction. The Enviropod, a two-camera system used to locate and document sources of environmental pollution, is designed to be strapped onto a Cessna 172 or 182 airplane without requiring any modifications to the aircraft. It thereby provides a low cost, rapid response to environmental emergencies or routine monitoring.

High-altitude aircraft can be used to track and evaluate weather systems. In 1958 a U-2 reconnaissance aircraft photographed Typhoon Ida about 750 miles off the coast of the Philippines. Ida contained surface winds of more than 200 knots and the lowest sea level atmospheric pressure ever recorded.

In order to acquire aerial stereo photography which allows an interpreter to see the ground in three dimensions, a series of overlapping photographs must be taken along a designated flight line. One of the major uses of stereo aerial photography is the gathering of military intelligence. Usually unsung and unrecognized, photo-interpreters and reconnaissance pilots play a vital role, often spelling the difference between success or failure, accuracy or blunder, readiness or surprise.

In August, 1914, at the outbreak of World War I, aerial reconnaissance was still primarily visual, but in March, 1915, a trench map based primarily on information gained from aerial photographs was used "with great success by Sir Douglas Haig in the attack at Neuve Chapelle," Constance Babington-Smith, a photo-interpreter with the RAF during World War II, noted in *Air Spy*, her book about the development of photo interpretation,

and from then on there was a continual urgent demand for photographic reconnaissance, both for making maps and for checking enemy activity. . . . By mid-1915 both sides were hard at it, and both sides were realizing that steps must be taken to prevent the enemy from recording their secrets from above. This need stimulated the rapid development of aircraft equipped with guns, for the work of the reconnaissance planes was so vital that they had to be protected—by an escort of specialized fighting aircraft. . . .

"By 1918, photographic reconnaissance was being used to a lavish extent," Babington-Smith continued:

There had been great advances in camera design and photographic techniques, as well as in the methods of deriving information from the pictures—methods which soon became known as photographic interpretation. A great revolution had taken place in the whole field of military intelligence. The traditional methods of obtaining information—the reports of secret agents, censors and interrogators—were not superseded, but they were supplemented, in the same revolutionary manner that the traditional methods of communication had been supplemented by the telephone and wireless telegraphy. By the time of the Armistice, photographic intelligence had indeed proved itself, and was recognized on every hand as the indispensable eye of a modern army.

—*Air Spy*, by Constance Babington-Smith

Photo reconnaissance aircraft during World War II were primarily stripped-down fighter aircraft: Spitfires, Lightnings, Mosquitoes, Mustangs, and the like. But it was not that way in the beginning. Recon missions were flown by twin-engined Bristol Blenheims, a medium bomber hastily converted from a civilian transport design and pressed into production in 1935. The lumbering Blenheims, however, proved no match for the German flak and speedy Messerschmitt fighters. Flying Officer Maurice "Shorty" Longbottom, assigned to the RAF special unit formed to provide photo reconnaissance, wrote a memo on the feasibility of using high-speed aircraft instead for such work. His summary read as follows:

The best method appears to be the use of a single small machine, relying on its speed, climb and ceiling to avoid destruction. A machine such as a single-seater fighter could fly high enough to be well above Ack-Ack fire and could rely upon sheer speed and height to get away from the enemy fighters. It would have no use for armament or radio and these could be removed to provide room for extra fuel, in order to get the necessary range. It would be a very small machine painted so as to reduce its visibility against the sky.

By January, 1940, statistics showed the value of Longbottom's thinking. The RAF, flying Blenheims and single-engined, high-winged Lysanders, had photographed 2,500

317

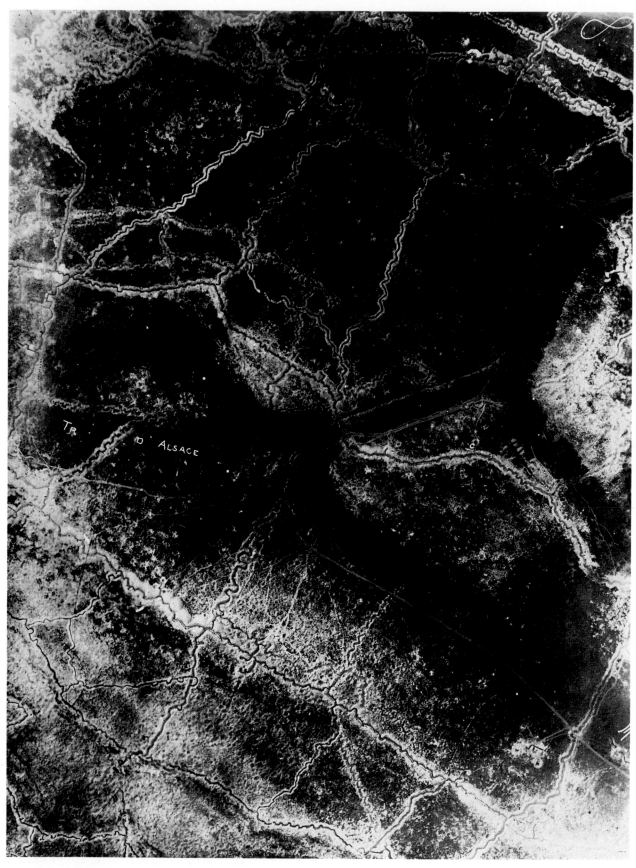

In August, 1914, at the outbreak of World War I, aerial reconnaissance was still primarily visual, but in March, 1915, a trench map based primarily on information gathered from aerial photographs greatly helped British General Sir Douglas Haig prepare for his attack at Neuve Chapelle.

square miles of enemy territory with a loss of 40 aircraft; and the French had photographed 6,000 square miles with a loss of 60 aircraft. But flying out of the commercial airport at Heston, England, the secret British reconnaissance detachment, of which "Shorty" Longbottom was a part, managed to photograph 5,000 square miles of enemy territory without losing the one and only Spitfire that had done the whole job.

When Sidney Cotton, leader of the reconnaissance unit, presented these dramatic statistics to the Air Ministry on January 26, 1940, in defense of his need for the specially redesigned and equipped fighter planes, he pointed out, "The Messerschmitts haven't a chance against the Spitfire because of its speed and altitude. And all that can be seen from the ground is an occasional condensation trail very high up."

Not all that many years would pass before an aircraft was developed that was capable of flying at altitudes so high no tell-tale vapor trails would form. That aircraft was the U-2—one of which hangs not far from the DH-4 in this gallery.

"Whatever else it has done and ever will do," wrote Clarence L. "Kelly" Johnson director of Lockheed's legendary "Skunk Works" where the U-2 was born, "the U-2 is indelibly identified in the public mind as the 'Spy Plane' in which Francis Gary Powers was shot down over Russia on May Day of 1960 while on a photo reconnaissance mission for the CIA."

During World War II it was still possible for a reconnaissance plane to outrun other fighters. But just as the long-bow had rendered the spear obsolete, ground-to-air and air-to-air missiles rendered the conventional reconnaissance aircraft obsolete. As in the past, the need was to build a higher and safer platform far removed from the enemy's reach.

"Kelly" Johnson noted the new requirements:

In 1953, we at Lockheed had been made aware of this country's desperate need for reconnaissance of Soviet missile and other military capabilities. A requirement existed for an airplane that could safely overfly the USSR and return with useful data. . . .

. . . The airplane would have to fly at an altitude above 70,000 feet so vapor trails would not give away its presence, have a range better than 4,000 miles, have exceptionally fine flight characteristics, and provide a steady platform for photography with great accuracy from this high altitude.

It would have to be able to carry the best and latest cameras as well as all kinds of electronic gear for its own navigation, communication, and safety.

—Kelly: More Than My Share of It All, by Clarence L. "Kelly" Johnson with Maggie Smith

The U-2's first flight was in 1955; nine months later, the slim, 80-foot-wingspan aircraft was operational. By the late spring of 1956 U-2s were regularly flying over the U.S.S.R., or near enough to its borders for reconnaissance. Such flights continued until Powers was shot down.

"[Powers] had taken off from Pakistan and was to fly over Russia to Norway," "Kelly" Johnson wrote,

He was flying a U-2C, with a new improved Pratt & Whitney engine that gave it 3,000 to 5,000 feet more altitude than the original U-2A. Photos from earlier overflights had shown as many as 35 Russian fighters trying to climb up to intercept the U-2.

. . . Between what we had deduced and what Gary told us, it appeared that an SA-2 missile had knocked off the right-hand stabilizer while he was at cruising altitude. The airplane then, predictably, immediately went over on its back at high speed and the wings broke off in downbending. Gary was left sitting in the fuselage with a part of the tail and nothing else. He did not use the ejection seat, but opened the canopy to get out.

With the airplane spinning badly and hanging onto the windshield for support, he tried to reach the destruct button to destroy the airplane. It was timed to go off about ten seconds after pilot ejection. But he could not reach the switch. We simulated the situation and it just was not possible with the forces on his body. He had to let go. His biggest worry then was that the fuselage and flailing tail would descend through his chute. But he landed uninjured in a farming area and was captured almost immediately.

—*Kelly: More Than My Share of It All*

Powers was returned to the U.S. two years later in exchange for the Soviet spy Rudolph Abel.

The next logical step from aircraft-borne reconnaissance systems has been satellite reconnaissance which provides the technology to look at Earth from space. Orbiting satellites can view huge areas at a glance, scan whole continents, and trace the courses of immense rivers and mountain ranges.

The Landsat satellites have been monitoring the Earth since 1972. Landsat 4, launched in 1982, represented a new generation of Landsat spacecraft. Designed to complete an orbit every 100 minutes, Landsat 4 circled the Earth 14.5 times a day. Like its predecessors Landsat 1, 2, and 3, it carried a multi-spectral scanner to provide continuity with early images and allow comparisons of land changes in time. In addition, it was equipped with a new sensor, the Thematic Mapper, capable of resolving features nearly three times smaller than earlier Landsat instruments could record thereby providing scenes of the Earth in great detail. Landsat 4, which malfunctioned early in its mission, was replaced by Landsat 5.

Visible light is only one kind of electromagnetic radiation that satellites can monitor. Infrared and radar are also part of the electromagnetic spectrum. Thermal infrared imagery can reveal important information on the temperature and properties of surface materials; radar can "see" through clouds and does not require the light of day. Imagery can therefore be recorded through poor atmospheric conditions at any time of day or night. The radar image of any feature is dependent upon its physical properties—the roughness

Developed in the mid-1950s, by Clarence L. "Kelly" Johnson and his Lockheed Aircraft "Skunk Works" team, the U-2 was designed for high-altitude reconnaissance. Equipped with an 80-foot wingspan to aid in achieving maximum altitude, the U-2 at first could fly over the Soviet Union unharassed by Russia's jets and anti-aircraft missiles which were unable to match its performance. In 1960, however, the U-2 piloted by Gary Francis Powers was brought down during a reconnaissance mission in Soviet air space. Since that time, U-2s have played a vital role in reconnaissance of the Soviet missile buildup in Cuba in 1962, verification of nuclear testing in China, reconnaissance in Vietnam and the Middle East, civil disaster assessment, and environmental monitoring.

The Museum's aircraft is a U-2C painted with camouflage colors for a special Air Force project.

Meteorological satellites look at Earth to study the weather and atmosphere. The world's first weather satellite, TIROS-1 (Television Infrared Observation Satellite-1) was launched on April 1, 1960. It provided more than 22,000 pictures of the Earth from orbit. This new way to look at Earth's weather revolutionized the science of storm prediction.

The Museum's satellite is a TIROS-II prototype designed for ground testing.

Taken at 50,000 feet this aerial infrared photograph of San Francisco Bay reveals characteristics of sea and land not otherwise visible.

of its surface, its orientation, moisture content, composition, as well as the wavelength of the radar signal that "illuminates" the scene.

Often satellite images are not photographs at all. Complex satellite sensors do not record a scene on film but, instead, collect information that can be converted to computer images. These computer scenes are composed of mosaics of tiny rectangles called "pixels" or picture elements. By manipulating the image on the computer, different aspects of the terrain can be emphasized.

Seasat, launched in 1978, was the first satellite designed specifically to monitor the oceans with radar. The satellite operated for 98 days and acquired enough data to produce images of 100 million square miles of the Earth's surface. Seasat orbited the Earth at an altitude of about 500 miles and supplied radar images of many ocean features, including sea ice, currents, eddies, and internal waves. Cameras, however, are still used in space.

A 1,000-pound Large Format Camera flown on the Space Shuttle in 1984 provided more than 2,000 black-and-white, color, and infrared photographs of the Earth's surface. From an altitude of about 150 miles, a single Large Format frame will cover more than 23,400 square miles with enough resolution to distinguish individual roads and other man-made structures.

Meteorological satellites look at Earth to study the weather and atmosphere. Clouds are the most obvious features appearing on satellite weather imagery. Close-up views reveal complex and delicate patterns, from criss-crossing streaks and oceanlike waves to concentric storm systems and eddies in the wakes of islands. The world's first weather satellite, TIROS-1, was launched on

April 1, 1960, and provided more than 22,000 pictures of the Earth from orbit. This new way to look at Earth's weather revolutionized the science of storm prediction. Now even the most remote places on Earth could be monitored regularly. Ten TIROS (Television and Infrared Observation Satellite) satellites were launched before being superseded by TOS (Tiros Operational Satellite) and ITOS (Improved Tiros Operational Satellite). ITOS-1, launched in January, 1970, greatly surpassed the performance of the earlier satellites by providing both direct transmission and storage of television and infrared imagery. Later ITOS spacecraft also supplied vertical profiles of atmospheric temperature.

GOES (Geostationary Operational Environmental Satellite) provides continuous monitoring of the weather and atmosphere over the same region of the Earth. Placed in a geostationary orbit about 22,300 miles above the Equator, the satellite at this altitude orbits at the same speed as the Earth rotates and so appears to remain fixed over one spot on the ground. From this vantage point GOES provides intensive coverage of a region's daily weather developments, as well as warnings of severe storms to come.

With the completion of a permanent space station some time in the 1990s, we will not only have come a long way from the time when hunters and soldiers climbed to the tops of hills and trees to look down over the land, we will have also come full circle. A human observer on an orbiting observation platform is simply utilization of the latest technology to provide us with an even more lofty "tower" that will enable us, like our climbing ancestors, to make use of and take delight in our ability and need for looking at Earth.

As the visitor enters the South lobby of the Museum he passes between two giant murals on the side walls. Robert McCall's The Space Mural—A Cosmic View *is on the east side;* Eric Sloane's Earthflight Environment *is to the west.*

Earthflight Environment

Eric Sloane's L-shaped Earthflight Environment *mural on the west wall of the Independence Avenue lobby stretches 75 feet across the horizontal segment and 58 feet 6 inches upward. Sloane met many of the early transatlantic pilots when they would drop into the Half Moon Hotel near Floyd Bennett Field in Brooklyn in the late 1920s when he was painting murals. They asked him to letter their planes and after several flights with Wiley Post, Sloane began painting cloud formations. Amelia Earhart was his first "cloudscape" customer.*

Sloane's interest in clouds and weather led him to write his first book Clouds, Air, and Wind, *which subsequently became an Air Force weather manual. He also built the*

Abraham Lincoln said: "I cannot imagine a man looking at the sky and denying God." I feel the same way about clouds, air, and wind which have inspired me to specialize in painting skyscapes.

The sky is ageless; man has not left his imprint there. The sky above the Grand Canyon today is exactly the same as when that landscape was being formed. Only an occasional plane and condensation trail has been added. I am thankful for the opportunity to add to an awareness of weather and dedicate this work to America's spacious skies.

—Eric Sloane

ROCKETRY AND
SPACE FLIGHT

APOLLO TO
THE MOON

SPACE HALL

EXPLORING THE
PLANETS

STARS

Rocketry and Space Flight

The Rocketry and Space Flight gallery celebrates the realization of one of mankind's oldest dreams: to abandon his planet's confining sanctuary for voyages into space. It is ironic, however, that for much of the more than 2,000 years that man's dream was evolving, speculation on such travel could hold almost as many physical perils as the voyage itself.

When Galileo Galilei (1564 – 1642), the great Italian astronomer, mathematician, and physicist, focused his crude, newly invented telescope upon the heavens he discovered, among other things, how dangerous it could be to discuss what he had found. Mountains marching across a lunar landscape were fine. That the Milky Way was composed of thousands of millions of stars, that Saturn had rings, that the sun had spots, and that Venus went through phases like our Moon, all these discoveries by Galileo were safe, too. But when, on January 7, 1610, Galileo perceived four dimly lit bodies orbiting Jupiter, he had discovered a dangerous truth: that the speculations set forth a half-century earlier by the Polish astronomer Nicolaus Copernicus (1473–1543) in his *De Revolutionibus Orbium Coelestium* were correct. Copernicus' theory that the Sun was the center of a great system with the Earth and other planets revolving around it was so opposed to accepted beliefs and considered so "dangerous to the faith" that Copernicus dared not publish his treatise until near death. And even though Copernicus placatingly dedicated his work to the pope, Paul III, it was swiftly suppressed. Galileo's

discovery sixty-seven years later of Jupiter's four moons, however, not only showed the validity of the Copernican Theory by proving that all heavenly bodies did not revolve around the Earth, even worse it indicated that if Jupiter had four moons to the Earth's mere one, then our planet might not even be of major importance in the celestial scheme.

The year before Galileo had turned his telescope on Jupiter, Johannes Kepler (1571–1630), the German astronomer and mathematician, had published the Danish astronomer Tycho Brahe's (1546–1601) precise and exacting calculations of the orbit of Mars, thus providing additional proof of the Copernican Theory. Kepler's work contained two of the three laws that now bear his name, which he had formulated on the rules governing planets' orbits. The first of Kepler's laws—that the orbit of each planet is an *ellipse* with the center of the Sun being one of its foci— would have surprised even Copernicus, who had supposed, as had the other astronomers, that a planet's orbit about the Sun would be a perfect circle. However, the theological grip on celestial science was still so strong that in 1633 Galileo was summoned to Rome, tried before the Inquisition, and imprisoned until forced to renounce upon oath any beliefs and writings that did not hold the Earth to be central to the universe with the Sun, planets, and stars in orbit about us.

A generation later Sir Isaac Newton (1642–1727), the English physicist and mathematician, came along "banishing,"

as Arthur C. Clarke has written,* "the last traces of metaphysics from the heavens, and turning the solar system into one vast machine whose every movement is explained by a single all-embracing law— the Law of Universal Gravitation." Newton's proof that no distinction exists between the rules governing the movement of the earth and those obeyed by every other celestial body destroyed forever the last vestiges of what Clarke calls "that closed and tidy medieval cosmos which contained only Heaven, earth and Hell like a three-story building." Astronomy ceased being a theological and philosophical science; it became instead an extension of the realm of mathematicians and geographers. Still, it would be almost three hundred more years before scientists and engineers would give serious thought to the realities of space flight. And when they did, they returned once more to Sir Isaac Newton, who, in 1687, had discovered and formulated the principle which accounts for how all rockets operate:

> To every action there is always opposed an equal reaction; or, the mutual actions of two bodies upon each other are always equal, and directed to contrary parts.

Visitors to the Rocketry and Space Flight gallery find rocket operation explained with illustrations from everyday life: when a frog jumps to shore from a chip of wood, the chip is propelled in the opposite direction. The force of the air escaping through the neck of an inflated balloon causes the balloon to dart about the room. A rocket is simply a device that creates a steady supply of gas to eject through a nozzle like the air from a balloon.

A rocket—a reaction-propelled device that carries both its own supply of fuel and the oxygen necessary to support combustion in airless space—is the only vehicle capable of

carrying man beyond the earth's atmosphere. And yet for many centuries the desire to travel in space and the technical development of the rocket engine proceeded along entirely separate paths, which did not converge until the twentieth century, when fantasy merged with technological fact. The divergence of these paths is illustrated as soon as the visitor enters the gallery where exhibits introduce both the history of the black-powder rocket and the dream of the first known science-fiction writer, the second-century A.D. Greek Lucian, whose book, *True History,* is the story of a fifty-man ship's company whose bark meets with a fierce Atlantic storm's whirlwind which picks them up and a week later deposits them on the Moon. There they encounter the cavalry of the Moon King, who ride into battle on three-headed buzzards; Windrunners, who are propelled into battle by the wind trapped within their great billowing shirts; salad birds, with lettuce leaf feathers, and so on. (Lucian was probably more a satirist poking fun at Homer's *Iliad* and *Odyssey* than a serious science-fiction writer.)

Exhibits trace China's discovery of gunpowder—an explosive mixture of potassium nitrate, sulfur, and charcoal, the earliest known recipe for which appeared in a Chinese volume written in A.D. 1040 — and its use in the thirteenth century in the first black-powder rockets. The evolution of the rocket from an incendiary arrow to a true war rocket is covered here. The second period exhibit deals with the fourteenth-century contributions to rocket technology from the Middle East. Included here is a model of the rocket-propelled "self-moving and combusting egg" first described by the Syrian scholar Hassan-er-Rammah in A.D. 1280. This weapon, powered by two black-powder rockets, contained an explosive or incendiary mixture in its flat pan and twin tails to direct it in a straight line through the water. It is uncertain whether such a device was ever built.

Another somewhat whimsical-looking

*Arthur C. Clarke and the Editors of "Life," *Man and Space,* Life Science Library, Time, Inc.

war machine appears in the third period exhibit: seventeenth-century Europe. It is Giovanni di Fontana's rocket-propelled ram for use against fortresses and other defensive works. A modified version was proposed for use against ships; neither was actually constructed. Fontana's *Bellicorum Instrumentorum Liber*, published in 1420, preceded by some fifty years Leonardo da Vinci's (1452–1519) famous war-machine sketches; and, in addition to a ram, Fontana proposed a rocket-propelled pigeon, hare, and fish. All three devices were incendiary weapons. Although the early use of rockets in the seventeenth century is covered in this exhibit, the technological advances made in rocketry were considerably less significant than the advances achieved in the sciences and, here, the gallery visitor is presented with the theories proposed by Copernicus, Brahe, Kepler, and Galileo.

Ironically, the rocket as a war weapon had, by the eighteenth century, been rendered obsolete by the development of more accurate and effective artillery. While improvements in rocket propellants and design continued, the emphasis was on firework displays and pyrotechnics, which had become increasingly popular forms of mass amusement.

The final exhibit in the Rocketry and Space Flight "corridor of history" concentrates on the re-emergence of the black-powder war rocket during the nineteenth century and its use in whaling and lifesaving. The nineteenth century is considered the "Golden Age" of black-powder rockets—although anyone familiar with *The Star Spangled Banner*'s "And the rockets' red glare, the bombs bursting in air, gave proof through the night that our flag* was still there..." knows that all that

---

*This very flag observed by Francis Scott Key, which flew over Baltimore Harbor's Fort McHenry throughout the British bombardment, is on exhibit in the Smithsonian Institution's National Museum of History and Technology.

glittered during that age was not gold. Sir William Congreve (1772–1828), having been impressed by the rockets used against the English in India, began a series of experiments in 1804 that led to the development of a metal-case stick-guided rocket capable of being fired in large barrages against enemy troop concentrations and fortifications. Congreve rockets played an important role during the War of 1812 and were used in many engagements against the Americans.

Toward the end of the nineteenth century science fiction and science fact began to merge. The French author Jules Verne, one of the most famous science-fiction writers of all time *(20,000 Leagues Under the Sea; Around the World in 80 Days; Journey to the Center of the Earth)*, wrote *From the Earth to the Moon* in 1865. Verne's tale, reflecting a careful blend of diligent scientific research, technical accuracy, prophetic vision, and sheer story-telling power, envisioned three men, a dog, and a couple of chickens, fired into lunar orbit from the Florida coast in a conical projectile by a 900-foot cannon. Their spacecraft was built for comfortable travel and decorated with the lavish appointments of the period.

Of course Verne's launching device with its 400,000 pounds of guncotton as a propellant charge would have eliminated his astronauts, but the author did correctly foresee the necessity of a 25,000-mph escape velocity to leave earth's gravity, that weightlessness would occur, that collisions with meteoroids were a possibility, and that their plush, upholstered "command module" could be steered by rockets in space. Two years later, in 1867, Verne retrieved his astronauts from permanent lunar orbit in the sequel, *Around the Moon*, and after bringing them through white-hot re-entry into earth's atmosphere, had their spacecraft splash down in the ocean.

By 1891, Hermann Ganswindt (1856–1934) was in Berlin drawing up

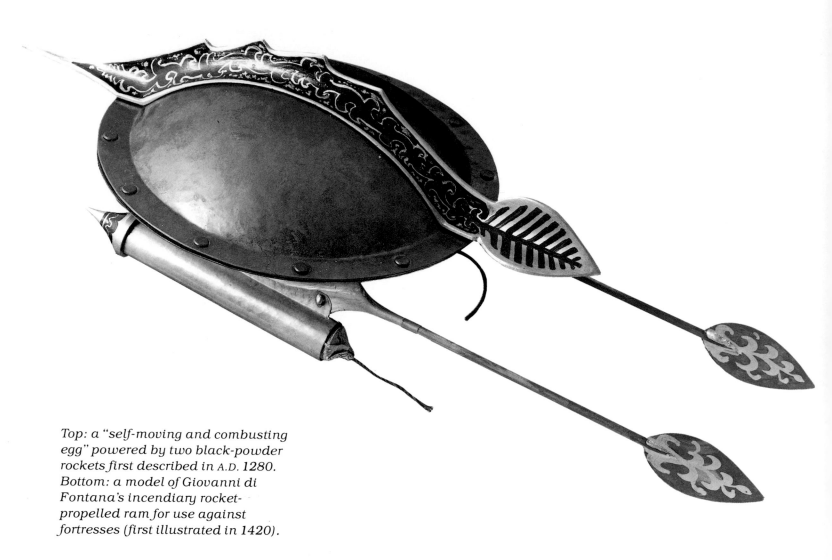

Top: a "self-moving and combusting egg" powered by two black-powder rockets first described in A.D. 1280. Bottom: a model of Giovanni di Fontana's incendiary rocket-propelled ram for use against fortresses (first illustrated in 1420).

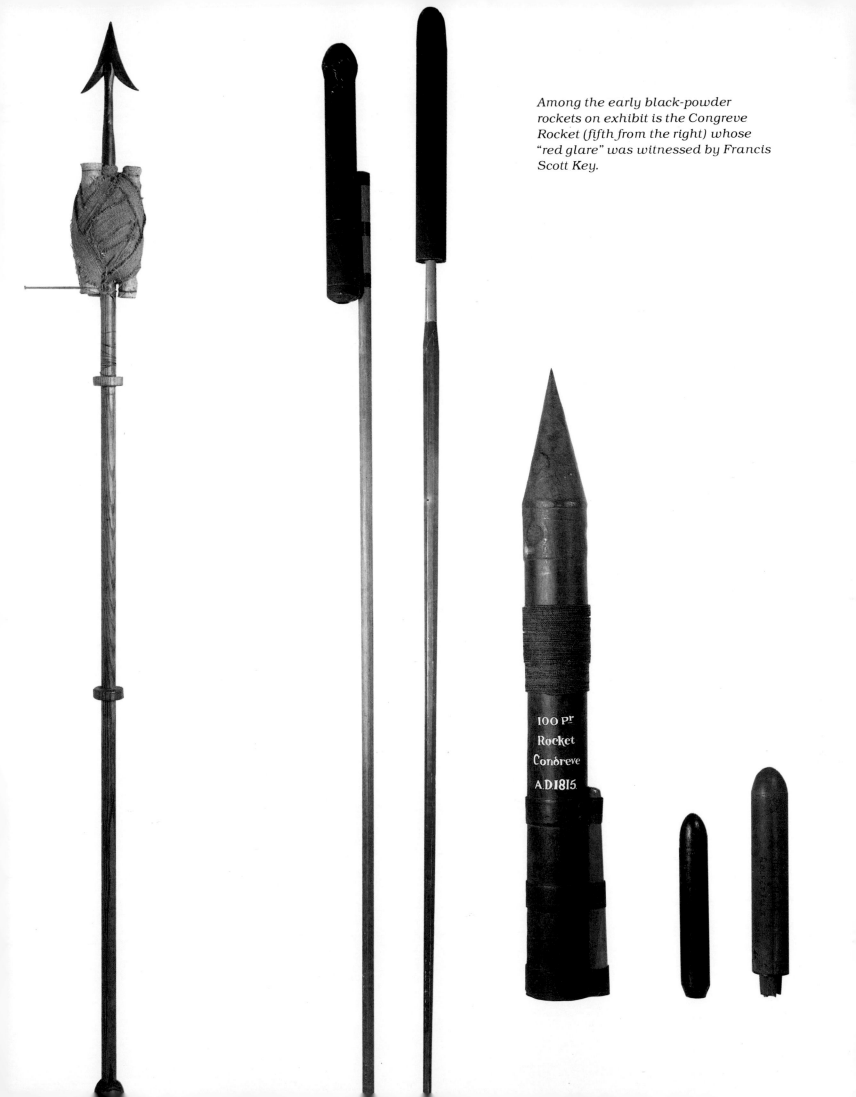

Among the early black-powder rockets on exhibit is the Congreve Rocket (fifth from the right) whose "red glare" was witnessed by Francis Scott Key.

100 Pr
Rocket
Congreve
A.D.1815.

Four hundred thousand pounds of guncotton within a 900-foot cannon fired Jules Verne's fictional 1865 spacecraft with its cargo of astronauts, chickens, dogs, and a small cask of cognac from the Earth to the Moon.

Right: A model of K.E. Tsiolkovsky's 1903 proposed spaceship had the cosmonauts survive the G-forces of launch and reentry by immersing themselves i bathtubs.

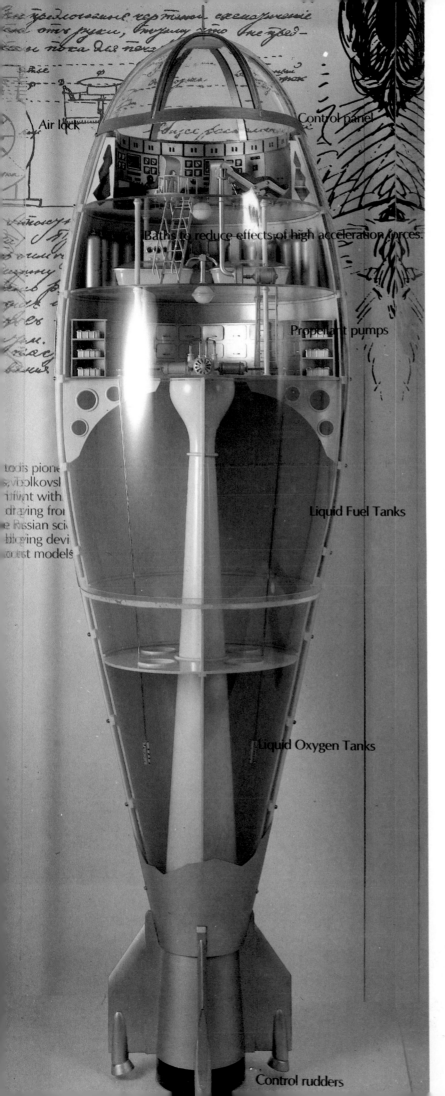

Air lock

Control panel

Baths to reduce effects of high acceleration forces.

Propellant pumps

Liquid Fuel Tanks

Liquid Oxygen Tanks

Control rudders

the first designs for a spaceship and demonstrating that solid-propellant rockets were the only means of getting man into space. In 1898, H.G. Wells published his classic *War of the Worlds*, which described a Martian invasion of Earth. This story formed the basis of Orson Welles' legendary October 31, 1938, Mercury Theater of the Air broadcast, which so terrified radio listeners that hundreds fled their homes believing the Martians had actually landed.

As the NASM visitor leaves the "corridor of history" the two divergent threads—the science and technology of rocket flight, and the desire to travel in space—at last converge and the twentieth century commences with Tsiolkovsky, Oberth, and Goddard, the three great pioneers of astronautics who laid the foundations for space flight.

Konstantin Eduardovitch Tsiolkovsky (1857–1935) was a self-educated Russian schoolteacher who, in 1903, published his now-classic article, "Exploration of Space with Reactive Devices," a theoretical study of rocket fuels and rocket-motor efficiency. Although he did not experiment with rocket engines, Tsiolkovsky showed why rockets would be necessary for space travel and proposed liquid hydrogen and liquid oxygen as the most efficient propellants. He conducted the first studies that demonstrated that space travel was, at least, theoretically possible and advanced the concept of multistage rockets. By 1903 Tsiolkovsky had also given careful thought to manned-spacecraft design and his description of such a ship demonstrates how completely fact and fantasy had by this time merged:

> Let's imagine the following configuration: a metal elongated chamber (having forms of least resistance), provided with its own light, oxygen, with absorbers of carbon dioxide, noxious effluvia and other animal excretions, intended not only for the maintenance of various physical devices, but also to provide life support to the men controlling the chamber.... The chamber contains a large supply of

339

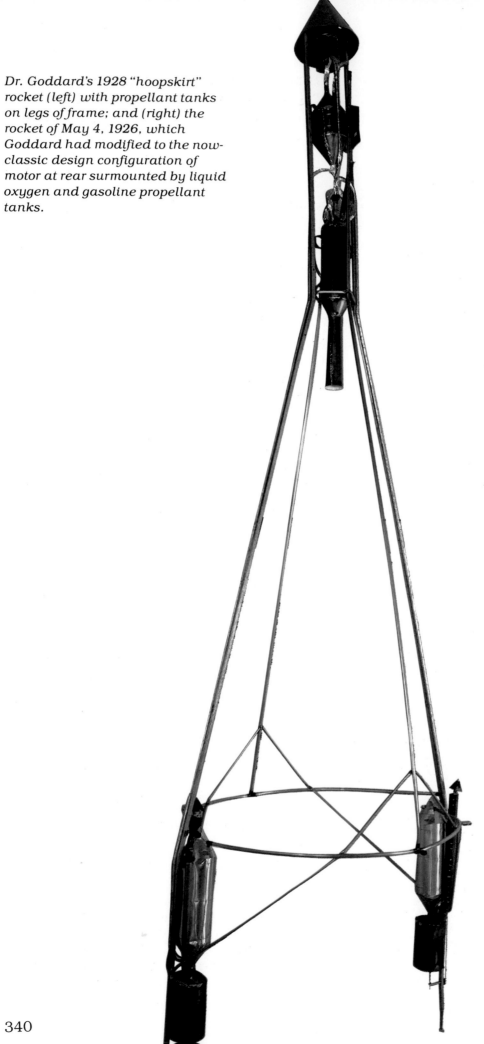

Dr. Goddard's 1928 "hoopskirt" rocket (left) with propellant tanks on legs of frame; and (right) the rocket of May 4, 1926, which Goddard had modified to the now-classic design configuration of motor at rear surmounted by liquid oxygen and gasoline propellant tanks.

340

materials which, when combined, immediately form an explosive mass.

Tsiolkovsky proposed dividing his spaceship into three bays. The top bay in the rocket's nose housed the crew. Here would be the control panels, automatic instruments, decompression chambers, and comfortable couches. The second bay contained the oxygen supply and the "bathtubs" filled with water in which the crew members would immerse themselves to ease the powerful G-forces experienced during the rocket's launch and re-entry. The third bay housed the pumps necessary to move the propellants from the fuel tanks to the engine. The fuel tanks were below this third bay and would contain, Tsiolkovsky suggested, a liquid hydrocarbon and liquid oxygen.

To remain in orbit a spacecraft must achieve a speed of 18,000 mph, to escape gravity 25,000 mph. Even using hydrogen as a fuel, a rocket's exhaust could not exceed about 8,000 mph. By 1898 Tsiolkovsky had already asked the fundamental question upon whose answer all space flight depended: could one build a rocket that could fly faster than its own exhaust gases? Tsiolkovsky published his results in 1903, the year the Wright Flyer stumbled into the air under its own power. Tsiolkovsky's equation showed that up to the speed of light (670 million mph) there is no limit to the speed rockets can reach. The one major limiting factor was that at least 75 percent of a rocket's weight had to be its fuel. Nearly 400 years before Tsiolkovsky advocated the use of multistage rockets, an artillery officer, Conrad Haas, had proposed the same idea with black-powder rockets; Tsiolkovsky, however, suggested that the staging should be set up on a principle similar to an aircraft jettisoning its empty wing tanks. K.E. Tsiolkovsky died in 1935 an honored hero of the Soviet Union and was given a state funeral. One is impressed even today by Tsiolkovsky's vision of space exploration as an inevitable process that would transform and spread human life throughout the solar system.

"Earth is the cradle of the mind," Tsiolkovsky wrote, "but one cannot live in the cradle forever."

Hermann Oberth, another of the great astronautic pioneers, was, like Tsiolkovsky, a schoolteacher. Born in Transylvania (now central Rumania) in 1894, Oberth was fascinated by his childhood reading of Jules Verne's Moon books. In 1923, largely at his expense, Oberth published his own slim volume in Munich. This book, *The Rocket into Interplanetary Space,* was a serious attempt to demonstrate the theoretical possibility of space flight as well as to formulate its basic mathematics. In addition to proposing designs for man-carrying spacecraft and high-altitude research rockets, Oberth advanced the concept of orbital rendezvous for refueling and resupply by reviving the idea of orbiting a space station or large satellite—an idea first suggested in an 1870 *Atlantic Monthly* magazine fiction serial, "The Brick Moon," written by the Boston clergyman Edward Everett Hale.* Oberth was, like Tsiolkovsky, a theoretician, but his book excited the imagination of many young men who banded together to form rocket societies in America, Germany, and the Soviet Union and inspired that generation of engineers who actually built the rockets that would carry man into space. Oberth worked in the German rocket program during World War II, but came to this country afterward, and in 1955 joined the staff of America's Redstone Arsenal.

The only one of the three great pioneers to actually build and fly rockets was Robert H. Goddard, who was born in Worcester, Massachusetts, in 1882. Goddard was a quiet man who loved music, painting, and nature; a schoolteacher like Oberth and

---

*A prolific writer of magazine articles, Hale is perhaps best known for his short novel *The Man Without a Country.* From 1903 until his death in 1909 he was chaplain for the U.S. Senate. Hale's satellite was launched by a spinning flywheel.

The V-2 rocket engine made in the United States after World War II was much larger than any previously built here.

Robert H. Goddard, at the blackboard at Clark University, in 1924, outlines for his students the problems of reaching the Moon by rocket.

Tsiolkovsky, he not only independently worked out the physical principles and calculations on rocketry and space flight, but went on to construct the world's first working liquid-fuel rockets.

Every visitor to the National Air and Space Museum probably knows who Orville and Wilbur Wright were and has some idea of the effects their invention has had on everyday life, but few of them realize that as a result of Robert Goddard's genius an immeasurably greater impact will be made on the lives of their children and generations yet unborn. The Wrights' invention has enabled us to race across our planet; Goddard's has made it possible for us to speed into space. As rocket expert Jerome Hunsaker said, "Every liquid-fuel rocket that flies is a Goddard rocket."

Just as Hermann Oberth had been inspired by his childhood reading of Jules Verne, Goddard's childhood reading of H.G. Wells' *The War of the Worlds* had enormous influence on him. By the age of seventeen, Goddard was already giving serious thought to rocketry and space flight. Goddard's studies of various rocket fuels while majoring in physics at Worcester's Clark University led him to conclude that the most effective propellant would be a combination

of liquid hydrogen and liquid oxygen—neither of which was then commercially available. Upon completing his doctorate in physics, Goddard began teaching at Clark, where his lectures in conventional physics also contained speculations upon methods of traveling in space. His suggested use of rockets as a means of reaching the Moon was included in a monograph entitled "A Method of Reaching Extreme Altitudes," published in 1919 by the Smithsonian Institution. The Smithsonian had granted $5,000 to Goddard for rocket research in 1917.

Goddard was ridiculed by the press, which called him "The Moon Man," and the *New York Times* derided him in an editorial for lacking "the knowledge daily ladled out in high schools." Mary Pickford, then a twenty-six-year-old movie starlet known as "America's Sweetheart," asked to be able to send a message in that first rocket to be launched to the Moon.

The Smithsonian provided the primary funding for Goddard's research from 1917 through 1929, during which period Goddard, who had early on recognized that liquid fuels provided a higher exhaust velocity than solid fuels, had been concentrating on developing a liquid-fuel rocket capable of carrying meteorological instruments to altitudes higher than those achieved by balloons. On March 16, 1926, after several successful static-fire tests, Robert Goddard launched his—and the *world's*—first liquid-propellant rocket. In his report to the Smithsonian's C.G. Abbot, Goddard wrote:

> In a test made March 16, out of doors, with a model...weighing 5¾ lb empty and 10¼ loaded with liquids, the lower part of the nozzle burned through and dropped off, leaving, however, the upper part intact. After about 20 sec. the rocket rose without perceptible jar, with no smoke and with no apparent increase in the rather small flame, increased rapidly in speed, and after describing a semi-circle, landed 184 feet from the starting point—the curved path being due to the fact that the nozzle had burned through unevenly, and one side was longer than the

The LR-87 gimbal-mounted twin-chambered, liquid-propellant rocket engine used to power the Titan I ICBM developed 300,000 pounds of thrust.

Foreground: The H-1 liquid-propellant rocket engine was used in the 8-engine cluster of the first stage of the Saturn I and IB launch vehicles. This engine was used in the launch of the U.S. crew of the Apollo-Soyuz Test Project. Background: The LR-87.

Three of these gold-plated vernier engines were used to soft-land the Surveyor spacecraft on the Moon. The engine's thrust could be varied from 30 to 104 pounds.

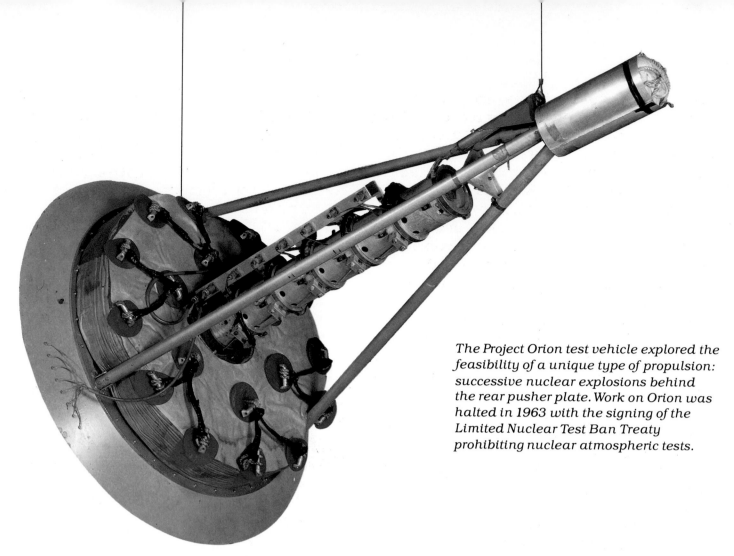

The Project Orion test vehicle explored the feasibility of a unique type of propulsion: successive nuclear explosions behind the rear pusher plate. Work on Orion was halted in 1963 with the signing of the Limited Nuclear Test Ban Treaty prohibiting nuclear atmospheric tests.

other. The average speed, from the time of the flight measured by a stopwatch, was 60 mph. This test was very significant as it was the first time that a rocket operated by liquid propellants travelled under its own power.

—"Robert H. Goddard and the Smithsonian Institution," by Frederick C. Durant III

The reason that Goddard's rocket lingered twenty seconds upon the test stand was that the rocket weighed more than its thrust of 9 pounds, and until the motor had expended enough fuel so that the rocket weighed less, its thrust could not lift it. The movie camera held by Goddard's wife, Esther, who was attempting to record this historic event for posterity, contained only seven seconds of film. The film, of course, had been expended before the rocket left the ground. "It looked almost magical as it rose without any appreciably greater noise or flame," Goddard enthusiastically reported, "as if it said, 'I've been here long enough; I think I'll be going somewhere else.'" Where the rocket went was 41 feet high into the air and, although it

traveled "down-range" only 184 feet, the flight of Goddard's rocket was every bit as significant as Wilbur and Orville Wright's first flight.

Throughout the 1920s Goddard continued rocket tests at his Aunt Effie's farm outside Worcester, but at times the Smithsonian could not hide its disappointment at his results. Once, upon receipt of a Goddard progress report in which the physicist had speculated about space travel, Abbot responded, "Interplanetary space travel would look much nearer to me after I had seen one of your rockets go up five or six miles in our own atmosphere." In May, after another flight with his March 16th rocket design, Goddard changed the configuration, placing the motor in the now-classic position at the rocket's base. This particular rocket, referred to as the May, 1926, device, is on display in this gallery. A replica of the original March 16, 1926, rocket in its first configuration is on display in the Milestones of Flight gallery.

Goddard was aware that he needed some

sort of spectacular success, and to achieve that he needed a much larger rocket. In 1927 he built a rocket with a 200-pound thrust, but it could not lift its weight. "Instead of a little flier," Esther Goddard said, "he had built a big sitter." Goddard decided to compromise. A medium-size rocket in the 40-pound-thrust range would reduce both construction effort and cost. Components could easily be designed and replaced. Goddard bought a secondhand windmill frame from a farmer and modified it as a launching tower for his 1928 "Hoop Skirt" rocket. Although the hoop was designed to add stability to the rocket's flight, the hoop kept getting caught in the launching tower. The rocket did finally fly, however, and reached an altitude of 90 feet. Flights were so noisy that in July, 1929, police, ambulances, and, unfortunately, reporters showed up at Aunt Effie's farm, where the tests were being carried out. One newspaper, with that blend of malice and glee which aviation pioneers had come to expect, headlined: MOON ROCKET MISSES TARGET BY 238,799½ MILES. Goddard was forbidden to make any more tests on the farm; but with the Smithsonian's assistance, he was able to work on his rockets at the United States Army artillery range at Fort (then Camp) Devens, Massachusetts.

In November, 1929, Goddard was visited by an important figure: Charles A. Lindbergh. Lindbergh, impressed by the potential of rocket power, suggested that his good friend and strong supporter of the fledgling aeronautic science, Daniel Guggenheim, help sponsor Goddard's work. Goddard soon received a two-year $50,000 grant from the Guggenheims and, at about the same time, the Smithsonian received $5,000 from the Carnegie Institute to further Goddard's research.

With the Guggenheim funding, Goddard was at last able to devote himself full-time to developing all the various elements of the sounding-rocket design he was working on. In 1930 he moved to Mescalero Ranch in Eden Valley near Roswell, New Mexico, where the climate, the landscape, and the privacy

were perfect for his work. There in Eden Valley, Goddard set up a permanent test facility and with his crew (which never numbered more than seven: Goddard's brother-in-law, Albert W. Kisk, and Henry Sachs, instrument makers; Charles Mansur and his brother, Lawrence, machinists; Nils T. Ljungquist, another instrument maker; Goddard himself, and his wife, Esther, who, in addition to being the official photographer, was also responsible for dousing fires caused by rocket exhausts) he conducted his research on rocket power plants, pumps, fuel systems and control mechanisms, which included gyrostabilization, steering jet vanes in the rocket exhaust, and aerodynamic flaps. The rocket had to be almost entirely handmade:

> Goddard ordered materials from large, hardware mail-order houses and his crew prowled through hardware stores, sporting-goods displays and auto-parts outlets. When they found something that might do a particular job—a child's wristwatch, a length of piano wire, an automobile sparkplug—they proceeded to use it to perform a function undreamed of by its manufacturers.
>
> A good deal of time had to be spent in the shop salvaging rockets that were successful—that is, they flew. A rocket that would not take off was a disappointment—but the rocket was usually left intact. A successful flight meant jubilation—and often carrying home a hunk of junk, all that remained after a crash landing. The smashed rockets could seldom be rebuilt, so Goddard designed and built a rocket recovery system with parachutes to ensure soft landings.
> —*Man and Space*, by Arthur C. Clarke and the Editors of "Life"

In 1932, two years after Goddard had set up his test facility in Roswell, New Mexico, the Guggenheim funds dried up because of the worldwide financial depression. Goddard returned to teaching at Clark University. He even had to ask the Smithsonian for $250 so that he could perform special tests concerned with reducing rocket weight.

In 1934, when the Guggenheim funding was resumed, Goddard still had to turn to the Smithsonian for assistance on specific problems and, in recognition of his very special relationship with the Smithsonian, Goddard, at the urging of Harry Guggenheim and Charles Lindbergh, sent the Institution a complete 1935 A-Series rocket—with the understanding that it not be exhibited until either he requested it or, in the event of his death, at the request of Lindbergh and Guggenheim. When the rocket arrived, its crate was bricked up inside a false wall within the Smithsonian and not exhumed until after World War II. This rocket is on display in the Satellites gallery.

Goddard was a gifted, ingenious engineer and an experienced and responsible scientist. During the Roswell years, 1930–40, and with but a small staff, Goddard built rocket engines and systems years ahead of their time. Among the many pioneering steps Goddard accomplished en route to successful liquid-propellant rocket-powered flight were the following: he obtained important performance data through static tests which enabled him to continually improve his designs; he pioneered gas-generator-powered, turbo-pump-fed rockets; he developed automatic launch-sequence control; he developed timed sequential actuation of tank pressurization, ignition, umbilical release, automatic shut-down, thrust determination, vehicle release, among others. Goddard also engineered on-board controls for guidance and engine shut-down, systems for parachute and payload recovery, and pioneered gyrostabilization and in-flight aerodynamic and rocket-exhaust deflection controls, as well as gimbal-mounted rocket motors. He developed recording and optical-telescopic tracking systems. And finally, he established a remarkable safety record throughout the Roswell period. Despite working with highly combustible and potentially dangerous propellants, there was not one serious accident during the innumerable static tests and thirty-one launches.

During World War II, Goddard left New Mexico for Annapolis, Maryland, to work for the government and developed Jet-Assisted Take-Off engines (JATO) for the Navy. In March, 1945, Goddard saw captured German V-2 Rocket parts for the first time. Although Goddard and the team of German scientists had worked separated not only by an ocean but by that theoretically most insurmountable of barriers, rigid wartime secrecy, Goddard saw that the V-2 rocket, though much larger than his latest model, used many similar systems. As Dr. Walter R. Dornberger, head of the German V-2 rocket team, explained, "That was the only way to build a rocket." But when, in 1950, Wernher von Braun, the German scientist who had presided over the V-2's development, examined the more than 200 patents Goddard was granted covering almost every aspect of liquid-fuel rockets, he said, "Until 1936, Goddard was ahead of us all."

Robert H. Goddard died in August, 1945. In his unpublished papers Goddard speculated about flights to the Moon, the planets, and beyond. He never outgrew the dream of that small boy who had read H.G. Wells' *War of the Worlds* and been struck with wonder at the challenge of the unknown. In 1932, Goddard wrote H.G. Wells a letter that said something important about his dream:

> How many more years I shall be able to work on the problem, I do not know; I hope, as long as I live. There can be no thought of finishing, for "aiming at the stars" both literally and figuratively, is a problem to occupy generations, so that no matter how much progress one makes, there is always the thrill of just beginning.
>
> —"Robert H. Goddard and the Smithsonian Institution," by Frederick C. Durant III

When the visitor leaves that part of the gallery devoted to Goddard's work he sees exhibits tracing the development of the different national rocket societies of the 1920s and 1930s, and from there he is

Right: Despite the problems of restricted mobility and poor stowage qualities, this RX-1 suit, on at least one occasion, kept a man alive and working on simulated tasks in a pressure environment like that of space.

Far right: This type of suit, worn by the astronauts on all Project Mercury flights, was designed to serve as an emergency backup system in case of cabin decompression.

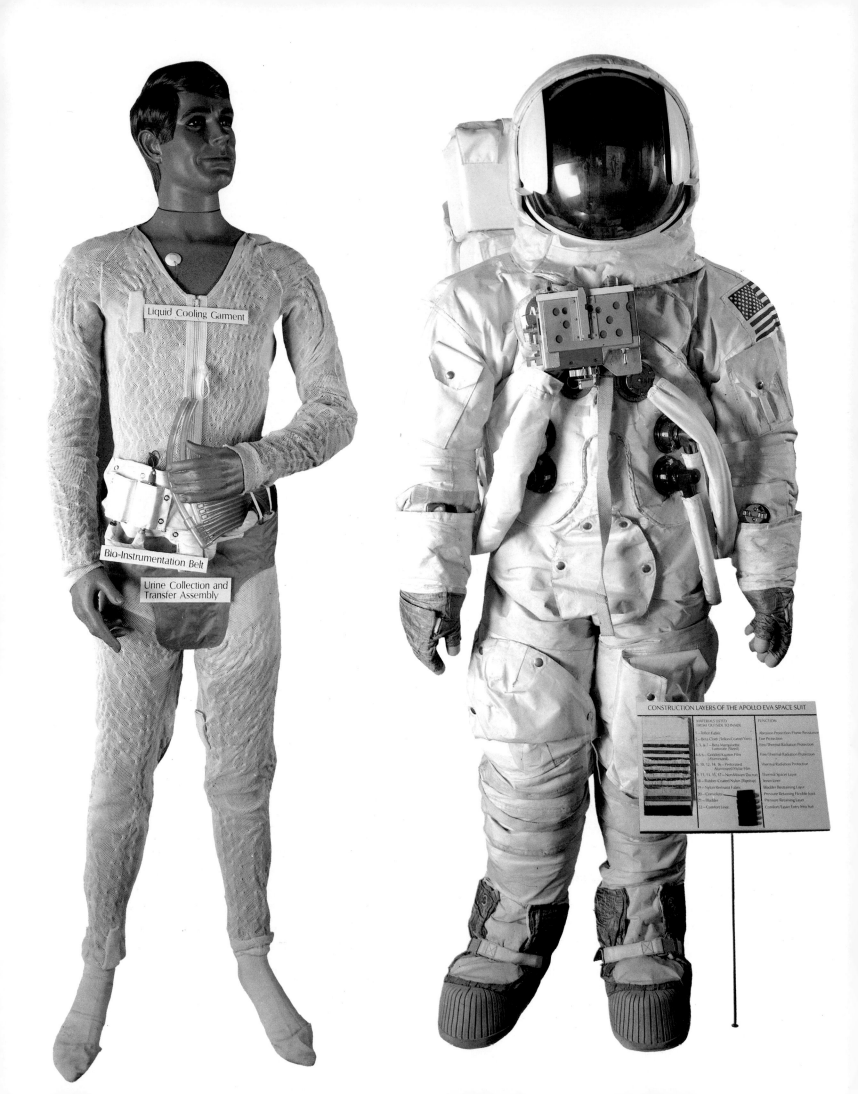

Liquid Cooling Garment

Bio-Instrumentation Belt

Urine Collection and
Transfer Assembly

CONSTRUCTION LAYERS OF THE APOLLO EVA SPACE SUIT

MATERIALS LISTED FROM OUTSIDE TO INSIDE	FUNCTION
1—Teflon Fabric	Abrasion Protection/Flame Resistance
2—Beta Cloth (Teflon-Coated Yarn)	Fire Protection
3, 5, & 7—Beta Marquisette Laminate (Sized)	Fire/Thermal Radiation Protection
4-6-8—Coated Kapton Film (Aluminized)	Fire/Thermal Radiation Protection
9, 10, 12, 14, 16—Perforated, Aluminized Mylar Film	Thermal Radiation Protection
11, 13, 15, 17—NonWoven Dacron	Thermal Spacer Layer
18—Rubber-Coated Nylon (Ripstop)	Inner Liner
19—Nylon Restraint Fabric	Bladder Retaining Layer
20—Convolute	Pressure Retaining Flexible Joint
21—Bladder	Pressure Retaining Layer
22—Comfort Liner	Comfort Layer Extra Into Suit

introduced to some of the larger rocket motors. The largest motors exhibited here are the RL-10 and the LR-87.

The RL-10 has an upper-stage propulsion system that can be stopped and restarted in space. RL-10s pioneered the use of liquid hydrogen as a rocket fuel and powered the Centaur launch vehicles that boosted crafts such as the Surveyor and Viking into space. A cluster of six RL-10 engines propelled the second stage of the Saturn I rocket. The RL-10 exhibited here has been cut away so that the visitors can see the "plumbing" inside and gain some understanding of how the rocket works.

The LR-87 was a twin-chamber liquid-propelled rocket engine developed for the Titan I intercontinental ballistic missile. The combustion chambers are gimbal-mounted so that the exhaust could be swiveled to direct the missile's trajectory during the powered phase of flight. One of the most fascinating exhibits here is the display area set aside for the sounds the various rockets made.

When the visitor completes his tour of Rocketry and Space Flight he enters the final area of this gallery, where he is introduced to the development of the space suit.

As man ventures out of his Earth's environment it is obvious that he must carry his "environment" with him. As his craft climbs away from Earth, the atmospheric pressure gradually diminishes until it disappears entirely. Beyond 40,000 feet, the limit for unpressurized flight, lungs no longer absorb sufficient oxygen to replenish the bloodstream; beyond 460,000 feet the atmosphere no longer provides protection against micrometeoroids. Were the pilot's pressure suit to be punctured by micrometeroids, death would rapidly follow;

Far left: A layered liquid cooling garment that provides for body cooling.
Left: Designed and created primarily for Moon walking, this suit, with its backpack, enabled the lunar astronauts to dispense with the tether-umbilical used on the Pioneer-Gemini "spacewalks" and to roam free over the lunar surface. The development of this space suit system was one of the most complex elements in the history of manned space flight.

therefore, several layers of plastic protect the pressure-retaining layer of the suit.

Pressure-suit development was an inevitable outgrowth of international competition for altitude and speed records. Since many of the problems confronting the man in the thin upper reaches of the atmosphere would be the same as for a man in the depths of the sea, the first pressure suits resembled deep-sea diving suits more than "space suits," a resemblance already familiar to the visitor who saw, in the Flight Testing gallery, the world's first practical flight pressure suit worn by Wiley Post. There is a difference between pressure suits and "space suits," however. Both space suits and pressure suits need to provide protection, mobility, comfort, and minimum bulk at light weight. In addition, the space suit must make provision for micrometeoroid protection, waste management, and the extreme temperatures of space. Consider, for example, the fact that the Apollo space suit worn during the walk on the Moon would be alternately exposed to temperatures on the surface of $+250°$ F. in the sun and $-200°$ F. in shadow. The astronaut's portable life-support system (PLSS) had to create and maintain a livable atmosphere inside the space suit. The PLSS could be worn for seven hours without being recharged. It supplied oxygen for breathing purposes, suit pressurization, communication, and ventilation. It also supplied cool water and oxygen for body cooling and removed contaminants from the oxygen circulating through the suit. Fully charged, the pack weighed 104 pounds. Fortunately, because of the lower gravity, the pack weighed but 17 pounds on the Moon.

The visitor to this gallery might be amused by the exhibits showing the at times startling similarity between the space suits worn by the astronauts and those proposed by cartoonists and early science-fiction writers; but he should also come away from this gallery with an appreciation that what was science fiction yesterday is fact today. And, in like manner, what is science fiction today will tomorrow be reality.

The fully charged 104-pound Apollo Portable Life Support System (PLSS) fortunately weighed but 17 pounds on the 1/6 Earth gravity of the Moon.

Apollo to the Moon

I believe that this nation should commit itself to achieving the goal, before this decade is out, of landing a man on the moon and returning him safely to earth. No single space project in this period will be more impressive to mankind or more important for the long-range exploration of space. And none will be so difficult or expensive to accomplish.

    —President John F. Kennedy, May 25, 1961

A videotape of the young President delivering this speech before a special joint session of Congress is one of the first images a visitor to the Apollo to the Moon gallery sees. President Kennedy's announcement that the United States intended to land a man on the Moon before the decade was out did more than define a new role for America's manned space flight program; it expressed in unequivocal terms this country's determination to win the space race, which had not only already clearly begun—but which also, just as clearly, America was already losing.

Three and a half years earlier, Sputnik's faint, otherworldly *beep-beep-beep* had stunned the American people and proven that the Soviet Union's scientific and technological capabilities were far advanced of what had been believed. Furthermore, the launching and successful orbiting of the 1,121-pound satellite Sputnik 2, several weeks later, indicated that the Soviet boast that they possessed intercontinental ballistic missiles capable of striking American cities was possibly true. To understand the significance of President Kennedy's challenge and how the lunar landing was to become a tangible symbol of this nation's resolve to restore its lost prestige, it helps to recall the events that led up to the President's speech.

One successful Soviet space launch after another had placed increasingly massive payloads into orbit while, in the most publicized failure in history, America's 3.3-pound grapefruit-sized Vanguard satellite had plaintively whistled even as its rocket consumed itself in flames. The American people's growing frustration and humiliation as they watched their rockets explode on their launch pads had resulted in grave doubts not only about this nation's technological prowess, but about the whole educational process in the United States. And then, on April 12, 1961, just before the President's speech, the Russians had orbited Major Yuri A. Gagarin. And, five days later, there occurred the debacle of the Bay of Pigs. American morale could not have been lower.

In the fall of 1958, NASA had begun the process of selecting the astronauts and establishing the criteria for the sort of men they needed: the men would have to be pilots, engineers, explorers, scientists, guinea pigs; they would have to be physically strong enough to endure the stresses and requirements of space flight and emotionally strong enough to withstand the pressures and demands made upon them before and after their return. The men would have to have daring and courage, but above all would have to remain cool and resourceful in the face of unforeseen emergencies or hazards. NASA set their top age at 40 (there have since been changes); their height was to be no more than 5'11"* and weight no more than 180 pounds. NASA further announced that no applicant would be considered who did not have a formal engineering degree or its equivalent. NASA immediately turned to the ranks of practicing military test pilots on the theory that these men already had the sort of experience and credentials future astronauts would need. At 2 PM, on April 9, 1959, the seven men who had been selected were introduced to the press. When one of the reporters asked the astronauts who, among them, would be willing to go into space right then and there, all seven raised their hands. They were M. Scott Carpenter, L. Gordon Cooper, Jr., John H. Glenn, Jr., Virgil I. Grissom, Walter M. Schirra, Jr., Alan B. Shepard, Jr., and Donald K. Slayton. Two years after our astronauts' introduction to the press, Russia's Yuri Gagarin was launched into space. Alan Shepard in *Freedom 7* would follow Gagarin three weeks later, on May 5, 1961.

The morning Shepard was scheduled to become the first American in space, the launch of his *Freedom 7* Mercury spacecraft had been delayed by cloud cover and problems with a small inverter that needed to be replaced. Shepard and Grissom had been killing time in the van trying to relax. Prior to Gagarin's orbit, the Soviets had launched seven dogs into space, Laika (11/3/57), Strelka and Belka (8/19/60), Pshchelka and Mushka (12/1/60), Chernushka (3/9/61), and Zvezdochka (3/25/61), four of whom had survived. Shepard was listing the desired qualities for being an astronaut: courage, perfect vision, low blood pressure, coordination. "And you've got to

have four legs," Shepard said.

"Why four legs?" asked Grissom.

"They really wanted to send a dog," Shepard replied, "but they thought that would be too cruel."

About four hours later, Shepard was launched in *Freedom 7* atop a Redstone rocket that generated 78,000 pounds of thrust:

Just after the count of zero Deke [Slayton, seated at the Capsule Communicator desk at Mercury Control Center] said, "Lift-off." Then he added a final tension-breaker to make me relax. "You're on the way, José," he said.

I think I braced myself a bit too much while Deke was giving me the final count. Nobody knew, of course, how much shock and vibration I would really feel when I took off. There was no one around who had tried it and could tell me; and we had not heard from Moscow how it felt....

There was a lot less vibration and noise rumble than I had expected. It was extremely smooth—a subtle, gentle, gradual rise off the ground....But there was no question that I was going....I could see it on the instruments, hear it on the headphones, feel it all around me.

It was a strange and exciting sensation. And yet it was so mild and easy—much like the rides we had experienced in our trainers—that it somehow seemed very familiar....For the first minute the ride continued to be very smooth. My main job just then was to keep the people on the ground as relaxed and informed as possible.... So I did quite a bit of reporting over the radio about oxygen pressure and fuel consumption and cabin temperature and how the G's were mounting slowly, just as we had predicted they would....

One minute after lift-off the ride did get a little rough. This was where the booster and the capsule passed from sonic to supersonic speed and then immediately went slicing through a zone of maximum dynamic pressure as the forces of speed and air density combined at their peak. The spacecraft started vibrating here. Although my vision was blurred for a few seconds, I had no trouble seeing the instrument panel. We had known that something like this was going to happen, and if I had sent down a garbled message that it was worse than we had expected and that I was really getting buffeted, I think I might have put everybody on the ground in a state of shock. I did not want to panic anyone into ordering me to leave. And I did not

---

*The reason for the height limitation was determined by the size of the Mercury spacecraft already on the drawing boards, whose dimensions, in turn, were dictated by the size of the available Redstone and Atlas boosters, which would launch the Mercury spacecraft into space. The diameter of the Mercury spacecraft at its base was 74", or 6'2". Once an astronaut was in his space suit and helmet, anyone taller than 5'11" simply would not fit.

want to leave. So I waited until the vibration stopped and let the Control Center know indirectly by reporting to Deke that it was "a lot smoother now, a lot smoother."

... The engine cutoff occurred right on schedule, at two minutes and 22 seconds after lift-off. Nothing abrupt happened, just a delicate and gradual dropping off of the thrust as the fuel flow decreased. I heard a roaring noise as the escape tower blew off...and then I heard a noise as the little rockets fired to separate the capsule from the booster. This was a critical point of the flight, both technically and psychologically. I knew that if the capsule got hung up on the booster, I would have quite a different flight, and I had thought about this possibility quite a lot before lift-off.... Right after leaving the booster, the capsule and I went weightless together and I could feel the capsule begin its slow, lazy turnaround to get into position for the rest of the flight. It turned 180 degrees, with the blunt or bottom end swinging forward now to take up the heat.... The capsule was traveling at about 5,000 miles per hour now.... All through this period, the capsule and I remained weightless. And though we had had a lot of free advice on how this would feel—some of it rather dire—the sensation was...pleasant and relaxing. It had absolutely no effect on my movements or my efficiency. I was completely comfortable, and it was something of a relief not to feel the pressure and weight of my body against the couch. The ends of my straps floated around a little, and there was some dust drifting around in the cockpit with me. But these were unimportant and peripheral indications that I was at Zero G.

... At five minutes and 14 seconds after launch, the first of the three [retro-] rockets went off, right on schedule. The other two went off at the prescribed five-second intervals. There was a small upsetting motion as our speed was reduced, and I was pushed back into the couch a bit by the sudden change in Gs. But each time the capsule started to get pushed out of its proper angle by one of the retros going off I found I could bring it back again with no trouble at all. I was able to stay on top of the flight by using the manual controls and this was perhaps the most encouraging part of the entire mission....

In that long plunge back to earth, I was pushed back into the couch with a force of about 11 Gs.... All the way down, as the altimeter spun through mile after mile of descent, I kept grunting out "O.K., O.K., O.K.," just to show them back in the Control Center how I was doing.... All through this period of falling the capsule rolled around very slowly in a counterclockwise

direction, spinning at a rate of about 10 degrees per second around its long axis. This was programed to even out the heat and it did not bother me. Neither did the sudden rise in temperature as the friction of the air began to build up outside the capsule. The temperature climbed to 1,230 degrees Fahrenheit on the outer walls. But it never got above 100 degrees in the cabin or above 82 degrees in my suit.... By the time I had fallen to 30,000 feet the capsule had slowed down to about 300 miles per hour. I knew from talking to Deke that my trajectory looked good and that *Freedom 7* was going to land right in the center of the recovery area.... At about 1,000 feet I looked out through the porthole and saw the water coming up towards me. I braced myself in the couch for the impact, but it was not bad at all. It was a little abrupt, but no more severe than the jolt a Navy pilot gets when he is launched off the catapult of a carrier. The spacecraft hit, then flopped over on its side.... One porthole was completely under water.... I could not see any water seeping into the capsule, but I could hear all kinds of gurgling sounds around me, so I was not sure whether we were leaking or not. Slowly but steadily the capsule began to right itself. As soon as I knew the radio antenna was out of the water I sent off a message saying that I was fine.

It took the helicopter seven minutes to get me to the carrier. When we approached the ship I could see sailors crowding the deck, applauding and cheering and waving their caps. I felt a real lump in my throat.

—*We Seven, by the Astronauts Themselves,*
by M. Scott Carpenter, et al.

Visitors to the Apollo to the Moon gallery can see Shepard's historic *Freedom 7* Mercury spacecraft. The engineers who went over the capsule after Shepard's flight decided that *Freedom 7* had come through its 15 minute 22 second, 536-mile trip in such good shape that it could be used again.

The next Mercury launch was Gus Grissom's ill-fated, nearly identical, sub-orbital ballistic flight aboard *Liberty Bell 7* *

Freedom 7 had been the seventh Mercury spacecraft built, hence the number 7. However, afterwards, the 7 was retained by the astronauts, who wished to indicate their sense of unity at being the first seven astronauts chosen.

The Freedom 7 *Mercury spacecraft in which, on May 5, 1961, astronaut Alan B. Shepard, became the first American to enter space.*

on July 21, 1961. Grissom's craft had a crack painted on it in honor of its namesake and when, upon landing, its hatch inexplicably blew, flooding the spacecraft and causing it to sink, they lamely joked at the Cape that it was the last time they'd launch a spacecraft with a crack in it. Despite the loss of *Liberty Bell 7*, Grissom's flight was judged so successful that no further Redstone launches were called for. The next mission was John Glenn's three-orbit flight in *Friendship 7*, the spacecraft on exhibit in the Milestones gallery. *Friendship 7* was launched on the morning of February 20, 1962, atop an Atlas launch vehicle whose 360,000-pound thrust accelerated the Mercury spacecraft to an orbital velocity of 17,540 mph in slightly more than 5 minutes. Shepard's and Grissom's suborbital flights had shown that the Mercury spacecraft was a safe vehicle for manned flight; Glenn's *Friendship 7* flight tested the performance of the pilot in a more extended weightless condition and how well the pilot could operate and interact with the various automatic systems in the spacecraft. When an attitude-control rocket malfunctioned during one of the orbits, forcing Glenn to take over manual control of his craft, the advantage of manned space flight was clearly proven. Scott Carpenter in *Aurora 7* duplicated Glenn's three-orbit flight on May 24, 1962. Despite constant problems with faulty instruments, an overheating space suit, and, during preparation for re-entry, misfiring of the rockets, which caused Carpenter to overshoot his target area by some 250 miles, *Aurora 7*'s flight, too, was a success. Wally Schirra in *Sigma 7* was next; his flight was virtually perfect and after six orbits he splashed down only four miles from his recovery ship. The last Mercury flight was Gordon Cooper's *Faith 7* on May 15, 1963. Cooper made 22 orbits and was aloft for 34 hours, 19 minutes, and 49 seconds, an American endurance record, during which he traveled some 583,000 miles.

The Russians, too, had been busy. In August, two months prior to Schirra's flight, the Soviets had launched their most spectacular manned flights. Vostok 3 with Andrian G. Nikolayev was launched August 11, 1962, and the following day Vostok 4 with Pavel R. Popovich was sent up to join him. The two Soviet spacecraft approached to within three miles of each other, but they did not attempt rendezvous. Popovich's 48 orbits and Nikolayev's 64 orbits continued to give the impression that the Soviet space program was more advanced than ours. Not until August 21, 1965, three years after the Vostok 3 and 4 launches, would an American Gemini spacecraft, Gemini 5, carrying astronauts Gordon Cooper and Charles Conrad, Jr., surpass the Soviet endurance record with a flight of 120 orbits lasting eight days. By this time, however, the Soviets had launched two of their Voskhod series spacecraft, which weighed 11,731 pounds, nearly as much as our Apollo spacecraft would weigh five years later. Voskhod 1, launched in October, 1964, had carried three cosmonauts. Voskhod 2, launched March 18, 1965, carried two, Aleksei Leonov and Pavel Belyayev, into a seventeen-orbit mission during which Leonov became the first man to "walk" in space. Difficulties with its automatic navigational system caused the Voskhod 2 to miss its planned landing site in the Ukraine. The spacecraft parachuted to earth far to the frozen north and it took several hours for the spacecraft to be located and nearly a day for a ground party to break through the forest to bring the cosmonauts out on skis. While waiting for rescue, the Soviet crew were forced to remain in their capsule out of fear of the lurking wolves. Ed White's 20-minute "space walk" in Gemini 4 took place not quite three months after Leonov's and was the highlight of Gemini 4's 62-orbit journey on June 3, 1965.

The Gemini 7 spacecraft in the Apollo to the Moon gallery was launched on December 4, 1965, and was followed by Gemini 6 on December 15. Gemini 6, containing astronauts Schirra and Stafford, had been

Suiting Up, *Norman Rockwell's 1965 painting of Virgil I. Grissom and John W. Young's Gemini 3 flight captures the intricacies of the preflight preparations.*

scheduled for launch October 25, 1965, to test rendezvous and docking procedures, but it had been delayed when their Agena target vehicle had exploded. Schirra and Stafford had watched the Agena's flawless launch from atop their own Titan II booster and then had begun to busy themselves with their own launch preparations when, six minutes into their target vehicle's flight, all telemetry ceased and the tracking radar at the Cape found itself following not one vehicle but five or six. Gemini 6's attempt at the first space rendezvous docking had to be put off until Gemini 7 could be pressed into service as the "target vehicle."

Gemini 6 was rescheduled for launch on December 12, eight days into Gemini 7's flight,

…and Launch Complex 19 at the Cape again had a Gemini-Titan poised for launch. Wally Schirra and Tom Stafford had no Agena to worry about this time, and all appeared normal during the countdown. In fact, they got engine ignition and an indication of lift-off—then sudden silence. They had about a second to review two scenarios: (1) the engines for some reason had shut down after lift-off, and they were now on the brink of disaster and would either settle back down or topple over, requiring immediate ejection to avoid the ensuing holocaust; (2) the engines had shut down the instant *before* lift-off, in which case they were still firmly bolted to the launch pad and could stay put unless some new danger developed. The design of the hardware spoke for option 1 (supposedly the lift-off signal in the cockpit was foolproof), but the seat of Wally's pants spoke for 2. It felt solid under him. A panicky type might have ejected anyway, and

*The cramped Gemini 7 spacecraft in which
Frank Borman and James Lovell, Jr.,
spent two weeks in December, 1965. It was
America's longest manned space flight until
the Skylab missions in 1973.*

December, 1965: Gemini 6 and Gemini 7 practiced rendezvous techniques with each other approaching at times as close together as 6 feet. Here Gemini 7 is seen from Gemini 6.

had Wally and Tom done so, certainly no one familiar with the hardware would have blamed them. But supercool Wally kept his head, picked the correct option, and Gemini 6 was saved to fly another day.

—*Carrying the Fire*, by Michael Collins

An electrical plug had vibrated loose just after engine ignition and a split second before lift-off, thereby sending a shut-down signal to the engines. It was later found that "a plastic dust cover carelessly left in a fuel line would have blocked the Gemini 6 launch even if an electrical plug had not dropped out of the tail and shut down the Titan II engines."* The device, installed at the plant, had not been removed due to "human error." The third launch attempt, on December 15, 1965, was successful, and after chasing Gemini 7 for four orbits, Gemini 6 caught up. The two spacecraft flew in close

---

*NASA SP-4006, "Astronautics and Aeronautics, 1965"

formation for five and a half hours, their crews talking back and forth, snapping pictures of one another, and at times approaching as close together as six feet at an altitude of 185 miles. And then, after almost 26 hours in flight, Gemini 6 was recovered while Gemini 7 remained aloft for two more days. On December 18, Frank Borman and James Lovell rotated their Gemini 7 into a retrofire attitude and prepared the electrical circuitry that would fire the retrorockets to bring them back down. These motors had been exposed to the cold vacuum of space for more than two weeks and there was some worry that they might not fire. Mission Control Center was relieved to hear Borman report a successful retrofire, and when the two astronauts were recovered and were seen on television walking confidently across the carrier deck with no apparent serious problems after their record-setting 330-hour endurance flight in space—nearly double the length of time required for a round-trip to the Moon— it now seemed clear that the United States had at last caught up with, if not actually overtaken, the Soviet space effort.

Only five more Gemini flights were planned, and the single major unanswered question that the Gemini program had been designed to resolve was the practicality of a lunar strategy requiring a space rendezvous for its successful completion. An actual docking had yet to take place, and the remaining missions were designed to perfect rendezvous and docking techniques and the sort of Extra-vehicular Activity (EVA) a Moon landing would require. Although Ed White's 22-minute "space walk" on Gemini 4 had been successful, there still did not exist enough practical experience to prove that a man could operate effectively outside of his spacecraft. Gemini 8, launched March 16, 1966, carrying Neil Armstrong and David Scott, completed the first space docking with an Agena target rocket, but when moments later a runaway thruster on the Gemini craft caused the linked vehicles to gyrate uncontrollably, they were forced to undock.

June, 1966: "We have an angry alligator on our hands," radioed Gemini 9's Thomas Stafford as they approached their still nose-shrouded target docking vehicle.

After undocking, the Gemini began to tumble at a rate which built up to 300 degrees per second. Armstrong had to activate the re-entry attitude-control system to bring the spinning Gemini back under control; but once that was done, they had no choice but to return to earth, 60 hours earlier than planned, after spending 10 hours and 42 minutes in space.

Gemini 9, scheduled for launch May 17,

Paul Calle. End of Mercury. *1963. Oil on canvas. Formerly on loan from the artist*

1966, had to be delayed when the Atlas booster that was to lift its Agena target vehicle into orbit went out of control, causing the rocket to plunge into the Atlantic. The launch, which was rescheduled for June 1, was delayed once more; but on June 3, Gemini 9 finally thundered into orbit only to discover that the nose shroud of their target vehicle had failed to separate and was instead still attached. Its appearance caused astronaut Stafford to report, "We have an angry alligator on our hands." Cernan's two-hour space walk during this flight was hampered by his exertion, which caused his visor to fog over. Gemini 10, launched July 10, 1966, rendezvoused with two different Agena

vehicles and, after successfully docking with one, astronauts John Young and Michael Collins used its power to propel them into a higher orbit. Collins transmitted a long series of three-digit commands to the Agena to commence its automatic firing sequence, and since they were docked nose-to-nose, the astronauts had the extraordinary experience of watching their own rocket fire up at them:*

---

*Collins' and Young's Gemini 10 was docked on the near end of their Agena, whose engine was mounted on the far end so that they couldn't quite see it. The Agena engine with its 16,000-pound thrust would take them to an altitude of 475 miles, higher than any manned spacecraft had gone before.

At the appointed moment, all I see is a string of snowballs shooting out of the back of the Agena in a widening cone. The unexpected white stream is quite pretty against the black sky. Aw shucks, I think, it's not going to light, when suddenly the whole sky turns orange-white and I am plastered against my shoulder straps. There is no subtlety to this engine, no gentleness in its approach. I am supposed to monitor the status display panel, but I cannot prevent my eyes from wandering past it to the glorious Fourth of July spectacle radiating out from the engine. Out of long habit, however, I check my instruments, and all seems to be going well inside the cockpit. We are swaying mildly back and forth...and as the clock passes through fourteen seconds, I send a command for this raucous engine to cease. At that very instant it has come to the same conclusion, and we are now jerked back into weightlessness and treated to a thirty-second barrage of visual effects even more spectacular than the preliminaries. It is nearly sunset with the sun directly behind us; it clearly illuminates each particle, spark, and fireball coming out of the engine, and there are plenty of them. Some of them are small as fireflies, others large as basketballs; some depart lazily, others zing off at great speed. There is a golden halo encircling the entire Agena that fades very slowly.
—*Carrying the Fire*, by Michael Collins

One of Collins' assigned EVAs was to retrieve a micrometeorite package from an Agena that had been exposed to space for four months:

...Gently, gently I push away from Gemini, hopefully balancing the pressure of my right hand on the open hatch with that of my left hand on the spacecraft itself. As I float out of the cockpit, upward and slightly forward, I note with relief that I am not snagged on anything but am traveling in a straight line with no tendency to pitch or yaw as I go. It's not more than three or four seconds before I collide with my target, the docking adapter on the end of the Agena. A cone-shaped affair with a smooth edge, it is a lousy spot to land because there are no ready handholds, but this is the end where the micrometeorite package is located and, after all, that is what I have come so far to retrieve. I grab the slippery lip of the docking cone with both hands and start working my way around it counterclockwise. It takes about 90 degrees of hand-walking in stiff pressurized gloves to reach the package. As I move I dislodge part of the docking apparatus, an electric discharge ring which springs loose, dangling from one attaching point. It looks like a thin scythe with a wicked hook, two feet in diameter. I don't know what will happen if I become ensnarled in it.... Best I stay clear of it. By this time, I have reached the package, and now I must stop. Son-of-a-bitch, I am falling off! I have built up too much momentum, and now the inertia in my torso and legs keeps me moving; first my right hand, and then my left, feel the Agena slither away, despite my desperate clutch. As I slowly cartwheel away from the Agena, I see absolutely nothing but black sky for several seconds, and then the Gemini hoves into view. John [Young] has apparently watched all this in silence, but now he croaks, "Where are you, Mike?" "I'm up above. You don't want to sweat it. Only don't go any closer if you can help it. O.K.?" "Yes."
—*Carrying the Fire*, by Michael Collins

Collins did succeed in controlling himself eventually and, with the help of a handheld "maneuvering gun" whose propellant gas would act as a tiny rocket in space, was able to retrieve the micrometeorite package and return to the Gemini.

The final two Gemini flights had astronauts Conrad and Gordon launched in Gemini 11 on September 12, 1966, where they linked up with an orbiting Agena and, using its power, kicked themselves up to a record height of 853 miles. Later, after undocking at 180 miles altitude, the astronauts created some gravity by spinning the two vehicles around each other at the end of a 100-foot rope. Gemini 12, with astronauts Lovell and Aldrin, launched November 11, 1966, concentrated on an extensive EVA, which included simple calisthenics, a space walk, photography, more rendezvous and docking; and the Gemini program was concluded. Borman and Lovell in Gemini 7 had proven that men could remain weightless for fourteen days without suffering any serious consequences; and a lunar landing mission would require only about half as much time.

Gemini 6, 8, 9, 10, 11, and 12 had successfully attempted a variety of rendezvous and docking techniques, proving that a lunar strategy calling for a space rendezvous was practical and that if it could

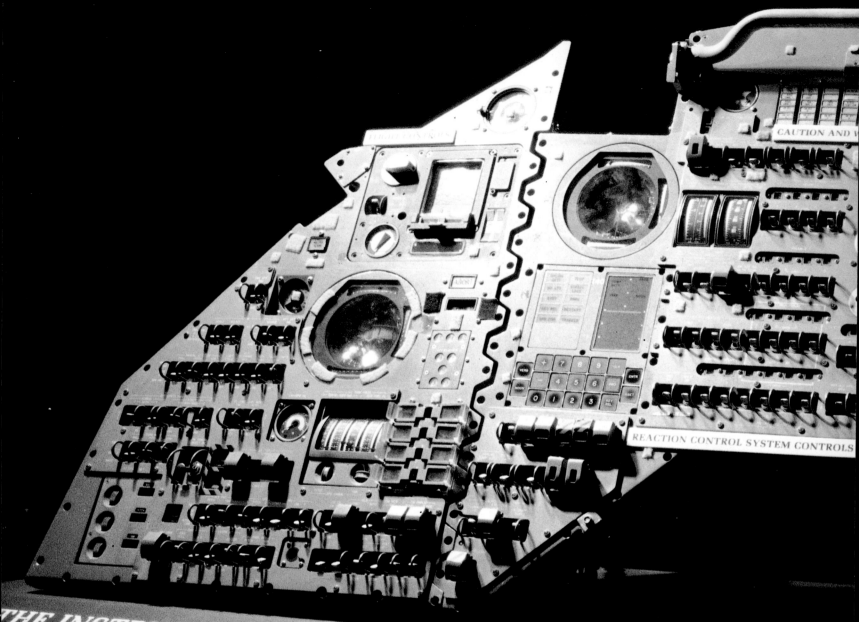

CAUTION AND W

REACTION CONTROL SYSTEM CONTROLS

THE INSTRUMENT PANEL

COMMAND MODULE M

display console faces the three crew
the Apollo Command Module. Ti
ches, dials, and meters used to control
nd monitor its performance.
rols for related sub-systems are
Flight controls are on the left
stems controls are on the

from a command module
for astronaut training.

SERVICE MODULE PROPULSION SYSTEM CONTROLS

ELECTRICAL POWER CONTROLS

COMMUNICATIONS CONTROLS

DISPLAY CONSOLE

be done in earth orbit, it could be done around the Moon.

And not the least important by any means was the training and experience gained not only by the astronauts but by the men on the ground; an intricately coordinated, competent testing, planning, and flight control team had been created and had shown its ability to minimize the hazards of sending men on missions a quarter of a million miles from earth and of returning them safely in spite of the problems that arise whenever men deal with complex, unproven machinery in the unknown. It was time to go to the Moon.

By this time tens of thousands of photographs had been taken by American Ranger, Lunar Orbiter, and Surveyor spacecraft and by Russian Luna and Zond vehicles. Both the Americans and Soviets had accomplished successful soft landings on the Moon's surface. Lunar Orbiter 1, on August 23, 1966, had returned hundreds of medium-resolution close-ups of potential Apollo landing sites, revealing that the *mare* areas, which appeared so smooth in telescopes, were actually pitted with craters. (Lunar Orbiter 2, launched in November, 1966, had confirmed the puzzling fact, first indicated by Soviet photographs, that there are no large *maria* on the hidden side of the moon.) And then, early in 1967, both the Soviet and the American manned space programs came to a tragic halt.

On January 27, 1967, Mike Collins as senior astronaut present had to attend the Friday staff meeting at Deke Slayton's office. There were very few people around; Slayton,

Apollo 10's Lunar Module Snoopy *returns from its trip to within 8.4 miles of the Moon's surface to rendezvous and dock with the Command Module* Charlie Brown.

too, was absent and his assistant Don Gregory was presiding. Collins relates what it was like:

> …We had just barely gotten started when the red crash phone on Deke's desk rang. Don snatched it up and listened impassively. The rest of us said nothing. Red phones were a part of my life, and when they rang, it was usually a communications test or a warning of an aircraft accident or a plane aloft in trouble. After what seemed a long time, Don finally hung up and said very quietly, "Fire in the spacecraft." That's all he had to say. There was no doubt which spacecraft (012) or who was in it (Grissom-White-Chaffee) or where (Pad 34, Cape Kennedy) or why (a final systems test) or what (death, the quicker the better). All I could think of was, My God, such an obvious thing and yet we hadn't considered it. We worried about engines that wouldn't start or stop; we worried about leaks; we even worried about how a flame front might propagate in weightlessness and how cabin pressure might be reduced to stop fire in space. But right here on the ground, when we should have been most alert, we put three guys inside an untried spacecraft, strapped them into couches, locked two cumbersome hatches behind them, and left them no way of escaping a fire…. As we sat there stunned, the red phone rang again and delivered additional details—rescue crew on the spot but unable to enter because of excessive heat… damage confined to command module alone—no word from the crew or sign of activity from within. Hell no, nor would there ever be—the only question was: How quickly, how quickly?
>
> —*Carrying the Fire*, by Michael Collins

The deaths of Virgil Grissom, Edward White, and Roger Chaffee brought the United States manned space program to a sudden halt. Not quite three months later, Vladimir Komarov, a veteran Soviet cosmonaut, was killed when his new Soyuz 1 spacecraft, which had been tumbling badly during a flight, was ordered to terminate the mission. During descent, the Soyuz became tangled up in its parachute lines and plunged to earth. The Soviets did not launch another manned spacecraft until October 26, 1968; the Americans, too, had had to extensively redesign their Apollo spacecraft. After two unmanned test flights, astronauts Schirra, Eisele, and Cunningham were launched on October 11, 1968 (twenty-one months after the fatal fire), in Apollo 7 to test the Apollo command module in the relative safety of earth orbit. Their Apollo flight lasted nearly eleven days. The next mission, Apollo 8, would send the Apollo command module and service module to the Moon atop the huge Saturn V rocket, which had not yet been used for a manned flight.

Apollo 8, carrying astronauts Borman, Lovell, and Anders, was launched from Cape Kennedy on December 21, 1968. The giant 364-foot-tall, 6-million-pound Saturn V dwarfed all previous launch vehicles. The main thrust was provided by five F-1 engines clustered together to create 7.6 million pounds of thrust. Visitors to this gallery can get some idea of how huge these engines are by seeing the exhibited F-1 mirrored to create the illusion of all five engines together. The second powered stage contained five J-2 engines, which created a million pounds of thrust; and the third stage contained one J-2 engine atop which was a truncated conical housing that protected and held the lunar module during the flight. Above the third stage was the service module, which held the fuel cells for electrical power in space, tanks, and supporting systems, and a single restartable engine capable of being repeatedly fired in a space environment. This engine propelled and maneuvered the spacecraft once the third stage had been jettisoned. Above the service module was the command module, the crew's quarters during the flight, and the escape tower which would provide their emergency exit in event of trouble on or near the launching pad.

Although Apollo 8's flight might pose fewer unknowns than had Columbus's voyage, as NASA's safety chief Jerry Lederer pointed out three days before the launch, the mission would "involve risks of great magnitude and probably risks that have not been foreseen. Apollo 8 has 5,600,000 parts and one and one half million systems, subsystems, and assemblies. Even if all functioned with 99.9% reliability, we could expect fifty-six hundred defects…." What made Apollo 8's flight different from the six Mercury, ten

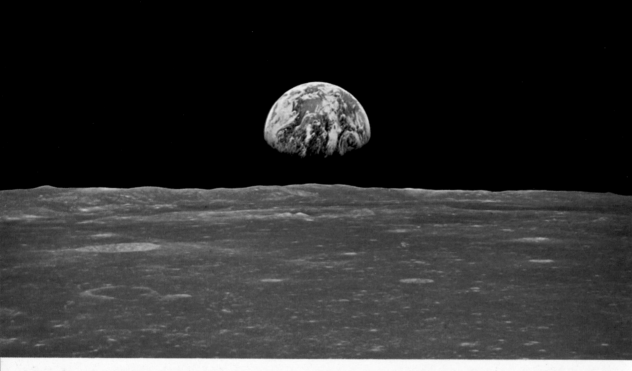

As Apollo 11 containing
the first men to walk
on the Moon swept over
the Moon's horizon,
the astronauts
photographed the
awesome majesty of an
earthrise from outer
space.

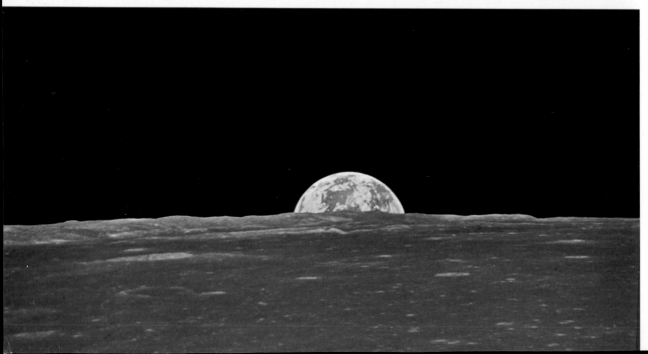

A large un-named crater, fifty miles in diameter, on the far side of the Moon photographed by the crew of Apollo 11.

Overleaf: Interior of the Apollo Lunar Lander
from which visitors can watch and hear
Apollo 17's landing.

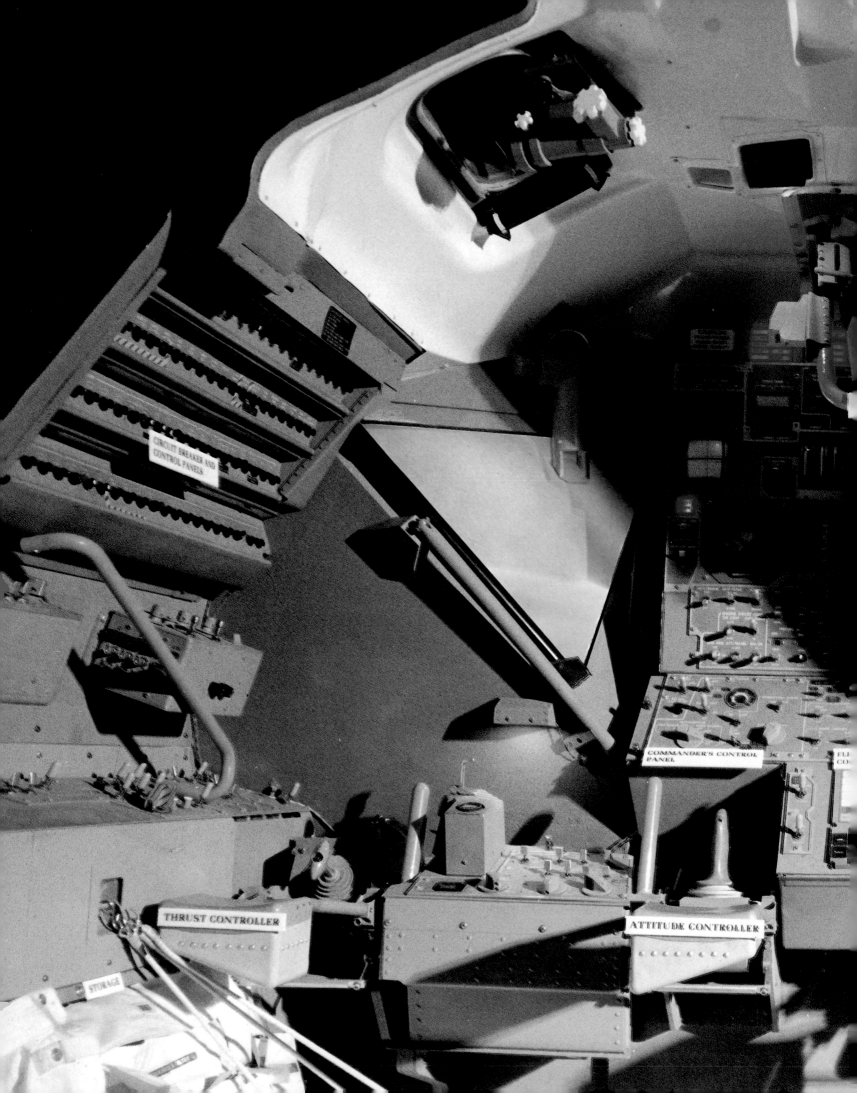

Lunar Module 2, on exhibit in the Museum's East Gallery, was built for an unmanned earth-orbital test flight. However, the mission of Lunar Module 1 was so successful that it was deemed unnecessary to fly the second craft. While being prepared for exhibition, the craft was outfitted to duplicate the Apollo 11 Lunar Module Eagle.

So that the Stars and Stripes might "wave" in the airless atmosphere of the Moon, the flag had to be stiffened with a brace.

and as they circled they took hundreds of photographs, made scientific observations, and shared with millions of people on earth, through live television broadcasts, the incredible scenes they saw. The astronauts found the Moon, "a very whitish gray, like dirty beach sand," and astronaut William Anders thought it "a very dark and unappetizing place." And then it was time for them to restart their SPS engine to come home.

Apollo 9, launched March 3, 1969, was the first manned flight of the lunar module. While Apollo 9 would be a less spectacular flight than the one that had preceded it in that Apollo 9 would never leave earth orbit, it was essential that the practicality of the rendezvous and docking of the Command Service Module (CSM) and the Lunar Module (LM) be tested. The "tissue paper spacecraft," as Jim McDivitt described the thin-skinned LM, was vital to the program and there was no need to fly it all the way to the Moon on its maiden flight. There was, however, the need to make sure everything worked. The Command Service Module had to be separated from the Saturn, turned 180 degrees, then driven back to where the Lunar Module was nestled within its protective sheath, and then docked. Once the CSM and the LM were docked nose to nose, the CSM would pull the LM free of the Saturn. On a lunar flight, this transposition and docking would be done only after the astronauts had been committed to leaving earth orbit and on course for the Moon. Fortunately, Apollo 9 went smoothly. The CSM pulled the LM free, and then they undocked from each other to perform a series of maneuvers which took them as far as 100 miles apart. They then duplicated as exactly as possible the sort of maneuvers and techniques required for a Lunar Module crew ascending from the Moon's surface to link back up with their Command Service Module.

Since Apollo 8's lunar flight had been successful and Apollo 9's had shown that the LM and CSM could transpose, dock, and rendezvous successfully, and since Apollo 10 was scheduled to be launched in just two more months, in May, why shouldn't Apollo 10's mission (which called for Tom Stafford and Gene Cernan to separate their LM from the CSM piloted by John Young and to descend to 50,000 feet above the lunar surface) attempt a lunar landing? There were many reasons, some having to do with the Moon being an entirely different environment in which to attempt a rendezvous, the lighting conditions, the orbital velocities, the ground tracking capabilities being different; an added complication was that the gravitational pull of the Moon was not evenly distributed and not enough information had been gained about where the concentrations of heavy-gravity spots were located to permit the tracking people with their computers to know exactly when and how they might affect an orbiting LM or CSM. The final argument was that Stafford's LM was some pounds overweight—which meant little in orbit, but which might be of immense significance when the time came to lift the LM off the surface of the Moon.

Apollo 10, launched on May 18, 1969, was the only full "dress rehearsal" for man's landing on the Moon. Astronauts Stafford and Cernan took their LM, *Snoopy,* down to 8.4 miles above the lunar surface to check out the Apollo 11 landing site in the Sea of Tranquility. After two passes and radioing back the description of the site's surface being "pretty smooth—like wet clay," they fired their ascent engine and rendezvoused with astronaut Young in the CSM, *Charlie Brown.* There was nothing left to do but to land men on the Moon.

On the morning of July 16, 1969, while astronauts Neil Armstrong, Michael Collins, and Buzz Aldrin were going through their final checklists prior to launch, Hermann Oberth was at the Cape to watch them go. He was seventy-five years old now, bent, white-haired, and the only one of the three great pioneers still alive. Tsiolkovsky had died in 1935, Goddard ten years later. Charles Lindbergh was there, too, along with T. Claude Ryan, who had built the *Spirit of St. Louis.* Everybody "who was anybody" or "had

Above: A medical kit with surgical scissors. Right: A tube of applesauce. Both of these items were carried by astronaut John Glenn on the Mercury mission of February 20, 1962.

Special tools designed for lunar exploration. Above, left to right: A contingency sampler with folding handle used by the astronauts to collect a soil sample immediately after they stepped from the Lunar Module to the surface. Scongs—a combination scoop-and-tong tool—for obtaining small samples of soil and rocks. A small scoop that allowed the astronauts to pick up small rocks or small quantities of lunar soil. Tongs provided to aid the astronauts in picking up rocks and soil since it was difficult for them to bend over in their stiff Apollo space suits. Right: Numbered bags in which the astronauts collected lunar rocks and soil while describing the samples to Mission Control Center.

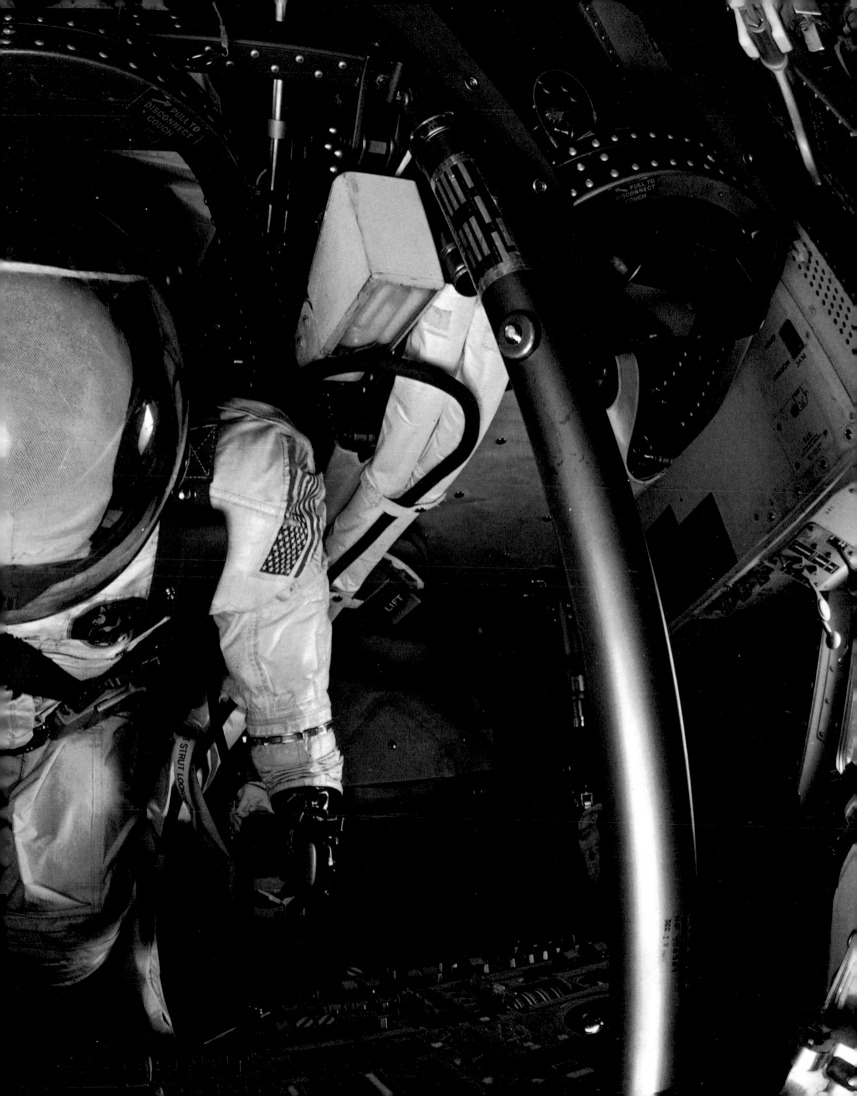

One of the vesicular basalt rocks formed by lunar vulcanism 3.7 billion years ago that was collected on the Moon and returned to this planet by astronauts during the six lunar landings of the Apollo program.

Preceding page: Interior of the Apollo 11 Command Module.

Chesley Bonestell. End of the World. *Oil on panel, 16¼ × 28¼". Formerly on loan from Alfred L. Weisbrich*

been anybody in the space program" was at the Cape to watch the launching of Apollo 11.

T minus sixty seconds and counting, Fifty-five seconds and counting. Neil Armstrong just reported back. It's been a real smooth countdown. We have passed the fifty-second mark. Our transfer is complete on internal power with the launch vehicle at this time. Forty seconds away...all the second stage tanks are now pressurized. Thirty-five seconds and counting. We are still go with Apollo 11. Thirty seconds and counting. Astronauts reported, feel good. T minus twenty-five seconds, guidance is internal. Twelve, eleven, ten, nine, ignition sequence starts. Six, five, four, three, two, one, zero, all engines running. LIFTOFF! We have a liftoff, thirty-two minutes past the hour. Liftoff on Apollo 11. Tower cleared.

> —*First on the Moon*, by Neil Armstrong, Michael Collins, Edwin E. Aldrin, Jr., with Gene Farmer and Dora Jane Hamblin

Michael Collins thought the first fifteen seconds "quite a rough ride":

...I suppose Saturns are like people in a way—no two of them are exactly the same....It was very busy. It was steering like crazy. It was like a woman driving her car down a very narrow alleyway. She can't decide whether she's too far to the left or too far to the right, but she knows she's one or the other. And she keeps jerking the wheel back and forth. Think about a nervous, very nervous lady. Not a drunk lady. The drunk lady would probably be more relaxed and do a much better job. So there we were just very busy, steering. It was all very jerky and I was glad when they called "Tower clear" because it was nice to know there was no structure around when this thing was going through its little hiccups and jerks.

> —*First on the Moon*, by Neil Armstrong, Michael Collins, Edwin E. Aldrin, Jr., with Gene Farmer and Dora Jane Hamblin

After a four-day voyage, the Lunar Module *Eagle* and the Command Service Module *Columbia* (named, in part, after the *Columbiad*, the cannon that fired a manned projectile to the Moon in Jules Verne's 1865 classic, *From the Earth to the Moon*) entered lunar orbit, averaging sixty miles above its surface so that they had a "noticeable sensation of speed." However, Collins noted:

> It's not quite as exhilarating a feeling as orbiting the earth, but it's close. In addition, it has an exotic, bizarre quality due entirely to the nature of the surface below. The earth from orbit is a delight—alive, inviting, enchanting—offering visual variety and an emotional feeling of belonging "down there." Not so with this withered, sun-seared peach pit out my window. There is no comfort to it; it is too stark and barren; its invitation is monotonous and meant for geologists only.
>
> —*Carrying the Fire*, by Michael Collins

The day before, as they had swung their spacecraft around to view the approaching Moon, Collins had sensed the Moon's hostile forbiddingness: "This cool, magnificent sphere hangs there ominously," he later wrote, "a formidable presence without sound or motion, issuing us no invitation to invade its domain." Neil Armstrong had commented then that it was "a view worth the price of the trip." Collins had added, "And somewhat scary too, although no one says that."
Now it is time to separate the *Eagle* from *Columbia*. And while *Columbia* with Collins

orbits overhead, the *Eagle* descends gradually to inspect its landing site more closely. On its second pass, a little more than 102 hours and 45 minutes after leaving Cape Kennedy, the following dialogue takes place. LMP = Lunar Module *(Eagle)* Pilot, Buzz Aldrin; CDR = Commander, Neil Armstrong; CMP = Command Module *(Columbia)* Pilot, Michael Collins; CC = Capsule Communicator, at Houston. 04 06 44 45 is the elapsed time of the flight: 4th day, 6th hour, 44th minute, 45th second.

04 06 44 45	LMP	100 feet, 3½ down, 9 forward. Five percent.
04 06 44 54	LMP	Okay. 75 feet, There's looking good. Down a half. 6 forward.
04 06 45 02	CC	60 seconds
04 06 45 04	LMP	Lights on…
04 06 45 08	LMP	Down 2½. Forward…Forward …Good.
04 06 45 17	LMP	40 feet, down 2½. Kicking up some dust.
04 06 45 21	LMP	30 feet, 2½ down. Faint shadow.
04 06 45 25	LMP	4 forward…4 forward. Drifting to the right a little. Okay. Down a half.
04 06 45 31	CC	30 seconds.
04 06 45 32	CDR	Forward drift?
04 06 45 33	LMP	Yes.
04 06 45 34	LMP	Okay.
04 06 45 40	LMP	CONTACT LIGHT
04 06 45 43	LMP	Okay. ENGINE STOP.
04 06 45 45	LMP	ACA—out of DETENT.
04 06 45 46	CDR	Out of DETENT.
04 06 45 47	LMP	MODE CONTROL—both AUTO. DESCENT ENGINE COMMAND OVERRIDE—OFF. ENGINE ARM—OFF.

The Apollo 11 Command Module Columbia, charred and flaked
from its fiery re-entry, being hoisted onto the deck of the
aircraft carrier Hornet 950 miles southwest of Hawaii. The
Columbia's flotation balloons are still attached.

Overleaf: Interior of the Apollo Command Module.

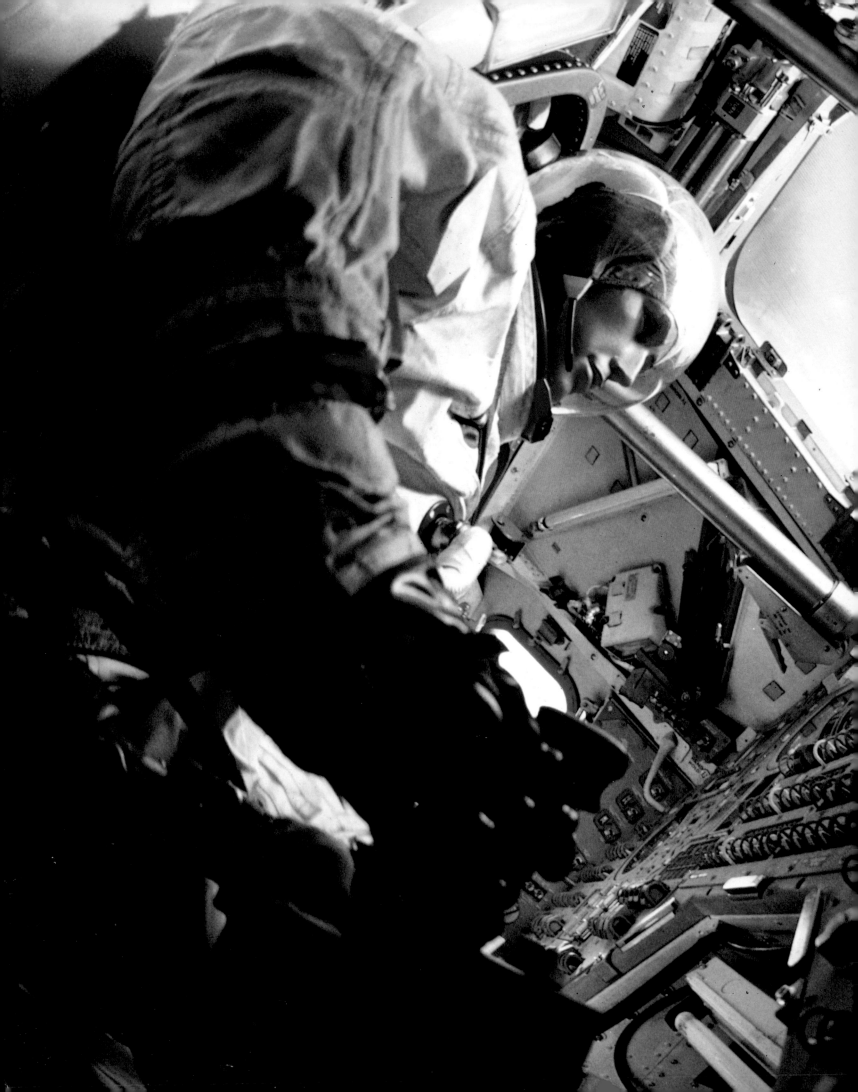

04 06 45 52	LMP	413 is in.
04 06 45 57	CC	We copy you down, Eagle.
04 06 45 59	CDR	Houston, Tranquility Base here.
04 06 46 04	CDR	THE EAGLE HAS LANDED.
04 06 46 04	CC	Roger, Tranquility. We copy you on the ground. You got a bunch of guys about to turn blue. We're breathing again. Thanks a lot.
04 06 46 16	CDR	Thank you.

Astronauts Armstrong and Aldrin were busy for a while, then Armstrong called Houston again:

04 06 55 16	CDR	Hey, Houston, that may have seemed like a very long final phase. The auto targeting was taking us right into a football-field-sized crater with a large number of big boulders and rocks for about...one or two crater diameters around it, and it required a...in P66 and flying manually over the rockfield to find a reasonably good area.
04 06 55 49	CC	Roger. We copy. It was beautiful from here, Tranquility. Over.
04 06 56 02	LMP	We'll get to the details of what's around here, but it looks like a collection of just about every variety of shape, angularity, granularity, about every variety of rock you could find. The colors—well, it varies pretty much depending on how you're looking relative to the zero-phase point. There doesn't appear to be too much of a general color at all. However, it looks as though some of the rocks and boulders, of which there are quite a few in the near area, it looks as though they're going to have some interesting colors to them. Over.

The relief and exhilaration expressed by

Mission Control over *Eagle's* safe landing was echoed all over the world, but, almost immediately, tension again built in anticipation of Armstrong and Aldrin's first steps on the Moon. After six and a half hours of preparation and as an estimated 600 million people throughout the world clustered around television sets, the hatch opened and slowly, very carefully Neil Armstrong began to climb out. The first thing anyone saw was Armstrong's leg:

> HOUSTON: Okay, Neil, we can see you coming down the ladder now.
> ARMSTRONG: Okay, I just checked—getting back up to that first step. Buzz, it's not even collapsed too far, but it's adequate to get back up...It takes a pretty good little jump...I'm at the foot of the ladder. The LM footpads are only depressed in the surface about one or two inches. Although the surface appears to be very, very fine-grained, as you get close to it. It's almost like a powder. Now and then, it's very fine...I'm going to step off the LM now... [*I had thought about what I was going to say, largely because so many people had asked me to think about it. I thought about that a little on the way to the Moon, and it wasn't really decided until after we got onto the lunar surface. I guess I hadn't actually decided what I wanted to say until just before we went out...*] THAT'S ONE SMALL STEP FOR A MAN, ONE GIANT LEAP FOR MANKIND.*

It was 9:56 PM in Houston, July 20, 1969. Man was standing on the Moon. President Kennedy's goal of landing a man on the Moon before the decade was out had been achieved. It had required eight years, an expenditure of $25 billion dollars, and the greatest technological mobilization the world has known. Visitors to the Apollo to the Moon gallery must remind themselves that when, on January 31, 1958, the United States placed its first satellite, the 31-pound *Explorer*, into orbit it was

---

*When Apollo 12 flight commander Charles "Pete" Conrad stepped on the moon on November 19, 1969, thereby becoming the third man to do so, he jubilantly paraphrased Armstrong, saying, "Whoopee, man, that may have been a small step for Neil, but that's a long one for me."

Astronauts stand before the seismic detector, its solar panels deployed, against a lunar landscape.

The first motor vehicle on the Moon was
Apollo 15's Lunar Roving Vehicle, which was
driven 17.3 miles at a top speed of 7-8
miles per hour. LRVs were also used on Apollo
missions 16 and 17 and made it possible for
the astronauts to carry heavy, bulky
equipment and scientific instruments to
locations distant from their Lunar Modules.

Enveloped in a cloud of blue vapor from his own Portable Life Support System, astronaut Alan L. Bean of Apollo 12 strides across the surface of the Moon.

The Apollo spacesuit worn by Apollo 17's Eugene Cernan had to withstand temperatures ranging from +250°F. in the sun to −250°F. in shadow.

399

Norman Rockwell. Apollo 11 Team. 1969. Oil on canvas, 28½ x 66″. Gift of the artist

considered an outstanding technological achievement. And yet within eleven years the Saturn V rocket, which was capable of orbiting 140 *tons*, would carry men to the Moon—and back. In 1958, to put one pound in orbit cost half a million dollars; by 1968 the Saturn was orbiting a pound for $500. Nevertheless, as Arthur C. Clarke pointed out, "It is impossible to tolerate indefinitely a situation in which a gigantic, complex vehicle like a Saturn V is used for a single mission, and destroys itself during the flight. The Cunard Line would not stay in business for long if the *Queen Elizabeth* carried three passengers—and sank after her maiden voyage."

The Apollo Program ended in December, 1972, after twelve astronauts* had walked on and explored the Moon and brought back 843 pounds of rocks. These rocks are made up of the same chemicals as earth rocks; but the proportions are different. Moon rocks contain more calcium, aluminum, and titanium than earth rocks and more rare elements like hafnium and zirconium. Other elements with low melting points, such as potassium and sodium, are more scarce in Moon rocks. The chemical composition of the Moon is different in different places: the light-colored highlands are rich in calcium and aluminum; the dark-colored *maria* contain less of those elements and more titanium, iron, and magnesium.

The inside of the Moon is not uniform; it is divided into layers. There is an outer crust probably composed of calcium and aluminum-rich rocks to a depth of about 37 miles. Beneath this crust is the mantle, a thick layer of denser rocks extending down to more than 500 miles. The deep interior is still unknown. The Moon does not have a magnetic field like the Earth's and yet magnetism has been discovered in many of the older Moon rocks that were brought back. Perhaps the Moon had an ancient magnetic field which disappeared after the old rocks were formed.

If, after viewing Alan Shepard's *Freedom 7* Mercury spacecraft, Frank Borman's and James Lovell, Jr.'s Gemini 7, the Lunar Roving Vehicle (three of these vehicles, used by the crews of Apollo 15, 16, and 17 are still on the Moon; the Museum's specimen was used in tests), the huge Saturn F-1 engine, the tools used by the astronauts, the Apollo Command Module from Skylab 4 (which ferried astronauts Gibson, Carr, and Pogue up to the Skylab for the last and longest, at 84 days, mission), and the various rocks and artifacts from the Moon, the visitor to the gallery is cynical enough to wonder whether all those billions of dollars spent were worth it, it would be nice for him to consider this editorial from the London *Economist* two days after Apollo 11 had safely returned to earth:

> And as the excitement dies and familiarity sets in, the voices that say the money could be better spent on ending wars and poverty on earth must gain converts.
>
> But this argument overlooks the factor in human make-up that sets us apart from the apes. When man first became a tool-maker, he ceased to be a monkey. The human race's way of sublimating its highest aspirations has been to build the greatest and grandest artifact that the technology of the time can achieve. Through the pyramids, the parthenons and the temples, built as they were on blood and bones, to the be-spired cathedrals conceived and constructed in ages of great poverty, the line runs unbroken to the launch pad of Apollo 11. Oddly—or perhaps not so oddly—the churchmen with their unstinting praise of the astronauts have recognized this where the liberally educated rationalists with their bored carping, and their ill-bred little jokes, have not. Spiralling to the planets expresses something in human nature that relieving poverty, however a noble cause that is, does not. And to the planets, sooner rather than later, man is now certain to go.

---

*The crews of the Moon landing missions were as follows: Apollo 11—Armstrong, Collins, Aldrin; Apollo 12—Conrad, Gordon, Bean; Apollo 14—Shepard, Roosa, Mitchell; Apollo 15—Scott, Worden, Irwin; Apollo 16—Young, Mattingly, Duke; Apollo 17—Cernan, Evans, Schmitt. (In each case, the second astronaut listed was the Command Module pilot, who did not walk on the Moon.)

Space Hall

Visitors enter the Skylab Orbital Workshop from the balcony overlooking Space Hall.

Skylab

The Space Hall is the third of the three huge exhibit halls facing the Mall (the other two being Milestones of Flight and the Hall of Air Transportation), and among its guided missiles, space launch vehicles, Skylab, the Apollo-Soyuz test project, Space Shuttle, and space missions of the future—exhibits which span the developments made in space flight from World War II through a hypothetical lunar base—are some of the largest artifacts in the Museum.

After World War II ended in 1945, Germany was between five and seven years ahead of all the other countries in rocket development—although it had not always been that way. When Dr. Wernher von Braun examined American rocket pioneer Robert H. Goddard's patents after the war, he had said, "Goddard was ahead of us all." Von Braun meant Goddard had been ahead until 1936, though even by then Colonel Walter Dornberger and Von Braun's team of engineers working for the German Army Weapons Department had already begun development of an engine for the A-4, a large rocket-propelled projectile, later designated the V-2. But while Robert Goddard and his crew of welders, instrument makers, and machinists (a team which never numbered more than seven persons, including Goddard and his wife Esther, who served as official photographer and volunteer fireman) conducted their liquid-propelled rocket experiments—a labor made possible by a $50,000 two-year Guggenheim grant—the Von Braun organization, numbering some 10,000, their research backed by millions of *Deutschmarks* funded by the German army and air force, worked at expressly constructed test facilities at Peenemünde, Germany. As Frederick C. Durant III, NASM's Assistant Director of Astronautics and one of the foremost authorities on Robert H. Goddard, explains, "That is why the Germans were literally years ahead of the Allies in 1945 in all areas of rocket and missile development. The V-2 represented

a quantum jump, a new plateau of technological capability in rocket and missile propulsion, guidance, aerodynamics, instrumentation, etc."

Components for nearly 100 captured V-2s were shipped to the White Sands Proving Grounds in New Mexico after the war and were used for training American personnel in the handling and launch operations of large rockets, as well as for lofting high-altitude, scientific data-gathering instrumentation. In 1946, sixteen V-2 flights were made, four of them to altitudes over 100 miles. It was to be months before the first American Aerobee and years before the first Viking sounding rockets were flown, and still longer before such altitudes were reached.

Whereas in the Satellite and Apollo to the Moon galleries the emphasis is primarily on payloads, here in the Space Hall's guided missile and space launch vehicle exhibit the focus is on the rockets that placed those payloads where they were supposed to go. The 46-foot-tall V-2, the world's first long-range ballistic missile, was seminal to all the large post-war launch vehicles designed to deliver intercontinental ballistic missile warheads and to place satellites in orbit, which led to men landing on the Moon and which send unmanned spacecraft to the outer planets. (Suspended above the V-2 are a number of rare German ground-to-air, air-to-air, and ground-to-ground missiles, among them the infamous V-1, known to wartime Londoners as the "Buzz Bomb.") The modern strategic missile systems that form the backbone of our national defense also made possible our entry into space. Nowhere is this lineage more evident than with the Jupiter-C rocket that stands in the pit near the V-2. The Jupiter-C, which launched America's first satellite, Explorer 1, on January 31, 1958, is nothing more than a modified Redstone Ballistic Missile, a 500-mile-range rocket developed for the United States Army in the early 1950s by Dr. Wernher von Braun and his team,

Skylab is exhibited minus one of its stubby solar wings—the actual vehicle lost one of its solar wings during a launch mishap.

LIGHTS
WARDROOM
1&3   2&4   WMC
OFF   OFF   OFF
630

many of whom had worked on the V-2.

The 68-foot-tall Jupiter-C shares the pit with a 70.8-foot-tall Vanguard (which launched our second satellite), a 73.8-foot NASA Scout (the only all-solid-propellant satellite launch vehicle), and a 60.4-foot United States Air Force Minuteman 3, a contemporary ICBM and one of our major strategic defense missiles. A Poseidon C-3, the two-stage United States Navy Fleet Ballistic Missile that is designed to be launched underwater from nuclear submarines, is displayed horizontally nearby.

It may help the Museum visitor to grasp the enormous size of an actual Saturn V launch vehicle by looking at the 48-foot-long, 80,000-pound Skylab Orbital Workshop and its 17-foot-long, 45,000-pound Multiple Docking Adapter and Airlock Module that dominate the Space Hall. The entire Skylab cluster consisted of four parts: the Orbital Workshop (OWS), the Airlock Module (AM), the Apollo Telescope Mount (ATM), and the Multiple Docking Adapter (MDA). In orbit the Skylab cluster was 118.1 feet long and weighed 199,750 pounds. The Saturn V rocket that launched these huge devices into orbit stood 364 feet tall—as high as a 36-story building, three-fifths the height of the Washington Monument, and more than half as long as the National Air and Space Museum itself.

When the scheduled number of lunar landings was reduced, the Saturn V and IB launch vehicles no longer needed for Apollo were utilized by Skylab.

Skylab was a space station* launched into

---

*It can be argued that Skylab was not, strictly speaking, a "space station," which is usually defined as a structure that can be indefinitely resupplied with consumable items. Skylab could be resupplied with some items, but there were no fittings to enable the station to be resupplied with nitrogen, oxygen, or water, and, after its effective life, it would necessarily have to be abandoned.

Earth orbit on May 14, 1973, and manned on three different occasions by three-man Skylab astronaut teams. The first Skylab crew, consisting of Charles Conrad, Jr., Joseph P. Kerwin, and Paul J. Weitz, were in space from May 25 through June 22, 1973, a period of 28 days. The second crew, Alan L. Bean, Owen K. Garriott, and Jack R. Lousma, was launched a month later, on July 28 and spent 59 days, until September 25th, aboard the space station. The third crew, Gerald P. Carr, Edward G. Gibson, and William R. Pogue, arrived on November 16, 1973:

> On that day, the space station—gold, white, and silver—was hard to make out against the swirling white clouds of the earth 269 miles below. Skylab resembled a huge, squat helicopter with a tower overhead surmounted by what looked like a big four-bladed rotor, but which was in fact an array of solar panels for generating electricity. It was somewhat battered now. After six months in space, the white paint had browned slightly, and some of the gold had baked and blackened. It was minus one pair of stubby, winglike solar panels that had broken off shortly after it was launched from Cape Kennedy [on May 14, 1973]; in the mishap, insulation protecting it from the sun had been shredded from its surface. The first crew [which had arrived twelve days later, on May 26, 1973] had erected a protected parasol over the space station; and the second crew [which had arrived on July 28] had spread a huge awning over that. Consequently, with its awning and its parasol, its pinwheel of solar panels overhead and its single remaining wing, Skylab looked less now like a space station than like Uncle Wiggily's airship.
> —*A House in Space*, by Henry S.F. Cooper, Jr.

In some ways Skylab was a far more ambitious and critically important program than Apollo to the Moon despite the "poor cousin" impression one might gather from its having been knocked together out of surplus Apollo parts. But if Skylab was not as dramatic a program as sending men to the Moon, it had a wholly different character and intent. Just as Gemini 7's fourteen-day flight had proven men could endure the prolonged weightless state required by a trip

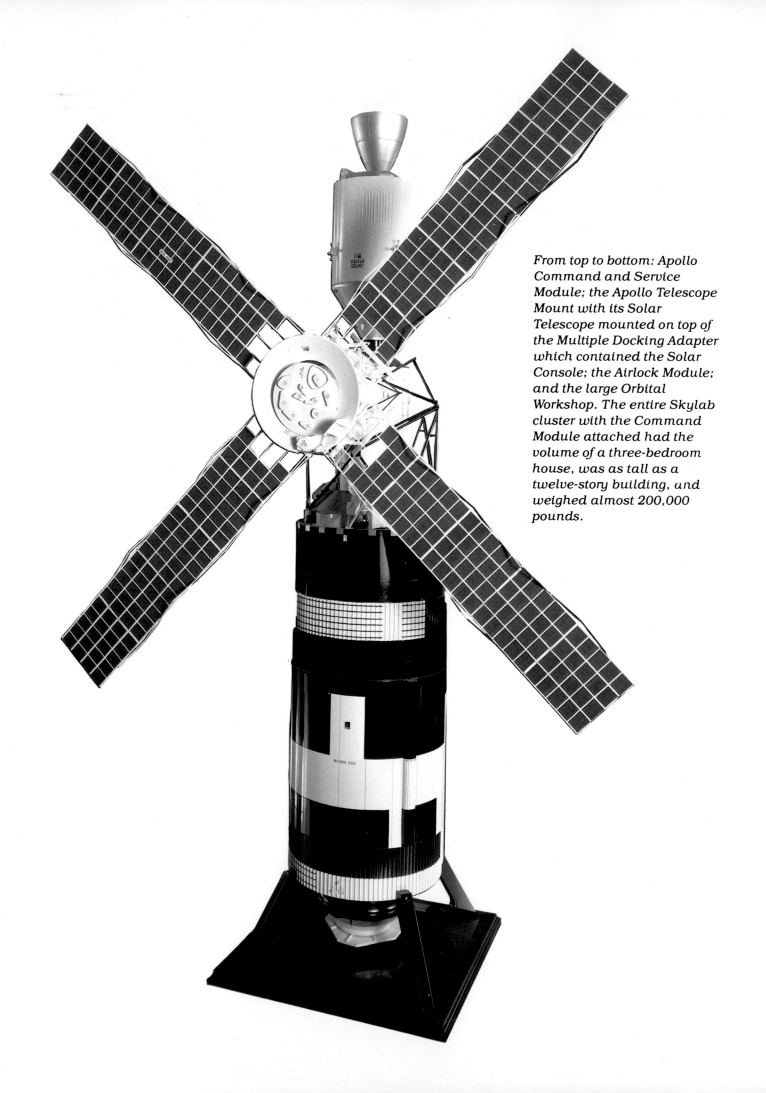

From top to bottom: Apollo Command and Service Module; the Apollo Telescope Mount with its Solar Telescope mounted on top of the Multiple Docking Adapter which contained the Solar Console; the Airlock Module; and the large Orbital Workshop. The entire Skylab cluster with the Command Module attached had the volume of a three-bedroom house, was as tall as a twelve-story building, and weighed almost 200,000 pounds.

WEIGHTLESSNESS
IN SKYLAB

Skylab's Multiple Docking Adapter and Airlock Module.

to the Moon and back, the eighty-four-day mission of Skylab's third crew proved that men might physically endure a flight to Mars. And, even though during the Skylab program the astronauts conducted various scientific experiments and performed a variety of tasks with an array of sophisticated equipment, the major scientific experiment—if not in fact the chief purpose of the Skylab program—was to see what happened to the astronauts themselves.

From the standpoint of time spent in space, as well as from the standpoint of the large size of the space station, both of which imposed special sets of circumstances, Skylab extended man's experience in a new way. The difference between it and earlier missions was between going on a quick trip through space in a vehicle the size of a car,* and moving there to stay a while in a house with all its rooms and corridors. The Skylab astronauts were the first to live in weightlessness. Like many long visits, the ones aboard Skylab were almost like a state of mind. It was not suspenseful, like the expeditions to the moon, but a steady, continuous experience, like life anywhere.

On a normal day during the third mission, the three astronauts were awakened by a buzzer at six o'clock in the morning, Central Standard Time, the time at the Johnson Space Center outside Houston, from which the flight was controlled. They felt refreshed; they logged much more sleep than the Apollo astronauts, who had been under more strain and had not had their own individual bedrooms.... Before they could get out of their beds, which were sort of sleeping bags hung vertically against the walls of their bedrooms, they had to unzip a light cloth that had kept them from floating off during the night. Then they soared upward out of bed as if by magic—as though they were genii escaping from bottles. They moved about without using their feet; and even if they made their feet go through the motions of walking, and then stopped these motions, they kept right on going until they hit something. If they thrust out an arm in one

direction, they moved back in the other. They could not have experienced these phenomena as readily in the smaller craft used by earlier astronauts; by its size, Skylab added a dimension to weightlessness. The astronauts, and everything they handled, moved as though they were underwater, in a sort of dreamlike, disembodied way—or as though they were in a magical place. They *were*, of course, if by magic is meant a suspension of natural laws familiar on earth. For two billion years, life on this planet has been conditioned by those laws, and all evolution has been determined by them. In gravity, where everything has weight, skeletons are needed for rigidity and leverage, muscles are needed for any sustained motion, and a circulatory system is required to pump blood against gravity. Arms, legs, fins, and cilia have all been developed for locomotion in gravity. Now, in weightlessness, all the effects of gravity vanished, and with them many of the reasons men are the way they are.

Before Skylab, nobody had known much about the long-term effects of weightlessness on man, and it was the purpose of the project to find out what they are—together with assessing what sort of useful work man might do in space, and how he might comfortably live in his new environment.
—*A House in Space*, by Henry S.F. Cooper, Jr.

Several disturbing symptoms had arisen in astronauts who had spent any extended period of time weightless in space. When Frank Borman and Jim Lovell returned from their two-week orbital flight aboard Gemini 7, flight surgeons detected a loss of muscle tissue (which led to diminished physical strength), a loss of calcium from their bones (diminished strength), and a sizable loss in the amount of bodily fluids, including blood which now contained fewer red cells. The astronauts' cardiovascular systems (heart and blood vessels) had weakened, and their body fluids had been redistributed to their upper from their lower parts. There was no way for the doctors to know whether these adverse changes in the astronauts' bodies were indications of unalterable courses— that the longer an astronaut was weightless, the worse his condition became—or whether these adverse changes progressed to a certain point and then ceased, or whether the body eventually compensated for the lack of gravity and

---

*Michael Collins said of the Gemini, "It's so *damned* small, smaller than the front seat of a Volkswagen, with a large console between the pilots, sort of like having a color TV set in a VW separating two adults.... The cockpit was tiny, the two windows were tiny, the pressure suits were big and bulky, and there were a million items of loose equipment which constantly had to be stowed and restowed." [from *Carrying the Fire*]

About Skylab's bathroom, which had a sheet metal floor instead of the more easily gripped triangular gridwork, one astronaut complained, "You just ricochet off the wall like a BB in a tin can."

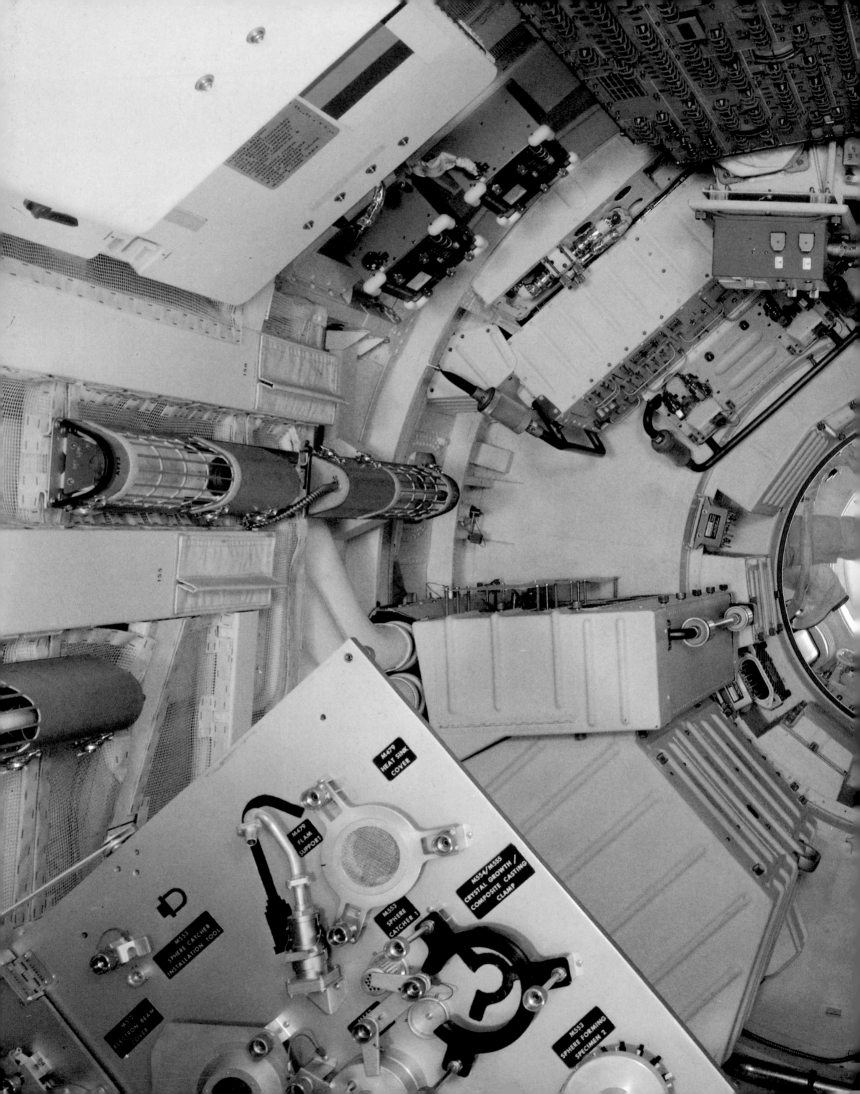

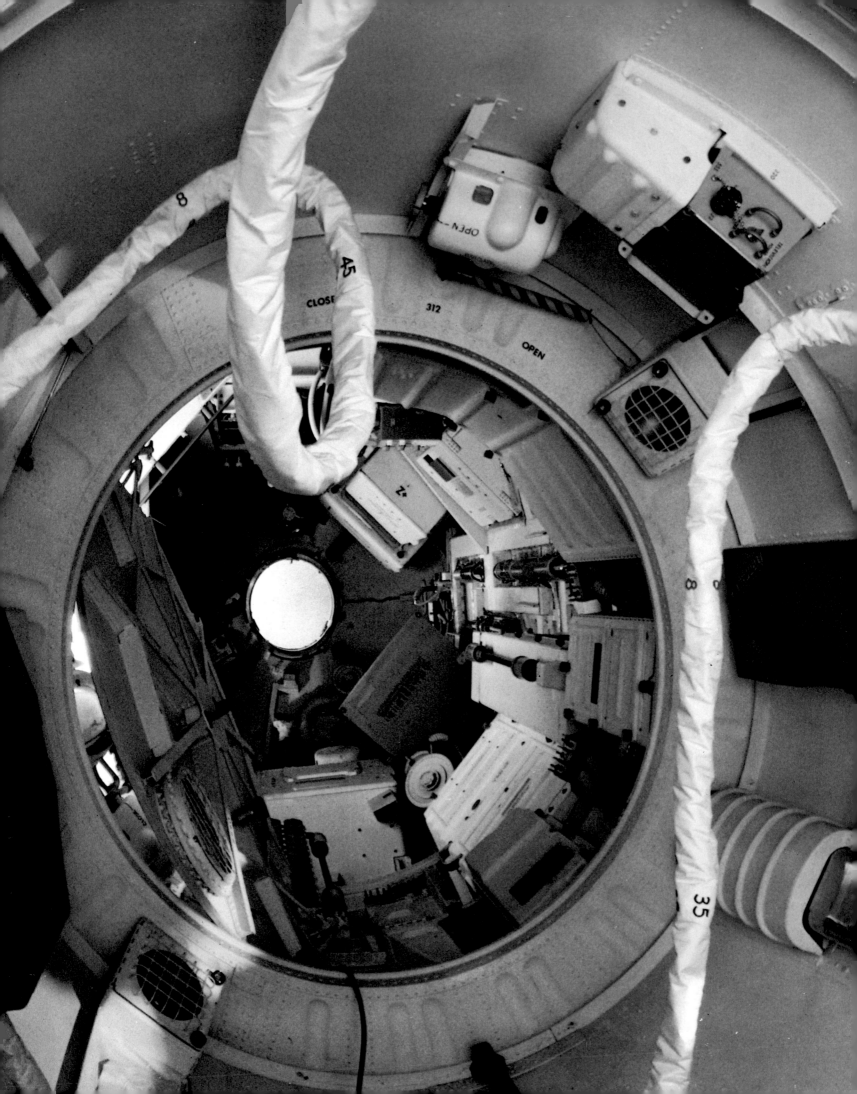

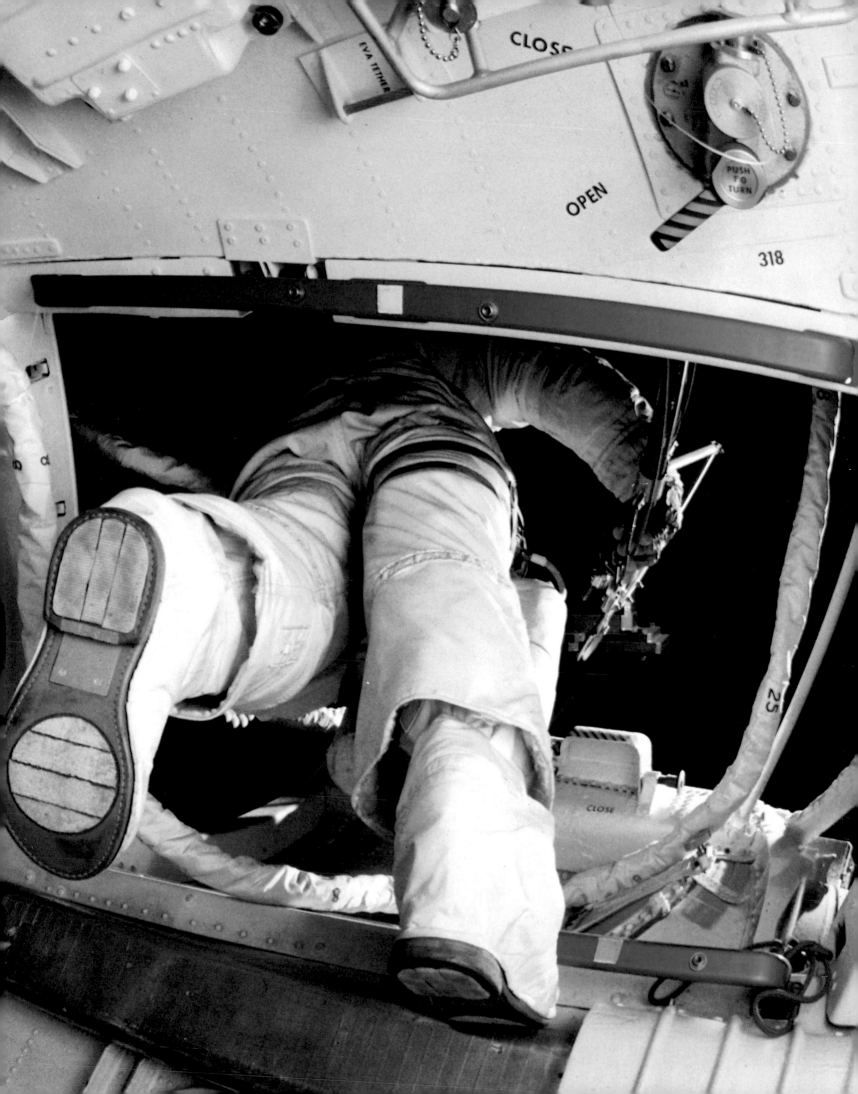

Preceding pages: In this sequence which takes the viewer through the Multiple Docking Adapter, which a Skylab astronaut described as "a very good example of how not to design and arrange a compartment," we continue into the Airlock Module, then seemingly follow an astronaut out of the Airlock Module and into space.

began to restore the muscle tissue, calcium, red corpuscles, and so on, it had lost. What was known so far was only that the more time an astronaut spent in space, the poorer the condition he seemed to be in upon his return. NASA flight surgeons were also aware that in 1970 two Soviet cosmonauts had been so weakened after spending eighteen days in orbit that they had to be carried from their spacecraft on stretchers upon landing. Even more ominous was the mystery surrounding the three other Soviet cosmonauts who had spent twenty-five days in space in 1971—longer than any other men. All three of the cosmonauts had died upon returning to Earth, as a result of a vent valve having popped open. Despite this, Skylab's first crew was scheduled to exceed by three days the dead cosmonauts' record duration flight.

The Skylab crews variously described their space station as "the cluster" or "the can"— what it was, actually, was a cluster of cans. The biggest "can," Skylab's Orbital Workshop (OWS), is on display in the Space Hall. It is a back-up vehicle for the one that was orbited. The OWS, with its huge single solar wing protruding from it like a giant sail, can be reached from the balcony overlooking the Space Hall. The visitor enters the lower level of the two-deck workshop, which contains the crew quarters, food preparation and dining area, washroom, and waste processing and disposal facilities. The upper portion contains a large work-activity area, water-storage tanks, food freezers, the film vault, and experimental equipment. The upper and lower decks were divided by an open-grid partition. The astronauts tended to prefer the lower deck since it contained the most domestic-seeming rooms. The wardroom, situated here, was the astronauts' favorite place, because it contained Skylab's only large window.

> ...When the astronauts were [at the big wardroom window] the earth appeared dynamic and alive. It was the view from [there] that convinced all nine Skylab astronauts that the

earth had to be observed directly, as any living object should be, with all the flexibility and intelligence that a man could provide....Part of the earth was always framed now in the round window, as though the astronauts were looking through the aperture of a microscope at a living tissue—all greens, blues, yellows, and browns. "I gained a whole new feeling for the world," Gibson [Edward G. Gibson, Third Skylab crew science pilot] told a visitor after he came back. "It's God's creation put before us, and whether you are looking at a bit of it through a microscope, or most of it from space, you still have to see it to appreciate it." Like the sun, the earth was an ever-changing kaleidoscope. [William R.] Pogue [Third Skylab crew pilot] said when he was back in Houston, "Every pass was different. It was never the same orbit to orbit. The clouds were always different, the light was different. The earth was dynamic; snow would fall, rain would fall—you could never depend on freezing any image in your mind."

The most direct view of all was from *outside* the space station, where an astronaut felt there was nothing between him and what he was looking at—as though he had slipped down the barrel of the microscope and was walking about the slide, magnifying glass in hand. "Boy, if this isn't the great outdoors!" Gibson said the first time he went out. "Inside, you're just looking out through a window. Here, you're right in it." And [Jack R.] Lousma [Second Skylab crew pilot] had said after his return, "...When you're inside looking out the window...it's like being inside a train; you can't get your head around that flat pane of glass. But if you stand outdoors, on the workshop, it's like being on the front end of a locomotive as it's going down the track! But there's no noise, no vibration; everything's silent and motionless; there are no vibrations going through your feet, no wires moving, nothing flapping." Skylab was moving down the track so fast that Lousma had actually *seen* the earth roll slowly beneath him; it was so big, though, that he could barely make out its curvature unless he was looking at the horizon.

...All the astronauts who went up [the scaffolding of the telescope tower to what the astronauts called the sun end] agreed that the sun end was the most exhilarating place aboard Skylab. "To be on the end of the telescope mount, hanging by your feet as you plunge into darkness, when you can't see your hands in front of your face—you see nothing but flashing thunderstorms and stars—that's one of the minutes I'd like to recapture and remember forever," Lousma had said afterward. It was a

The Skylab 4 Apollo Command Module in which Skylab's third crew, Gerald P. Carr, Edward G. Gibson, and William R. Pogue, returned after spending 84 days in space from November 16, 1973, to February 8, 1974.

During re-entry Skylab 4's exterior was subjected to temperatures of 2800°C. (5000°F). To protect it, the Command Module was covered with a 3,000-pound ablative heat shield composed of a phenolic epoxy resin in a Fiberglas honeycomb structure. The heat was carried away from the spacecraft as the heat shield charred and vaporized in friction with the earth's atmosphere.

little unnerving, too. When Pogue went up, he had the uneasy feeling that comes with being in the crow's nest of a ship. The telescope tower didn't sway like a ship's mast; it was just that an astronaut up there was far enough away from any large structure that he no longer felt part of the space station.

—*A House in Space*, by Henry S.F. Cooper, Jr.

At the forward end of the Skylab Orbital Workshop was a shorter, narrower "can" approximately the size of an Apollo spacecraft; this was the Airlock Module (AM). Because of the size of the Skylab "cluster of cans," the Airlock Module is displayed in the Space Hall horizontally next to the Orbital Workshop. In its orbit configuration, the AM rests atop the OWS, and further above the AM came the Multiple Docking Adapter, at the end of which the Apollo spacecraft ferrying up the crews would dock.

The Airlock Module (the back-up module for the one used) can be viewed by the visitor from all sides. The AM made it possible for Skylab astronauts to go outside the OWS without dispersing the Skylab's interior

atmosphere into space. The astronaut would put on his space suit, enter the AM, and close the hatch connecting the module to the OWS. He would then vent the Airlock Module's atmosphere and, when pressure in the airlock reached zero, he could open the hatch and float out into space, as the mannequin in this exhibit appears to be doing. The Airlock Module contained the control panels for the atmosphere and temperature for the entire Skylab; it also distributed electrical power throughout the Skylab cluster and supported communications and data handling. The importance of the AM was proven with the first Skylab crew, who had to carry out vital emergency repairs on the exterior of the Skylab after one of the huge solar wings and insulation had been lost during the launch. The Skylab astronauts regularly emerged from their spacecraft to replace or adjust equipment, change film, or carry out other extra vehicular activities.

The Multiple Docking Adapter (MDA) provided docking facilities for the Apollo Command and Service Modules (CSM) carrying the three-astronaut crews. In addition to the docking hatch at the end, a side docking hatch was provided in case of emergency so that two CSMs could dock simultaneously. The design of the Multiple Docking Adapter infuriated the third crew's Pogue; in fact, of the nine astronauts, only Gibson actually liked to be in the MDA. Because the MDA was a long tunnel with consoles, instruments, and boxes radiating from its cylindrical walls, there was no clear vertical and the astronauts had a hard time orienting themselves inside it. Pogue exploded, "Well, all I gotta say is, if you want a very good example of how not to design and arrange a compartment, the docking adapter is the best example. Boy, it's so lousy, I don't even want to talk about it until I get back down to the ground, because every time I think about how stupid the layout is in there, I get all upset!" Gibson's attraction for the MDA was that it was where the solar console was. After receiving his doctorate in physics from Cal Tech and writing a book about the sun, Gibson had become an astronaut because he knew that that was the only way he would see the sun from above theEarth's atmosphere. Sun watching was one of the astronauts' favorite pastimes. And the Apollo Telescope Mount's eight astronomical instruments, which were designed to observe the sun over a wide spectrum from visible light to X-rays, fed their images to the control and display console in the Multiple Docking Adapter.

All three Skylab missions were highly successful. The astronauts' tasks—to observe the Earth, using a variety of techniques designed to further knowledge of natural resources and the Earth's environment; to observe the Sun for increased understanding of solar processes and influences on Earth's environment; to conduct experiments in processing materials under the unique conditions of weightlessness and the vacuum of space— were carried out with such gratifying results that the data and photographs obtained will be under analysis for years.

Most important, however, was the successful experiment conducted by and on the astronauts themselves, for they proved that man:

> ...is a more adaptable creature and space a more suitable home for him than anyone had previously expected.

> The chief worry, though, was one trend that had not stabilized in space at all: the slow, steady loss of calcium from the astronauts' bones, which had gotten progressively worse the longer they had been weightless. This was the only area in which the two later crews were worse off than the first, and it was the trend that took the longest to right itself after their return.

> After the astronauts' bone calcium had returned to normal, there was no way a flight surgeon could tell by any clinical test that any of them had ever been in space. Yet as much as six

or eight weeks after their return, the astronauts' wives reported, they stumbled at night in the dark, evidently requiring a visual clue to the room's vertical even though their sense of balance had completely returned to normal. And for a long time afterward some of the astronauts kept trying to float things around them as they had done in the space station. One morning when he was shaving, Lousma tried to leave his can of shaving cream hovering in midair. It crashed to the floor.
—*A House in Space*, by Henry S. F. Cooper, Jr.

The third and last Skylab crew departed from it on February 8, 1974, after occupying it for 84 days. Astronauts Carr, Gibson, and Pogue; Bean, Garriott, and Lousma; and Conrad, Kerwin, and Weitz had lived inside the space station for a total of 171 days, orbited the Earth 2,476 times, and traveled some 70,500,000 miles. As their Apollo Command Module, Skylab 4, undocked and pulled away from Skylab abandoning it forever, Gibson radioed, "It's been a good home."

Five years after the last Skylab mission the 77.5-ton space station's orbit began to deteriorate faster than expected due to unexpectedly high sunspot activity; and on July 11, 1979, those few parts of the Skylab that did not burn up in the atmosphere or plunge into the Indian Ocean came crashing down in Western Australia near Perth. Sheep rancher John Seiler and his wife, Elizabeth, were jolted awake after midnight at the isolated sheep ranch 480 miles east of Perth by a great noise and rushed outside. Seiler reported, "It was an incredible sight— hundreds of shining lights dropping all around the homestead."

In 1962, when John Glenn, the first American to orbit Earth, passed over Perth, the city had turned on all its lights to greet him. Ironically seventeen years later, NASA's plunging Skylab unintentionally returned the compliment.

Apollo–Soyuz Test Project

On July 17, 1975, a few minutes after noon in Washington, D.C., and seven PM in Moscow, a 32-foot long silver Apollo Command and Service Module with an especially constructed ten-foot long Docking Module at its nose nudged gently closer to the smaller, pale-green Soviet Soyuz-19 spacecraft drifting in orbit 140 miles above the west coast of Portugal over the Atlantic. The Soyuz had been launched two days earlier from Tyuratam about 2,000 miles southeast of the Soviet capital at 3:20 PM Moscow time with cosmonauts Aleksey A. Leonov and Valeriy N. Kubasov inside. About seven-and-a-half hours later, at 3:50 PM Washington time (EDT), a Saturn 1B carrying the Apollo Command and Service Module plus Docking Module and with astronauts Thomas P. Stafford, Donald K. ("Deke") Slayton, and Vance D. Brand aboard had thundered aloft from Cape Kennedy, in Florida.

"Less than five meters distance," Apollo commander Stafford announced now in Russian. Although he was speaking to the Soviet cosmonauts inside their spacecraft, observers around the world could listen in and watch on their television screens as the Apollo made its slow and delicate approach. Against the black background of space, were it not for the Russian spacecraft's extended solar panels, * the Soyuz appeared like a flying eggcup. "...Three meters," Stafford said in

---

*The top of the Soyuz "eggcup" was the Orbital Module; 8.7 feet long and 9 feet at its widest, the Orbital Module was used by the crew for work and rest. The 7.2-foot long inverted cup-shaped Descent Module, which came next, contained the main controls and couches used by the crew during launch, descent, and landing. The final section, the 7.5-foot long Instrument Module, held the subsystems for power, communications, propulsion, and other functions. The 28-foot long solar panels extended from this module and were used to convert sunshine to electricity to recharge the Soyuz spacecraft's batteries. The Apollo's electrical systems were powered by fuel cells which generated electricity by chemical means.

Russian, "...one meter...." The two spacecrafts' reciprocal docking mechanisms touched. "Contact!"

"Capture!" a cosmonaut said in English.

The docking latches on the Apollo and Soyuz automatically began to hook and close; the flower petal-like guide plates in each spacecraft intertwined, clasped like fingers.

"We also have capture," confirmed Stafford, his Russian colored by an Oklahoma twang.

Cosmonaut Leonov added, "Soyuz and Apollo are shaking hands now." The two spacecraft were locked together as one while Western Europe slid beneath them.

"Close active hooks," Stafford ordered, "Docking is completed." Then, in English for the benefit of the Apollo ground controllers at the Johnson Space Center, Stafford repeated, "Docking is completed, Houston."

From within the Soyuz, Houston could hear a cosmonaut's accented English, "Well done, Tom. It was a good show. We're looking forward to shaking hands with you aboard Soyuz."

The successful linkup of the two space rivals' workhorse spacecraft, the Apollo and the Soyuz, high above Earth, was the direct result of an accord reached more than three years earlier when, on May 24, 1972, then-President Richard M. Nixon and Aleksey Kosygin, Chairman of the USSR Council of Ministers, signed an agreement in Moscow "concerning cooperation in the exploration and use of outer space for peaceful purposes." This agreement establishing the Apollo-Soyuz Test Project that was designed to develop and fly a standardized docking system "to enhance the safety of manned flight in space and to provide the opportunity for conducting joint missions in space," was the result of meetings and discussions already carried out by representatives of the two nations for several years. There had then followed three years of intense training, and the setting up of an

elaborate communications netw
it possible for Houston and Kali
controllers to be in constant tou
spacecraft and each other. Engi
country and the Soviet Union d(
built the necessary docking har
would enable the two incompat
spacecraft with different atmos
pressures to become one. The a
and the cosmonauts, engineers
technicians and interpreters fro
spent hundreds of hours in join
and in attempts to learn each ot
language. Two-fifths-size scale r
docking systems developed by b
were constructed on wheeled tal
pushed together to see if they jo
did. And the compatible dockin
now made possible a variety of p
involving manned spacecraft fro
Union and the United States su
internationally manned space s
addition to showing that an inte
rescue operation could be carrie
one nation have a stranded spac
orbit and no rescue vessel from i
country ready to be launched to
an imperiled crew.

About three hours after the or
and Apollo had docked, persons
Apollo Control in Houston and S
at Kaliningrad watched their tel
monitors. On their screens the p
interior of the Docking Module e
still-closed hatch connecting the
spacecraft. Two American astro
shoulders of their brown-gold sp
"fatigues" occasionally crowding
screen, were continuing down tl
checklist:

"Soyuz, our step 23 is completed,"
the astronauts into the intercom. H
informing the two-man crew in the i
Soyuz of his progress. "We are now v
step 24." He spoke in Russian.
"We're through with that, too," sa
astronaut. "Right here 25. Let me ge
on....Camera. Yep, we're right on s
Okay there."

nand Module in which
its lived. It was the only
turn from space; it
ie spacecraft's controls.
Exhaust cone of main

ng down the Apollo's tail pipe
uz' extended solar "wings."

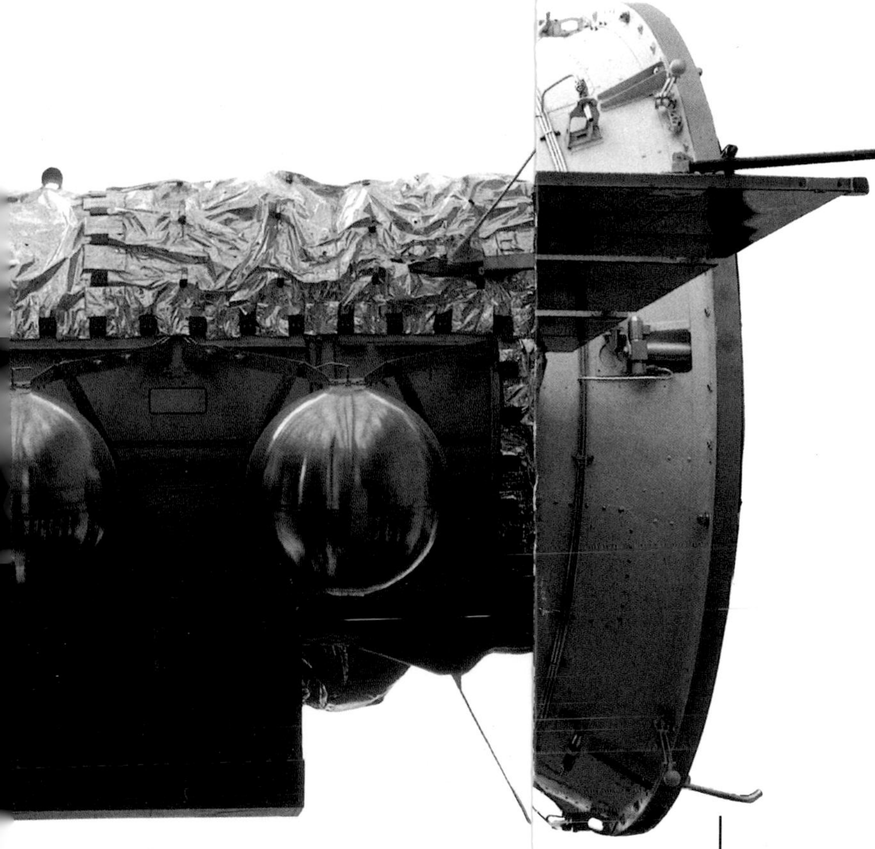

—— 66 feet ——→

e Docking Module carried into space
 the Apollo crew through which the
nerican and Soviet crews could pass
each others' spacecraft.

r power
The solar
to

UNITED STATES

1980 SPACE
TRANSPORTATION
SYSTEMS

1976 APOLLO-SOYUZ
TEST PROJECT
(ASTP)

so far to have flown men in space*—were working together rather than separately as they had through most of the period since the space age opened nearly 18 years earlier.

—Apollo Soyuz, by Walter Froehlich

For the next two days the crews of the Apollo and Soyuz shared meals, exchanged gifts and mementoes, conducted scientific experiments, but most of all provided the world with a vivid demonstration of detente on the way to achieving a series of "firsts": in addition to conducting the first space flight and docking of spacecraft and crews of different nations, Apollo-Soyuz detected the first pulsar outside our own galaxy, and the first stars emitting extreme-ultraviolet radiation (one of which is the hottest "white dwarf" known); the first separation of live biologic materials in space by electrophoresis was achieved; and for the first time communications between a manned orbiting spacecraft and ground controllers was carried out via an orbiting unmanned communications satellite.

Although among the millions of viewers in many different parts of the world who watched the live telecasts there were doubtless many who shrugged the Apollo-Soyuz project off as "nothing but a glittering soap bubble"—as did the West Berlin newspaper *Spandauer Volksblatt,* or regretted, like the Copenhagen *Jyllands-Posten,* the joint project's 450-million-dollar cost which prompted "reflection on how much the world's two biggest superpowers could achieve if they agreed to mobilize their strength and technology in joint service of more close-to-earth causes,"—no one would disagree with the London *Sun's* perception that "It was the world's most expensive

---

*Even as Soyuz and Apollo met, two Soviet cosmonauts were aboard a Soviet Salyut space station. Salyut 6, a space station launched in September, 1977, has now been visited by cosmonauts from Czechoslovakia (Vladimir Remek, March, 1978), Poland (Miroslaw Hermaszewski, July, 1978), and East Germany (Sigmund Jahn, August, 1978).

handshake, but it will not have cost a dollar or ruble too much if it is a handclasp for peace."

At the very least, the Apollo-Soyuz Test Project was a dramatic demonstration on the part of both the United States and the Soviet Union of a willingness to relax tension. Perhaps when spacecraft in the future establish the first extra-terrestrial colony, when international laboratories are orbited, when colonization of space occurs, Apollo-Soyuz will be remembered as a footnote: the first truly international manned space venture. As NASA's director for Apollo-Soyuz, Chester M. Lee said at the time, "Space is going to be explored. It's man's inherent nature to do that, and we might as well do it together."

The Apollo-Soyuz flight was the last Apollo launch and the last American manned space flight using a disposable spacecraft and launch vehicle.

Space Shuttle

At a little before 7 A.M. on April 12, 1981, at Cape Canaveral, Florida, space shuttle commander John W. Young and pilot Robert L. Crippen were waiting inside the cockpit of the *Columbia,* the first U. S. manned spacecraft ever to be launched without having undergone prior unmanned test flights in space. Already two years behind schedule and 3.6 billion dollars over budget, the *Columbia* was to carry the first U. S. astronauts into space since the Apollo-Soyuz link-up six years before.

The *Columbia,* the world's first reusable manned space vehicle, was the most complex flying machine ever built. Its three main space shuttle engines would have to provide 375,000 pounds of lift, deal with unprecedented pressures, and operate at greater temperature extremes than any mechanical system in common use today. In early tests of the space shuttle engines, seals had burst, propellant lines ruptured, valves had blown, faulty welds gave way,

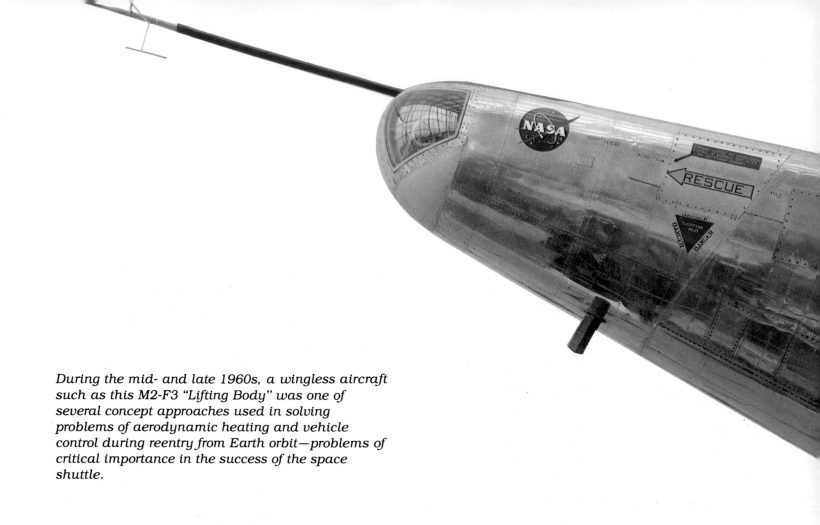

During the mid- and late 1960s, a wingless aircraft such as this M2-F3 "Lifting Body" was one of several concept approaches used in solving problems of aerodynamic heating and vehicle control during reentry from Earth orbit—problems of critical importance in the success of the space shuttle.

turbine blades cracked, ball bearings splintered, engines had exploded. The three main space shuttle engines develop just over 37-million horsepower and release an energy equivalent to that of 23 Hoover Dams. During launch and ascent these engines consume a half-million gallons of liquid-hydrogen fuel in about eight and a half minutes. It was not until 1980 that the space shuttle main engines, under development since 1972 (when an unenthusiastic President had asked a skeptical Congress to authorize inadequate funds for the space shuttle program), were pronounced ready for flight. By then it had already been discovered that thousands of the 30,761 tiles—each eccentrically shaped and unique, ranging in size from six-by-six inches to the palm of one's hand—might not withstand the rigors of a 17,100 mph reentry from space and would have to be reattached. In fact, some 7,500 tiles had been lost or damaged during the comparatively smooth flight of the *Columbia* when it was piggybacked from the West Coast to the Kennedy Space Center on top of

a special Boeing 747. The loss of any single tile during reentry could lead to a general irreversible burnout and the deaths of the crew. It took two years for the problems with the tiles to be resolved.

Despite these difficulties, however, there was reason on launch day to be optimistic. The *Columbia* was the most versatile and ambitious spacecraft ever developed and it stood against its launch gantry on Pad 39A glistening with promise. Before the space shuttle, everything was carried into space by expendable, one-time-only launch vehicles. The *Columbia*, which took off like a rocket, operated in orbit like a spacecraft, and landed like a glider, was designed to make space flight routine.

NASA officials envision the shuttle as a back-and-forth pick-up truck whose 65,000-pound cargo hold can put a space lab into position, manned and ready, or orbit a 22,500-pound space telescope as well as carry satellites and other payloads for science, the military, and industry. The shuttle is expected to be the first practical step toward the industrialization of space.

But, as April 12, 1981, dawned with astronauts Young and Crippen sitting strapped in *Columbia*'s cockpit watching the pelicans fly up and down the beach while the countdown for launch continued, all the hopes and expectations for America's shuttle program were compressed into the here and now: would the launch be scrubbed? Would the mission, once launched, have to abort? Would there be problems in orbit? Would the tiles withstand reentry?

In order for the mission to succeed, all three main elements of the space shuttle with its thousands of parts had to function together and in their proper sequence. There was the *Columbia* Orbiter, a delta-winged aircraft-spacecraft about the length of a twin jet commercial airliner, but far bulkier and built to be reused for at least 100 flights; the dirigible-like expendable External Tank; and, attached to the sides of the External Tank, two reusable Solid Rocket Boosters, each longer and fatter than a railway tank car. By T-minus-five minutes the astronauts were too busy with last-minute launch preparations to enjoy the Cape Canaveral dawn. Here, in the astronauts' own words, is how the first shuttle flight went. Pilot Robert L. Crippen speaks first; the story is then picked up by shuttle commander John W. Young.

CRIPPEN: It wasn't until we hit T-minus-27 seconds and nothing had gone wrong that I made up my mind we were really going to do it. That's when my heart rate went up to 130. I'm surprised it wasn't higher. John, I guess, was calmer. He had been into space four times and walked on the moon.* Maybe that's why his heart rate was only about 85.

YOUNG: I was excited too. I just couldn't get my heart to beat any faster. I was pretty impressed anyway, when at T-minus-five seconds I heard the three main engines start up with bangs. Then the two solid rocket boosters, which were strapped onto our big white external tank, exploded to life.

---

*Young had been aboard *Gemini 3*, *Gemini 10*, *Apollo 10* and, as part of *Apollo 16*, had been the ninth man to walk on the Moon.

Within three-tenths of a second we were off the ground. I saw pictures later of the conflagration all those engines and boosters made. I'm sure glad we didn't have rearview mirrors.

Looking out the side windows, I watched the vehicle go by the tower. There was a little vibration at first, and much less noise than I expected. Basically it was smooth, like riding a fast elevator.

. . .

We launched with *Columbia*'s tail facing south. Immediately after clearing the tower, we did this roll, pitch and yaw maneuver to get ourselves headed east-northeast toward Gibraltar.

We were getting more thrust from those solid rockets than we expected. When we jettisoned the solids, for instance, we were supposed to be at 164,000 feet, but were already at 174,000.

Those solids were putting out a tail of flame that was more than 600 feet long and 200 feet wide. Photographs from the chase planes show the flames were so hot that the back of the external tank was glowing bright white. We were about to lose the solids, however. We could feel a slight deceleration; their fuel was just about spent. Then two minutes and 11 seconds after lift-off when we were 29 nautical miles high, there was bright yellow-orange flame all across our windows. Six-tenths of a second later it was gone, and so were the solid rockets. Eight booster separator motors had flared up and fired the solids off into the Atlantic. That was some flash. We weren't expecting it to be so breath-taking, but for six-tenths of a second you don't have time to get nervous.

CRIPPEN: About two minutes later we were feeling good. All our testing data had indicated that if the engines were going to fail; they'd have done it already. Then we got the call from Mission Control in Houston we'd been waiting for: "Press to MECO." MECO, or main engine cutoff, occurs just before we reach orbital velocity. If we were cleared to go that far, we knew we weren't going to have to turn around and come back.

At this point, right on schedule, *Columbia* suddenly pitched over to level off our trajectory. As it did, I saw earth from space for the first time. "What a view! What a view!" was all I could get out. Seeing that curvature of the earth against the black of space, the multihued ocean, and the vivid blue shimmer at the top of the atmosphere grabbed my breath. "It hasn't changed any," said John.

YOUNG: Neither of us could sightsee, however. In less than four minutes we had to jettison the external tank. Pieces of white insulation from the

Seconds after 7 A.M. on April 12, 1981, Columbia, *America's first space shuttle was launched carrying commander John Young and pilot Robert Crippen.*

An artist's drawing showing the shuttle orbiter preparing to touch down on a runway following a mission into space.

tank drifted by our windows. They looked spectacular, like chunks of ice.

We were flying upside down underneath the tank, to make getting away from it easier. The main engines cut off, and the computers activated a 16-second separation sequence. Our umbilical lines were pulled out of the tank back into the orbiter. Then explosives blew the bolts fastening us to the tank, and *Columbia* was flying free. We couldn't see the tank, but knew it was up above us and would soon begin to drift down on a trajectory that would take whatever pieces survived the heat of entering the atmosphere into the Indian Ocean.

Up in the cockpit the only way we knew we had separated from the tank was that three red lights on the panel in front of Bob Crippen went out. There was no motion, no sense of the explosives firing. I took the stick and began manually flying off to the side to guarantee that we wouldn't run into the tank as it fell.

CRIPPEN: Actually, it felt like we were walking away from the external tank. The orbiter has 44 reaction control engines. These thrusters, which you can fire one at a time or in tandem, let us control the direction and attitude of *Columbia* most of the time we were in space. They were very physical and really shook the vehicle. They sounded like muffled howitzers right outside our door. Later,

when we fired them at night, we could see 30-foot-long tongues of fire leaping out from them. It was a little uncomfortable initially, and we never entirely got used to it. When one of the bigger ones fired, it was like something had hit the vehicle.

To get away from the tank, John flew the vehicle off to one side, and the computers fired one reaction jet on the nose and one aft. The one on the nose cut off intermittently to hold attitude. That kept *Columbia* positioned properly, but it also made us feel as if we were sidestepping across the sky. We didn't anticipate that. . . .

Once we hit orbit, we unbuckled our seatbelts and went to work. We loaded the computers with the on-orbit programs they needed to control and monitor just about every aspect of our flight and environment. As I got up to go into the aft flight deck behind the cockpit, I really noticed my weightlessness.

Right after the main engines had cut off, we had seen a little debris—washers and screws—floating through the cockpit. But not until I unstrapped my seatbelt did I realize how spectacular weightlessness is. I felt like a bird learning to fly from its nest.

I floated to the aft deck to open the large doors that cover the payload bay. The payload doors must stay open most of the time in orbit. They have reflectors that radiate into space the heat that builds up from all the electronic equipment.

John was back there, too, feet up in the air, and getting ready to take pictures with three remotely controlled T.V. cameras located in the cargo bay.

As soon as we got the door open, I noticed some dark patches on the pod that houses the starboard OMS engine. "Hey, John," I said, "we've got some tiles missing."

We didn't know it but those missing tiles caused quite a commotion back on earth. We weren't worried. The pods weren't supposed to get that hot during reentry, and NASA was being conservative by tiling that area. . . .

YOUNG: We spent most of that first day making sure *Columbia* was working properly. We went around checking our systems and doing routine first-flight things like surveying noise levels in the cockpit or checking the hand controller. One of us always stayed on the flight deck, since that is where the alarms and controls are. Because of zero G it was more fun to zoom down below into the mid-deck to do a checkout. Bob Crippen thought so too. He was learning how to swim in space pretty quickly.

CRIPPEN: After I assured myself I wasn't going to get spacesick, I spent a lot of time enjoying zero G. You don't have to do fancy aerobatics. Just moving around is enough. At first I did things that surprised me, like shoving off from one side of the mid-deck a bit too hard and finding myself sprawled on the opposite wall. But I mastered it pretty quickly. Soon I felt graceful and could fully control my body and motion.

YOUNG: We had a complex flight plan, detailing what we were to do almost minute by minute. After we finished the first day's chores, Crip fixed us dinner. Mission Control told us it was bedtime and signed off for the next eight hours. Neither one of us slept well that first night. For one thing it was light out much of the time and far too beautiful looking down at earth.

Sixteen years ago on *Gemini 3* we didn't have any windows to speak of. There was one porthole in front of me and one in front of Gus Grissom. The only way we could take pictures was to point a camera straight at something or open the hatch. (We didn't do that much.) On Apollo we were on our way to the moon. We didn't have much chance to look back and take pictures. We were moving too fast anyway.

CRIPPEN: The shuttle has those wraparound windows up front. But the best views are from the flight deck windows, looking out through the payload bay when you are flying upside down with the doors open, which we were doing most of the time. You see the whole earth going by beneath you.

I remember one time glancing out and there were the Himalayas, rugged, snow covered, and stark. They are usually obscured by clouds, but this day was clear and the atmosphere so thin around them that we could see incredible detail and vivid color contrast. The human eye gives you a 3-D effect no camera can. Sights like the Himalayas and thunderstorms, which we later saw billowing high above the Amazon, are especially dramatic.

YOUNG: I wasn't ready to go to bed that first night at quitting time even though we had been up for 18 hours. I slept only three or four hours. Crip did a bit better. When we did turn in, we just fastened our lap belts and folded our arms. We could have gone down to the mid-deck and just floated around, but I like some support. Anyway you do it, sleeping in zero G is delightful. It's like being on a water bed in three dimensions.

CRIPPEN: We were busy most of our second day, April 13, doing burns with the reaction control jets, going into different attitudes and performing maneuvers. We needed to understand how well the computer autopilot can control the vehicle. Could we make fine maneuvers? Houston wanted to see how well the crew could coordinate with the ground in positioning the orbiter.

YOUNG: I just kept feeling better and better about that vehicle. After we launched and got it into orbit, I had said to myself, "Well, that went pretty good." Then the vehicle worked so well the first day I had said, "We'd better take it back before it breaks." The second day it worked even better and so I thought, "Man, this thing is really good. We'd better stay up here some more to get more data." But Mission Control made us come back the next day.

CRIPPEN: We both slept soundly that second night. I was really sawing the Zs when an alarm started going off in my ears. I didn't know where I was, who I was, or what I was doing for the longest time. I could hear John saying, "Crip, what's that?" It was a minor problem, fortunately. A heater control in one of our auxiliary power units quit working. We just switched on an alternate heater and went back to sleep.

It was about 2:30 A.M. Houston time when flight control greeted us with a bugle call and some rousing music. John fixed breakfast that morning although usually I took care of the chow. Then we checked out the flight control system one last time and stowed everything away for reentry. We strapped on biomedical sensors to keep the doctors happy, and got back into our pressure suits. We programmed the computers for reentry and closed the payload bay doors.

The first step toward getting home was to deorbit. We had tested all our engines and were very confident they were working. We were really looking forward to flying reentry. Bringing a winged vehicle down from almost 25 times the speed of sound would be a thrill for any pilot.

We were orbiting tail first and upside down. We fired the OMS engines enough to feel a nice little push that slowed us down by a little less than 300 feet per second. That is not dramatic, but it did change our orbit back to an ellipse whose low point would be close to the surface of the earth.

When we finished the OMS burn, John pitched the vehicle over so it was in the forty degree nose-up angle that would let our insulated underbelly meet the reentry heat of the atmosphere.

YOUNG: We hit the atmosphere at the equivalent of about Mach 24.5 after passing Guam. About the same time we lost radio contact with Houston. There were no tracking stations in that part of the Pacific. Also, the heat of reentry would block radio communications for the next 16 minutes.

Just before losing contact, we noticed a slight crackling on the radio. Then, out of the sides of our eyes, we saw little blips of orange. We knew we had met the atmosphere. Those blips were the reaction control jets firing. In space we never noticed those rear jets because there were no molecules to reflect their light forward. Those blips told us that *Columbia* was coming through air— and hence, plenty of molecules to reflect the thrusters' fire.

That air was also creating friction and heating *Columbia*'s exterior. About five minutes after we lost contact with Houston, at the beginning of reentry heating, when we were still flying at Mach 24.5, we noticed the reddish-pink glow. Bob and I put our visors down. That sealed our pressure suits so that they would automatically inflate if somehow reentry heating burned through the cabin and let the air out. Other than the pink glow, however, we had no sense of going through a hot phase.

CRIPPEN: *Columbia* was flying smoother than any airliner. Not a ripple!

As we approached the coast of northern California we were doing Mach 7 and I could pick out Monterey Bay. We were about to enter the most uncertain part of our flight. Up to this point,

Columbia's course was controlled largely by firings of its reaction control thrusters. But as the atmosphere grew denser, the thrusters became less effective. *Columbia*'s aerodynamic controls, such as its elevons and rudder, began to take over.

We had more and more air building up on the vehicle, and we were going far faster than any winged vehicle had ever flown. Moreover, the thrusters were still firing. It was an approach with a lot of unknowns. Wind tunnels just cannot test such complex aerodynamics well. That was the main reason John took control of the flight from the automated system at a little under Mach 5. We had been doing rolls, using them a little like a skier uses turns to slow and control descent down a mountain. The flight plan called for John to fly the last two rolls manually. He would fly them more smoothly than the automatic system, helping to avoid excessive sideslipping and ensuring that we would not lose control as we came down the middle of our approach corridor.

YOUNG: It turned out to be totally unnecessary for me to manually fly those last two roll reversals. *Columbia* had been flying like a champ. It has all those sensors: platforms for attitude control, gyroscopes, and accelerometer. Its computers take all the data, assimilate it instantly, and use it to fire thrusters, drive elevons, or do anything needed to fly the vehicle. They are much faster at this than any man. The orbiter is a joy to fly. It does what you tell it to, even in very unstable regions. All I had to do was say, "I want to roll right," or "Put my nose here," and it did it. The vehicle went where I wanted it, and it stayed there until I moved the control stick to put it somewhere else.

CRIPPEN: Flying down the San Joaquin Valley exhilarated me. What a way to come to California! Visibility was perfect. Given some airspeed and altitude information, we could have landed visually.

John did his last roll reversal at Mach 2.6. The thrusters had stopped firing by then, and we shifted into an all-aerodynamic mode. We found out later that we had made a double sonic boom as we slowed below the speed of sound. We made a gliding circle over our landing site, runway 23 on Rogers Dry Lake at Edwards Air Force Base.

On final approach I was reading out the airspeeds to John so he wouldn't have to scan the instruments as closely. *Columbia* almost floated in. John only had to make minor adjustments in pitch. We were targeted to touch down at 185 knots, and the very moment I called out 185, I felt us touch down. I have never been in any flying vehicle that landed more smoothly. If you can imagine the smoothest landing you've ever had in an airliner, ours was at least that good. John really greased it in.

"Welcome home, *Columbia*," said Houston. "Beautiful, beautiful."

"Do you want us to take it up to the hangar?" John asked.

—"Our Phenomenal First Flight,"
by John Young and Robert Crippen
in *National Geographic*

In time, the *Columbia* was joined by three other shuttles: *Discovery, Challenger,* and *Atlantis.* Shuttle flights became so routine that television ceased its live coverage of launchings and landings; and newspaper accounts were relegated to inside pages. But then at 11:39 A.M. on January 28, 1986, after 24 successful space shuttle missions that had logged more than 50 million miles without injury, disaster struck. One minute, 15 seconds into the launch of the space shuttle *Challenger,* during those critical moments when the two solid-rocket boosters were firing along with the shuttle's main engines, the spacecraft exploded in a ball of flame about ten miles above the Earth killing all seven astronauts on board. They were Francis R. (Dick) Scobee, Commander; Michael J. Smith, Pilot; Dr. Judith A. Resnik, Mission Specialist; Ellison S. Onizuka, Dr. Robert E. McNair; Gregory B. Jarvis, Payload Specialist; and Christa McAuliffe, a teacher from Concord, N.H., who was to have been the first ordinary citizen to go into space.

As with all voyages of discovery and adventure there are risks. After the fatal flash fire that killed Apollo astronauts Virgil Grissom, Edward White, and Roger Chaffee, NASA investigated the accident, corrected the problem, and went on to the triumphant landing on the Moon. NASA's response to the *Challenger* has not been any different. The remaining shuttle fleet has been equipped with improved solid-rocket boosters, and the astronauts who will fly them are determined to keep the dream alive.

Two TDRSS (Tracking and Data Relay Satellite System) satellites like the one shown here, plus an in-orbit spare, will replace the large network of ground stations traditionally operated around the world to support manned spaceflight and earth-orbiting satellites. From geosynchronous orbits, the satellites will provide almost continuous tracking services.

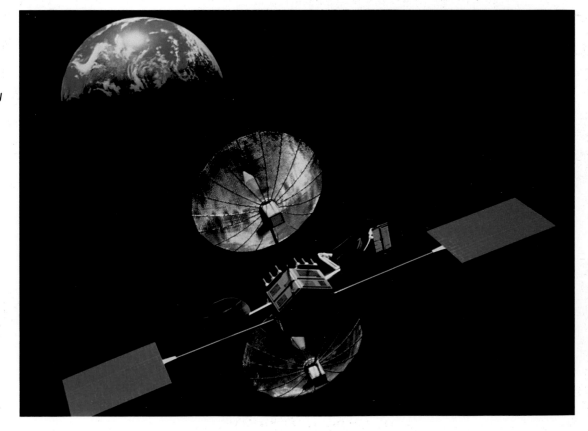

Exploring
the
Planets

THE SOLAR SYSTEM

DR. WATSON
INSTRUCTS
SHERLOCK HOLMES

After ancient man learned to search both philosophically and actually beyond the Sun and Moon which dominated his heavens, he discovered that five particular objects in the nighttime sky moved against the seemingly fixed patterns of the stars and were, therefore, in some fundamental way different from them. Once identified, they named these objects after gods: Mercury, Venus, Mars, Jupiter, and Saturn. Because of their comparative nearness, these planets were thought to move around the Earth like the Sun and the Moon. Ptolemy (c.A.D. 90-168), who perfected a system to describe their movements, placed the Earth at the center of the universe with the bodies in the following order moving in perfectly circular orbits about it: the Moon, Mercury, Venus, the Sun, Mars, Jupiter, and Saturn. And to account for observable irregularities in their orbits, each body also performed an epicycle–a small circular orbit within the circumference of the large orbit. Although several early Greek philosophers, Aristarchus (310-230 B.C.) in particular, had suggested that the Earth moved about the Sun, and not the other way around, this heliocentric theory was rejected until the Polish astronomer Nicolaus Copernicus (1473-1543) published as he lay dying his *De Revolutionibus Orbium Coelestium* containing his argument that the Sun was the center of the universe and all the planets revolved around it. So controversial and dangerous was this concept that it prompted the German theologian and leader of the Protestant Reformation Martin Luther (1483-1546) to protest, "The fool will turn the whole science of astronomy upside down!" Pope Paul III swiftly proclaimed Copernicus' theory to be heresy; but the courageous and inventive Italian scientist Galileo Galilei (1564-1642), his new telescope's discovery of Jupiter's four largest moons providing what he considered to be proof, published his support of the Copernican view of the universe in *Sidereus Nuncius*.

The telescope helped transform astronomy from a science dominated by theologic and philosophic reservations and reasoning to a pure science within the expanding realms of mathematicians and geographers. For two thousand years the only known bodies in the Solar System had been the Earth, its Moon, the Sun and the five nearest planets; now, in the seventeenth century, through Galileo's invention nine new bodies were added: the four moons of Jupiter that Galileo had discovered, and five moons of Saturn found by others.

During the next century a sixth planet, Uranus, was discovered along with two of its moons, and Saturn's sixth and seventh satellites.

On January 1, 1801, the planetoid Ceres was discovered where a planet "should" have been: in the vast empty space between the orbits of Jupiter and Mars. Within the next six years three more planetoids were discovered in what came to be known as the asteroid belt. The nineteenth century abounded in discovery; in 1846 Neptune was found and immediately thereafter one of its two moons. In 1848 Saturn's eighth satellite was discovered, in 1877 two tiny moons orbiting Mars, in 1892 a fifth satellite of Jupiter, in 1898 Saturn's ninth.

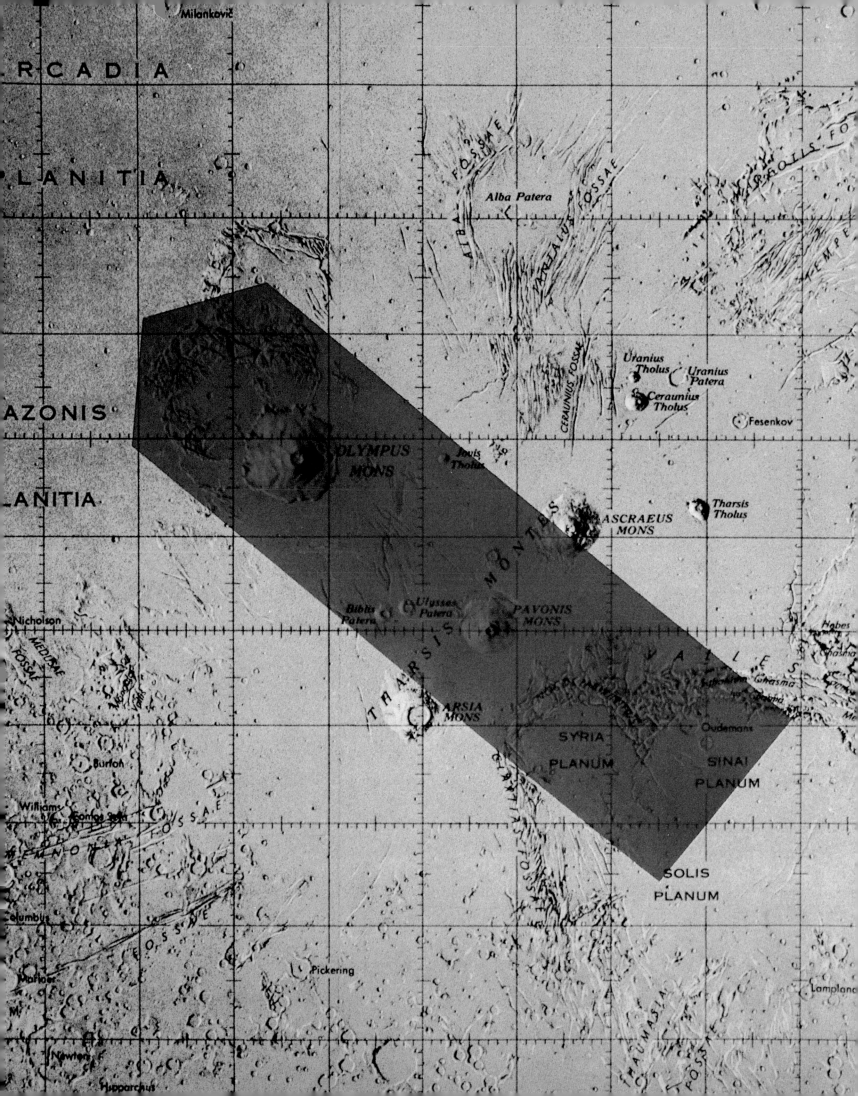

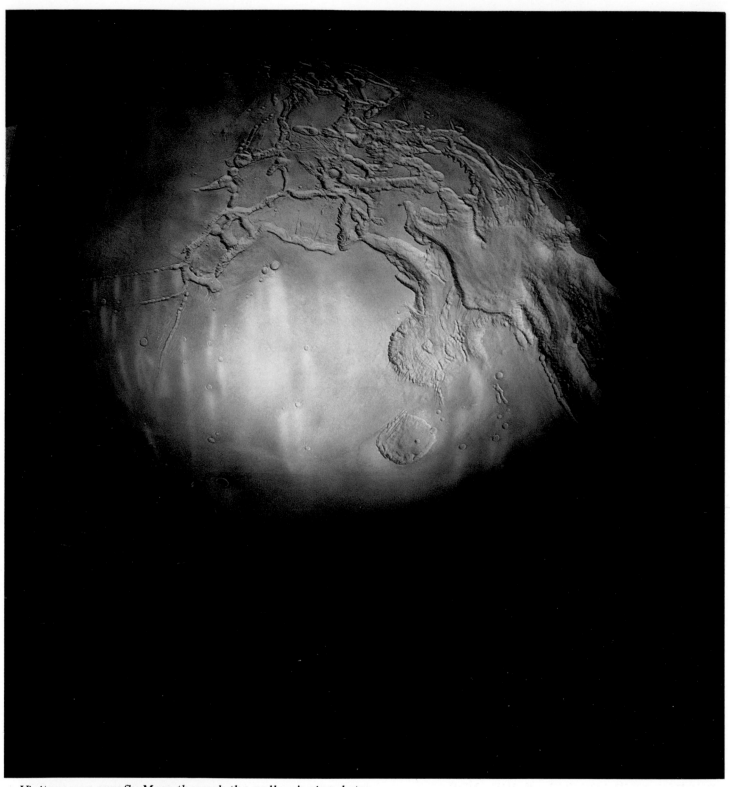

Visitors can overfly Mars through the gallery's simulator.

So far in the twentieth century seven more satellites of Jupiter have been discovered; in 1948 a Dutch-American astronomer Gerard P. Kuiper discovered a fifth and innermost satellite of Uranus and in 1950 a second moon of Neptune. In 1930, with the help of Percival Lowell's computations of where just such an object might be found, Clyde W. Tombaugh discovered the ninth planet in our solar system and called it Pluto since, like the Underworld god after whom it was named, it was the farthest removed from the light.

The first exhibits in Exploring the Planets, NASM's dramatic gallery, introduce this earthbound view of our solar system and show how Renaissance and Modern Man's expanding knowledge based on observable fact forced fantastic and superstitious explanations of natural phenomena to give way. Astronomers in the centuries since the telescope's invention have been able to detect objects so distant that their light began its journey to the telescopes' lenses before the Earth was even formed. Still, the great mysteries remain unsolved: How was the universe formed? Is it expanding from its creation? How long will it last? When did it begin? Are there millions of planets circling millions of stars? This last question suggests the greatest riddle of them all: Are we alone? No telescope can detect anything so small as a planet orbiting even the nearest star.

Now, as we near the last decade of the twentieth century, we look back on almost thirty years of space exploration that have increased our knowledge of the Sun, the Moon, the planets and their satellites. It was an era that began, really, with Russia's 1959 Luna 1 mission past the Moon and continues with Voyager 1's stunning photographs of Saturn. The knowledge we have gained and the wonderful devices that have helped us achieve it, are at the core of the Exploring the Planets gallery.

Visitors are shown the possible types of missions or trajectories used: flyby, impact, orbit, landing, and landing with return.

A series of photographs taken by Ranger 9 demonstrates the flight attitude, descent, and crash of that craft on the Moon's surface in March, 1965. A model Surveyor sits surrounded by a lunar landscape photo mosaic created from the pictures sent back by the 1968 Surveyor 7. Here, too, is Surveyor 3's camera which the crew of Apollo 12 brought back with them from the surface of the Moon.

Visitors are shown how Mariner 9 (1971) mapped the surface of Mars and discovered that the erosional and volcanic landforms on that planet far outscale anything to be found on Earth: the great Martian volcano, *Olympus Mons*, for example is over 81,000 feet high; Mt. Everest, by comparison, is but a puny 29,028 feet. That planet's enormous equatorial canyon, *Valles Marineris*, stretches nearly a third of the way around Mars. And photographs from the Viking 1 and Viking 2 Mars landers (1976) revealed a ruddy, rocky surface and a pinkish sky.

Mariner 10's triple rendezvous with Mercury (twice in 1974, once in 1975) showed that planet to be as barren and cratered as the Moon; Russia's Venera 9 (1975) soft-landed on the planet Venus and sent back photographs of a hostile, rockstrewn landscape beneath a thick blanket of clouds. Pioneer 10 (1973) and Pioneer 11 (1974) sent back the first good photographs of Jupiter. But now NASM visitors can see the spectacular photographs taken recently by Voyager 1.

A model of Voyager 1 dominates the gallery with its huge dish antenna and experiments boom. This marvelous spacecraft passed by Jupiter in March, 1979, and stunned the world with its wondrous photographs of the giant planet's red spot, its four largest moons, and even caught the first view of Jupiter's ring and the bursting of a volcanic explosion on airless Io as it continued on.

In 1986, five years after Voyager 1 discovered that Saturn was girdled by *thousands* of rings, Voyager 2 transmitted back data revolutionizing our knowledge of Uranus. Between them, these spacecraft

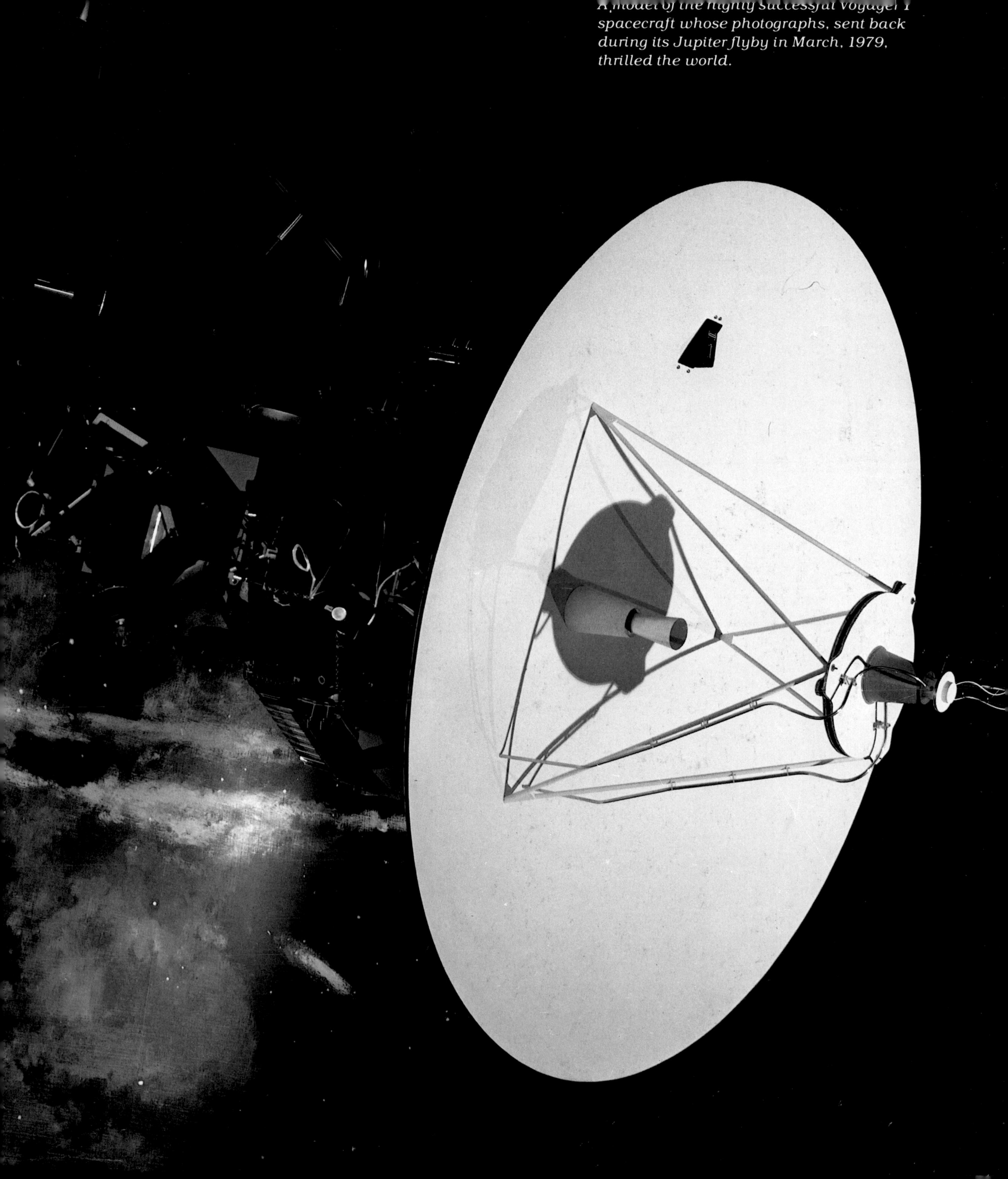

A model of the highly successful Voyager 1 spacecraft whose photographs, sent back during its Jupiter flyby in March, 1979, thrilled the world.

Children can monitor their spacecraft's descent onto the red-hot surface of Venus.

have returned to Earth information equivalent of about 100,000 encyclopedia volumes.

What then? Perhaps exploration of Titan, Saturn's largest moon, and the only moon in the solar system with an atmosphere of its own. While planetary exploration has so far answered a great many questions about our solar system, it has succeeded in raising still more. The final exhibit unit in this gallery is "Unanswered Questions," which leads us back to the biggest riddle of them all and reminds one of the probably apocryphal story about the renowned astronomer who received the telegram "IS THERE LIFE ON MARS. CABLE THOUSAND WORDS. HEARST." and sent back the following message, "NOBODY KNOWS" repeated five hundred times. He might have added the basic and crucial question, "What do you mean by Life?"

Stars

What is a star? Is our sun a star like other
stars, or is it different and somehow
special? How far away are the stars? How
large are they? Are they all the same size?
What are they made of? Why do they shine?
When we look up into the sky, how many
stars can we see? How many can we not
see? Have the stars always been there, or do
new stars appear and old ones die out?

Such questions are sometimes asked by
the youngest child. They are asked often by
visitors to the National Air and Space
Museum's Stars gallery. And they are asked
always by the most learned astronomers.
But even hundreds, possibly even
thousands of years ago, before there was a
science called "astronomy," these questions
were asked. Back then, however, we believed
we knew the answers—answers that sprang
from our deepest and earliest thoughts
about nature. For the stars and, above all,
the Sun have been a part of our culture for
as long as mankind has possessed collective
thought.

We have always used the Sun and stars to
find out what time of day it is, what time of
year it is, to know when is the proper time
to harvest and to sow. Neolithic people
would determine the time of day and time of
year by the length and direction of the Sun's
shadow made by a tree, a rock, or a stick.
Later, cultures throughout the world marked
time by the Sun. For instance, the Netchilli
Eskimos of the Boothia Peninsula in the

Northwest Territories, Canada, divided the
year into twelve parts to which they gave
such names as "The sun disappears," "It is
cold, the Eskimo is fishing," "The Sun is
returning," "The Sun is ascending," "The
seals are shedding their coats," and so on.

The Sun has also been used with either
natural or artificial horizon markers on
calendars. Certainly one of the most famous
is Stonehenge: the complex of stone rings
built starting at about 2800 B.C. in
southern England. Modern archeologists
and astronomers have discovered that these
stones could have been used to illustrate the
passage of the seasons. A 1/50-scale model of
Stonehenge is located near the entrance to
this gallery.

Eventually, calendars were created that
described astronomical cycles. Such
calendars became tools for survival. The
motion of Earth around the Sun changes
the Sun's apparent position with respect to
the stars. At different times of the year,
different stars appear opposite it at sunrise
or sunset. The Egyptians, for example, used
the first appearance of the star Sirius before
sunrise to mark the River Nile's seasonal
flood.

Thousands of years ago, the motions of
celestial bodies were thought to be the
determining agents of the course of one's life
as interpreted by astrological rules.
According to ancient astrologers, important
astronomical occurrences thought to affect
one's life were such events as eclipses,
"conjunctions" (planets moving close to each
other or to the Sun and Moon), and the
location and motion of the visible planets

On a clear, dark night in the country we can see a few thousand stars with our unaided eyes. Our galaxy contains about 100 billion stars. It has been estimated that the Universe contains as many galaxies as the Milky Way does stars. Beyond our galaxy lies the distant Trifid Nebula whose reddish glow is shown here clouded by streaks of interstellar dust.

(Mercury, Venus, Mars, Jupiter, Saturn) among the 12 constellations of the Zodiac. Today, astrology is still said to describe the celestial influences on our lives and actions; and horoscopes are among the most popular features of our daily newspapers, although considered by scientists to be a waste of time and money.

In markets, in the street, in our homes, in our everyday lives, we are inundated with images of the Sun and stars. Artists, industries, nations use the Sun and stars as metaphors for brightness, power, warmth, fate, the universe, eternity. From the Bible— "Canst thou bind the sweet influences of Pleiades, or loose the bands of Orion?" (Job 38:31), and Shakespeare, "The fault, dear Brutus, is not in the stars,/But in ourselves, that we are underlings." (*Julius Caesar*, Act I, Scene II)—to the nursery rhymer who wrote:

> Star light, star bright,
> first star I see tonight,
> I wish I may, I wish I might,
> have the wish I wish tonight.

we have looked to the stars for explanations of our earthly destinies.

From Sunrise, Alaska (and Sunrise, Florida), to Sunset, Maine (and Sunset, Louisiana and Utah, too); and in all the Sun Citys (Arizona, Kansas), Star Citys (Arkansas, West Virginia), Rising Suns (Maryland, Indiana), and Sunnysides (South Carolina, Washington) in between; be we in Sunnyland (Illinois), Sunnyview (South Dakota), Sunbeam (Colorado), or Sunbury (Pennsylvania); in North Star (Delaware), Lone Star (Texas), or Stella (Nebraska); we have sought succor, comfort, warmth and, arguably, prestige by ascribing the mysterious and wonderful qualities of the heavens to the products, places and players of life on our Earth.

The distances between objects in the Universe are so great that astronomers often describe them in terms of "light years." One light year is the distance light travels in one year—almost 6 trillion miles. Light travels at 186,000 miles a second. At this speed light would travel to Earth from the Moon in 1 and ⅓ seconds, from the Sun in about 8 minutes, and would take about 4.4 years to reach us from Alpha Centauri, the nearest bright star. Sirius, the brightest star in the night sky, is 8.6 years of light travel time; but most of the bright stars in the sky are dozens, hundreds, or even thousands of light years away.

On a clear, dark night away from city lights we can see a few thousand stars with our unaided eye, and they occupy only a very small part of our galaxy. Our galaxy contains about 100 billion stars; therefore, if we could shrink our entire galaxy down so that it was about 8 feet in diameter, most of the stars we see would fit within a space the size of an orange. Our galaxy is part of a small local cluster of about 25 galaxies. Andromeda, the nearest large galaxy in it, is about 2.2 million light years away.

When we look at the stars, we are looking into the past. For example, when we see the blue star Spica in the constellation Virgo we are seeing the light that left that star when the American Revolution began. Directing our telescopes toward the remote cluster of galaxies in the constellation Hydra, we collect light generated 3 billion years ago, when the first traces of life appeared on Earth. (The Earth itself formed as a solid body only about a billion years earlier.) And when we look even farther out into space, we see light that started toward us before the Earth existed, and before the Sun was shining in space.

To understand the relative sizes of the stars, if the Earth were a grain of buckshot, our Sun would be a tennis ball. And, at that same scale, whereas there are some stars the size of the Earth, there are also stars too large to fit into this gallery in the Museum.

Stars, self-luminous gaseous celestial bodies, are created when gas and dust clouds contract by their own gravity, raising the temperature and pressure inside. A star is born when fusion begins. All stars live by consuming matter and converting it into energy. When the temperature is high enough, hydrogen atoms fuse together in the stellar core liberating great amounts of energy that counteracts the gravitational collapse. When most of the available fuel for fusion is consumed, the core contracts, and a shell of fusing hydrogen grows outward. The star becomes a bloated "red giant." Heating eventually drives the outer shell into space, leaving the core which continues to contract and heat by compression. The collapsed, intensely white hot core is now visible as a "white dwarf" star. Eventually, this will probably happen to our Sun.

How each star lives, ages, and dies depends largely upon its mass. High-mass stars age quickly; low-mass stars age very slowly. Stars more massive than the Sun evolve into red super giants. When their cores heat up, helium atoms fuse, forming heavier elements like carbon. Helium fusion in massive stars is short-lived. During the helium fusion period, the star becomes unstable, swelling and shrinking. When the helium is exhausted, the star expands again and may begin to consume carbon, yielding iron in its core. The most massive stars go out with a flash. Stellar cores that try to fuse iron become highly unstable and require energy rather than provide it. This causes the star to explode as a "supernova." Part of the central core remains, intensely compressed into a "pulsar." A pulsar is an extremely dense rotating core of an exploded star that continues to collapse. It emits a focused beam of energy with each rotation not unlike the rotating beam of a lighthouse.

Unlike a pulsar, a "black hole" is so dense that nothing, not even light, can escape its gravity; therefore, it can only be detected by the radiation of energy emitted from material falling into it.

Galaxies come in a variety of shapes and sizes: small and large; spherical and elliptical; large tightly wound and loosely wound spirals; spirals with bars crossing their nuclei; irregular ones with no particular shape; and very peculiar ones suggesting unusual physical conditions such as distorted combinations of spirals and ellipticals. In addition, there are galaxies which appear to be in collision with each other and galaxies with enormous central explosions. One reason so many different types of galaxies exist is the different rotational energy each possesses.

Our galaxy's 100 billion stars are arranged in a series of spirals. Spiral galaxies contain a number of distinct regions. The *arms* contain young stars, gas, and dust. The central area, called the *nuclear bulge*, consists of old yellow stars, probably older than the Sun. The general plane of the galaxy—which is not necessarily confined to its spiral arms—is called the galaxy's *disc*. A spherical globe surrounding the galaxy and as large as the visible arms is called the galactic *halo*. This thinly populated region contains globular clusters—spherical cities—with up to one million stars each. All the objects in this region are very old. Beyond the halo, astronomers have recently found a vast *corona* of very hot, thin gas visible primarily in the X-ray region of the spectrum.

Galileo Galilei (1564–1642), the great astronomer, was the first to look at the Milky Way through a telescope and report on what he saw. He learned that it contains vast numbers of stars. Later, in 1755, Immanuel Kant was among the first to suggest that we live inside a "disc" of stars, the light of which blends to form the Milky Way. Astronomers have been able to map its local structure in some detail and have learned we live within, but not in the middle of, one of the galaxy's spiral arms. The Milky Way is brightest toward the constellation Sagittarius, the direction to the center of our galaxy. Toward this center, millions of stars and interstellar material block our view so that we cannot use ordinary light to study it. But infrared and radio energy reaching us from this region reveal intense sources of radiation possibly from supermassive stars surrounded by

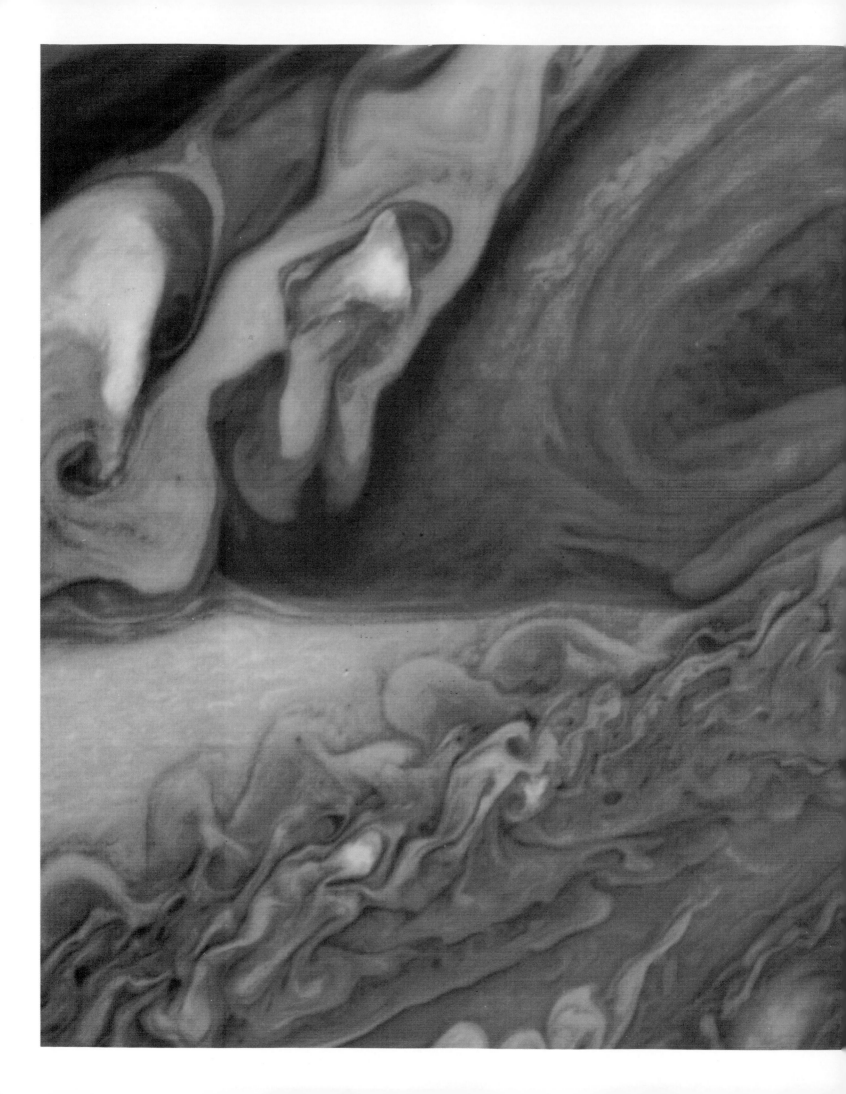

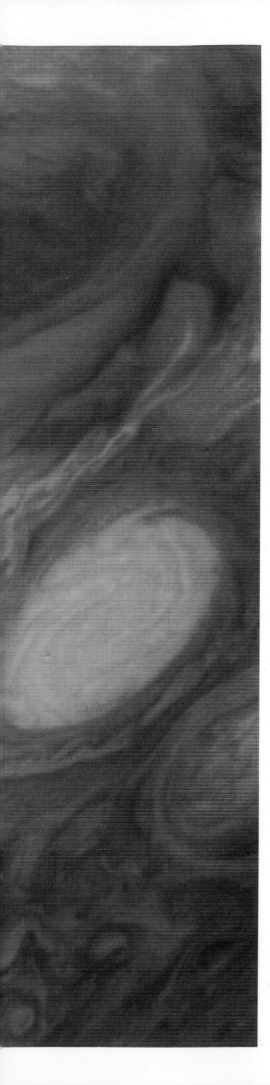

glowing gas, or even from a giant central black hole.

We have learned about the stars through observing them and by developing theoretical models to explain what we have observed. For us to understand the stars, theory and observation have to work together. Observations often require that theory be modified; and theory helps in understanding the observations that have been made. Over the years we have developed various devices with which we can better observe the stars; but it all began when Galileo focused his crude telescope on the heavens and saw mountains upon the face of the Moon.

The primary function of a telescope is to collect light. From our vantage point, except for the Sun, astronomical objects are very faint, and so, telescopes must be as large as possible to collect sufficient light for examination. Telescopes use either curved mirrors or lenses to gather and focus light. There are many types of optical telescopes with specific functions, and even more types of specialized telescopes like radio telescopes and X-ray telescopes. They all collect energy and bring it to a focus for analysis. Optical types include refractors (with lenses), reflectors (with mirrors), and combinations (with lenses and mirrors).

The first astronomers to use telescopes observed with their eyes and made notes and drawings of what they observed. Today, astronomers may use their eyes to find and set the telescope on an object, but rarely do they continue to use their eyes to actually examine celestial objects. Photographic and

Voyager 1 passed by Jupiter in March, 1979, and stunned the world with its wondrous photographs of the giant planet's red spot. Planets like Jupiter are condensed objects that do not have the mass to become stars.

457

video cameras, computers, and electronic sensors have for the most part replaced the eye at the telescope. This equipment can be used to record an image directly; or, in conjunction with other equipment, it can record the light of the object digitally for computer analysis.

A photograph provides a permanent record of an astronomical observation. Since the photographic emulsion can collect light over a long period and the eye cannot, photographs can record images too dim to be seen by the naked eye. Photography also provides permanent records of positions of stars, planets, asteroids, and systems of stars. The positions of these celestial objects can be determined with great precision and exposures separated by long periods can reveal changes. From measurements made of such photographs, distances to stars and their motions through space can be determined.

Visible light is only a tiny portion of the entire energy spectrum. Energy coming from objects in space ranges continuously from high energy X-rays and gamma-rays to ultraviolet, visible light, infrared, microwaves, and radio waves. Through spectroscopy, the study of the details of the energy distribution of light, we can learn a star's temperature, what it is made of, its motion through space, its rate of rotation, and the pressure and density of its atmosphere. Photometry, using light-sensitive photoelectric systems, can measure the rapidly changing brightness of an object.

The application of new technologies to open up new ways of examining the Universe is usually very costly, but scientists at the California Institute of Technology in the early 1960s managed to fabricate an economic, but highly effective telescope to find out just how many "invisible" infrared—or heat radiating—objects could be seen in the heavens. The Caltech Infrared Telescope designed by Caltech scientists R. Leighton and G. Neugebauer is on display in this gallery. All of its

components were constructed to permit reception of a portion of infrared energy that can penetrate the Earth's atmosphere. The results astonished astronomers. They had expected to discover only a few hundred infrared-emitting objects in the sky, but tens of thousands were found! These included glowing clouds of gas that might be precursors of very young stars, and the very intense nucleus of our Milky Way, as well as the nuclei of distant galaxies.

Beyond the infrared spectrum lies the radio spectrum. Radio waves also provide data about celestial objects and are used to observe galaxies in collision, stars exploding, and other violent processes in the Universe. And since they show the widely diffused hydrogen gas that lies between the stars, they also hint at the spiral arm structure of our galaxy.

The Earth's atmosphere distorts and blurs images of celestial objects and blocks most of their energy signatures. It is impossible to study objects from the ground that emit gamma-ray, X-ray, far ultraviolet, or most ranges of infrared energy. But as instruments travel up into and beyond the atmosphere, vast new areas of the spectra of celestial objects become accessible and more stars become visible, including X-ray and infrared stars.

X-ray astronomy reveals the violent nature of many celestial objects, from high temperature solar flares to blast wave radiation from supernovae. The first Small Astronomy Satellite, number 42 in the Explorer series, was launched from an off-shore platform in Kenya on December 12, 1970—the seventh anniversary of Kenyan independence. The satellite, named *Uhuru*, meaning "freedom" in Swahili, carried two X-ray telescopes pointing in opposite directions to scan the heavens as the satellite slowly rotated. In its three years of operation, *Uhuru* demonstrated the feasibility of X-ray astronomy in space by finding more than 200 X-ray sources and by producing an X-ray map that served as a guide for all later high-energy astrophysical

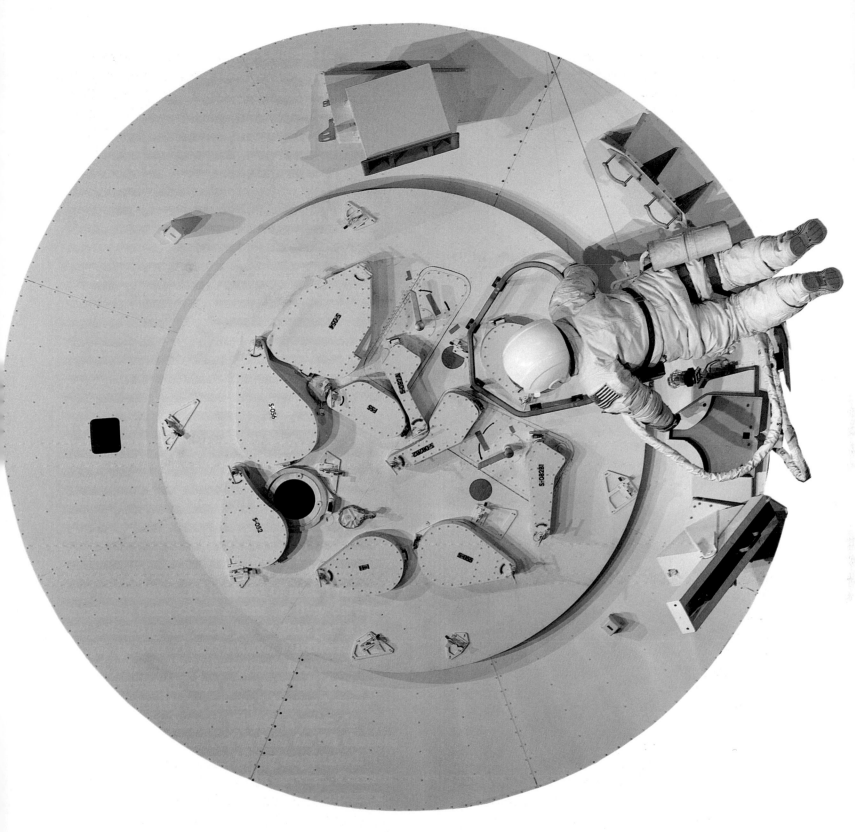

The Apollo telescope mount was a collection of eight very large, sophisticated solar instruments designed to examine the entire solar image in the high-energy range. It was launched atop a Saturn rocket in 1973 as part of the Skylab cluster.

"To be on the end of the telescope mount hanging by your feet as you plunge into darkness . . . you see nothing but flashing thunderstorms—that's one of the minutes to recapture and remember forever," said Jack R. Lousma, a member of Skylab's second crew.

459

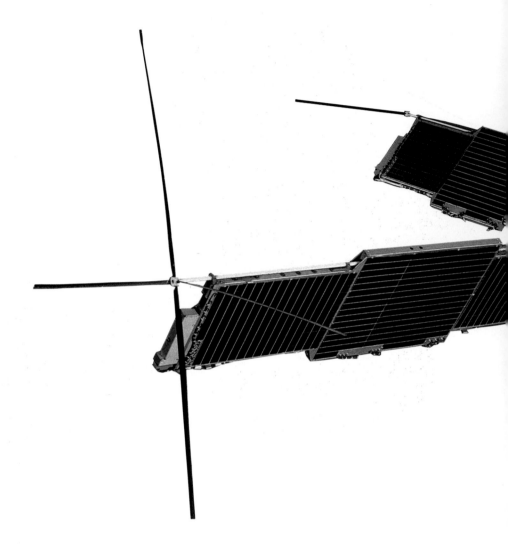

studies. *Uhuru* examined the centers of galaxies, quasars (quasi-stellar radio sources), and found that a pulsar is often a "partner" in a double-star system.

Lying between the X-ray and visible realms is the ultraviolet region. Most of the energy from within and around hot stars, and on the violent surface regions and in the atmosphere of our Sun, lies in the ultraviolet region. The Princeton telescope within *Copernicus*, the third Orbiting Astronomical Observatory launched in August, 1972, carried a 32-inch mirror, spectrograph, and ultraviolet detector called a photo-multiplier since it multiplies, or

amplifies, the energy that reaches it. The ultraviolet sensitive surfaces were exposed directly to portions of the spectra of celestial objects without passage through any optical material which might produce errors in the measurements. Over a nine-year period, *Copernicus* (the prototype of which is suspended in this gallery) observed the ultraviolet spectra of stars, galaxies, a nova, and the interstellar realm, which provided new information about how stars are formed, evolve, and die.

Several NASA aircraft, based at Ames Research Center, Moffet Field, California, fly at high altitudes to observe the infrared

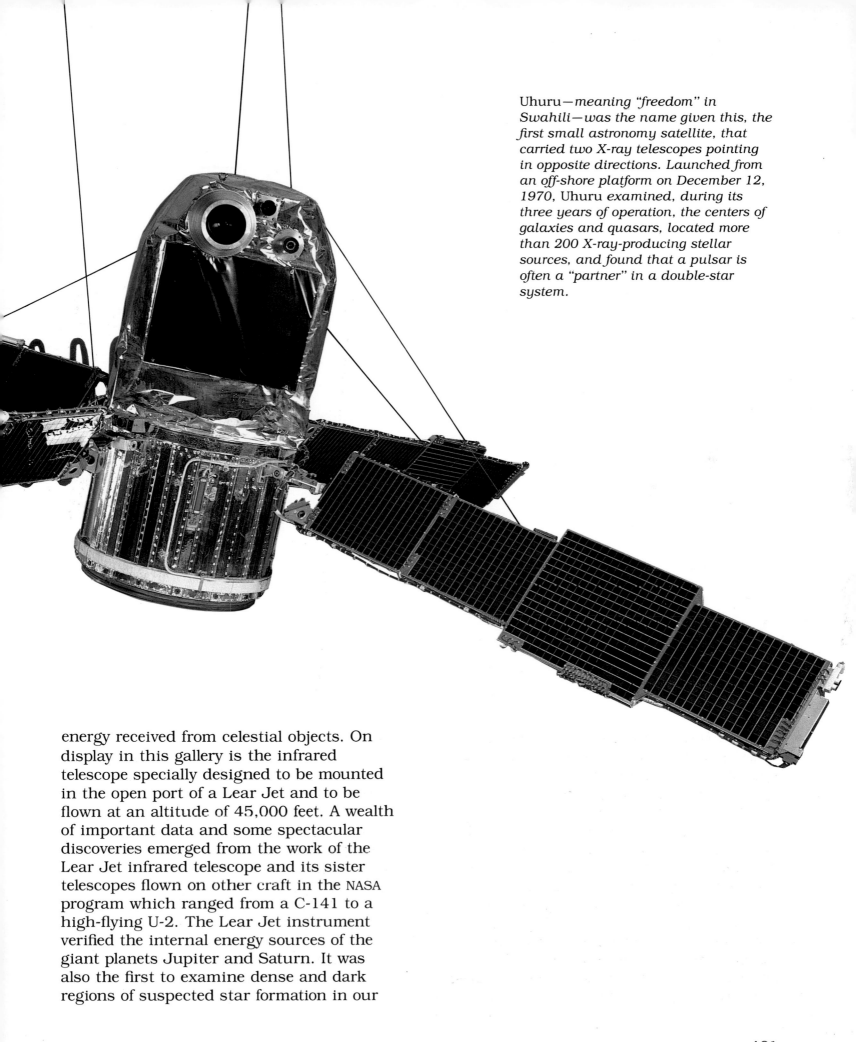

Uhuru—meaning "freedom" in Swahili—was the name given this, the first small astronomy satellite, that carried two X-ray telescopes pointing in opposite directions. Launched from an off-shore platform on December 12, 1970, Uhuru examined, during its three years of operation, the centers of galaxies and quasars, located more than 200 X-ray-producing stellar sources, and found that a pulsar is often a "partner" in a double-star system.

energy received from celestial objects. On display in this gallery is the infrared telescope specially designed to be mounted in the open port of a Lear Jet and to be flown at an altitude of 45,000 feet. A wealth of important data and some spectacular discoveries emerged from the work of the Lear Jet infrared telescope and its sister telescopes flown on other craft in the NASA program which ranged from a C-141 to a high-flying U-2. The Lear Jet instrument verified the internal energy sources of the giant planets Jupiter and Saturn. It was also the first to examine dense and dark regions of suspected star formation in our

Launched on January 25, 1983, the
Infrared Astronomical Satellite (IRAS)
carried a highly sensitive infrared
telescope and a unique liquid
helium cooling system to cool the
detectors to -455°F (-271°C). Such
temperatures allowed the telescope
to measure the heat from stars and
other celestial objects. By the time
its supply of supercold liquid helium
ran out 10 months later, IRAS had
surveyed 95 percent of the sky twice
in each of the four infrared bands.

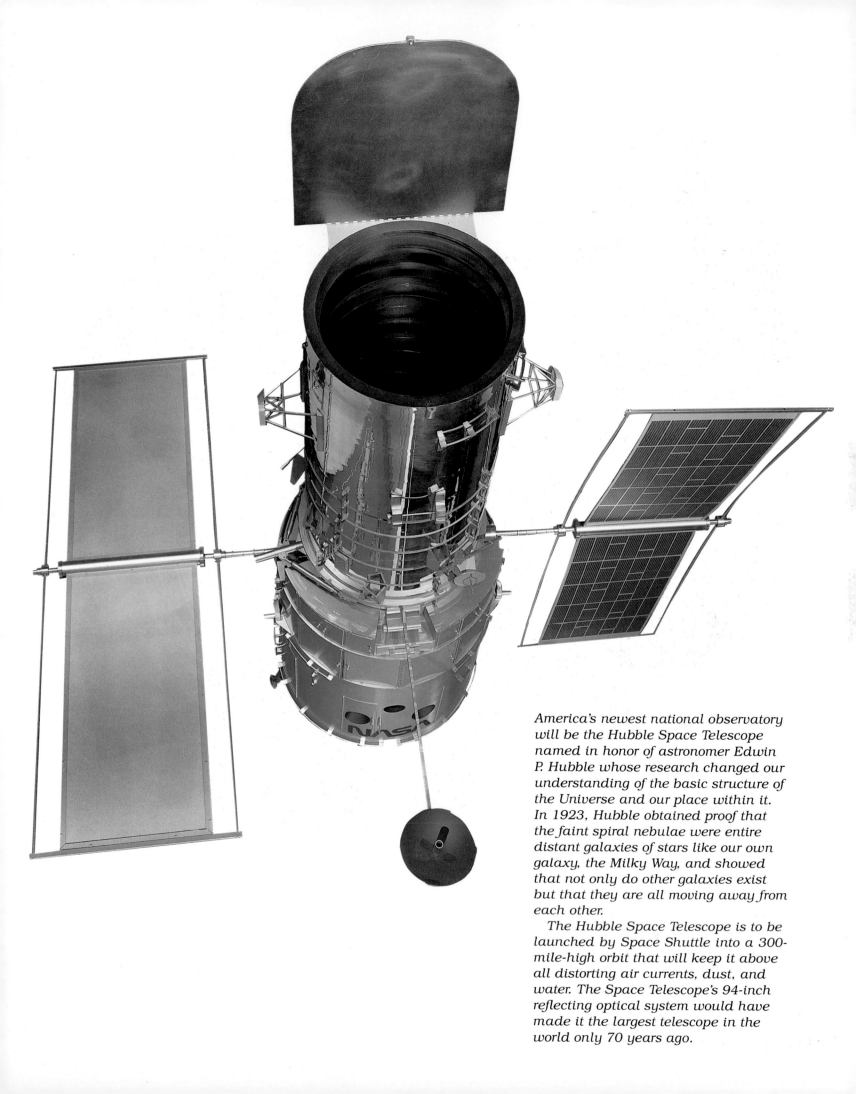

America's newest national observatory will be the Hubble Space Telescope named in honor of astronomer Edwin P. Hubble whose research changed our understanding of the basic structure of the Universe and our place within it. In 1923, Hubble obtained proof that the faint spiral nebulae were entire distant galaxies of stars like our own galaxy, the Milky Way, and showed that not only do other galaxies exist but that they are all moving away from each other.

The Hubble Space Telescope is to be launched by Space Shuttle into a 300-mile-high orbit that will keep it above all distorting air currents, dust, and water. The Space Telescope's 94-inch reflecting optical system would have made it the largest telescope in the world only 70 years ago.

galaxy. It examined the structure of giant molecular clouds in the galaxy, and was the first to peer into the infrared character of other galaxies.

The Infrared Astronomical Satellite (or IRAS), launched in 1982, was the first large space-based telescope devoted to infrared research. The 2,372-pound, 11.8-foot-long and—with its solar panels deployed—10.5-foot-wide IRAS telescope was developed and operated jointly by NASA, the Netherlands Agency for Aerospace Programs, and the United Kingdom's Science and Engineering Research Council. Previously, surveys of the infrared sky from high mountains and brief studies from balloons and aircraft had been done through the few narrow atmospheric "windows" that allow some infrared wavelengths to pass. From Earth orbit, the entire infrared spectrum is available.

Since IRAS was a survey instrument making new observations, discoveries came fast and furious. Perhaps the mission's most dramatic discovery came as a result of IRAS's examination of the star Vega, long used by astronomers as a brightness standard to calibrate instruments. But when IRAS mission planners observed this star, they were astonished at its excess of infrared radiation. This implied that Vega is surrounded by a swarm of particles of at least pebble size—the kind of dust rings out of which planets are thought to form.

IRAS also discovered several comets; and early in May, 1983, IRAS discovered a comet that later came only 3 million miles from Earth—closer than any comet since 1770. IRAS also collected data on Zodiacal Dust Bands, orbits into which much of the dust and small bodies of the inner solar system are grouped. Astronomers have known for some time that a great deal of dust lies in the Earth's orbital plane; but IRAS data showed that this dust is remarkably well organized into three distinct bands, two of them ten degrees above and below the plane. Since radiation pressure from the Sun and other forces should continually disperse this dust, the mystery of the source of its replenishment still remains to be solved.

Completely unexpected was the discovery in the IRAS data of faint, wispy clouds of material in all directions. Called Infrared Cirrus because of their resemblance to the cirrus clouds of Earth, the material seems to be 25 times as far from Earth as the planet Pluto, some of it within our solar system, and some of it associated with clouds of interstellar gas and dust.

Our newest national observatory will be the NASA Space Telescope which, in 1983, was named the Hubble Space Telescope in honor of Edwin P. Hubble (1889–1953). Working in 1923 with the 100-inch Hooker Telescope on Mount Wilson—then the world's largest telescope—Hubble was able to prove that the faint spiral nebulae were entire distant galaxies of stars like our home galaxy, the Milky Way, and that all the galaxies were moving away from each other: an indication that the Universe is expanding. Edwin P. Hubble's systematic investigation of extra-galactic nebulae won him great honor and changed our understanding of the basic structure of the Universe.

The Hubble Space Telescope, a multifunctional telescope with a 94-inch reflecting optical system, will be the largest telescope in space. Astronomers have long dreamed of having a large, general purpose observatory above the obscuring and filtering atmosphere of our planet. The Hubble Space Telescope, when launched by the Space Shuttle into its orbit free of all distorting air currents, dust, and water 300 miles above the Earth, will help to fulfill that dream. A ⅕-scale model of this telescope is suspended in this gallery.

Since the Hubble Space Telescope will be truly a national observatory, all the nation's astronomers will be able to use it to peer deeper than ever before into our fascinating and wondrous Universe.

SCIENCE, TECHNOLOGY, AND THE ARTS

Science,
Technology,
and the Arts

Samuel P. Langley Theater

"To Fly!" the first of the spectacular IMAX films to have been shown in the Museum's newly-opened 486-seat Samuel P. Langley Theater continues, today, to overwhelm each visitor who watches it. Photographed on 70mm film, a frame eleven times larger than the standard movie frame, and shown on a movie screen five stories high and seven stories wide, this breathtaking cinematic experience takes the viewer on a flying tour of America from its opening sequence of a post-Colonial balloon ascension (and the balloonist's near-collision with a church steeple) to the thundering lift-off of the last towering Saturn rocket—the one that, in July, 1975, carried an Apollo crew into orbit and link-up with the Soviet Soyuz space-craft. In between, a biplane flips upside down, the Navy's famed Blue Angels flight demonstration team screams across Arizona's deserts, a rainbow-striped hang glider soars off the coast of Hawaii, and "To Fly!" viewers cling white-knuckled to their armrests as if strapped into cockpits themselves.

In addition to "To Fly!" four other IMAX movies are shown in the theater: "On the Wing," "Living Planet," "Flyers," and "The Dream Is Alive."

"On the Wing" crisscrosses the globe, from Europe, China, and the remote mountains and jungles of Peru, to the Florida Wildlife refuge and the deserts of the American Southwest as the IMAX cameras record the wonder of flight in all forms—from birds and insects to kites and aircraft, with footage that includes the remarkable reproduction of a 65-million-year-old pterosaur come to life.

"Living Planet" is an armchair exploration of the great monuments of past civilizations: the Acropolis in Athens, the Cathedral of Chartres, and India's Taj Mahal. This bird's-eye view of the planet reveals the marvels of the world from the towers and canyons of New York City to the stunning natural beauty of some of the Earth's most remote corners in Africa and the Arctic.

"Flyers" takes visitors into the cockpit with two daring pilots as they swoop, soar, and dive over the land and sea in a series of spectacular aerial sequences that include an emergency landing on an aircraft carrier and a heart-stopping wingwalk on a biplane 14,000 feet above the Grand Canyon.

However, as spectacular as these films might be, none is more awesome than "The Dream Is Alive" which offers an insider's view of America's space program. This film features spectacular in-flight footage shot by 14 astronauts on three separate Space Shuttle missions. Viewers see astronauts at work both inside and outside the spacecraft, the deployment of scientific and communications satellites, the dramatic capture and repair of the "Solar Max" satellite, and the first space walk by an American woman astronaut, Kathy Sullivan. It is the next best thing to being in orbit oneself.

Albert Einstein Planetarium

After the exhausting excitement of visiting the Museum's gallery exhibits, the opportunity to merely rest in a comfortable armchair in the quiet, darkened, 230-seat Albert Einstein Planetarium while tranquil symphonic music washes over you through the planetarium's superb sound system might seem enough of a treat. But just as the Samuel P. Langley Theater's IMAX films give visitors the chance to view the marvels of the Earth from above, the Museum's planetarium provides visitors with the chance to view the marvels of above from the Earth.

The Albert Einstein Planetarium, named in honor of the man whose theories of relativity have altered our concept of the Universe, is one of the most advanced planetariums in the world. Its major piece of equipment, the Zeiss Model VI planetarium

An artist's view of the Zeiss planetarium instrument, a gift of West Germany for our Nation's Bicentennial. In a typical Planetarium scene, the planet Saturn is shown rising above the horizon as seen from one of its moons.

instrument (a Bicentennial gift to the American people from the people of the Federal Republic of Germany), projects simulated images of the Sun, the Moon, the five planets visible to the naked eye, and about 9,000 stars on the 70-foot-diameter overhead dome.

The planetarium uses a variety of vivid special effects to take visitors on a tour of the Universe. Viewers find out what we know about the world beyond—from ancient beliefs and superstitions to the latest predictions based on scientific theory. One sees stars and quasars, galaxies and black holes. The discoveries of past centuries unfold before viewers' eyes. A supernova explodes. Viewers travel through the Milky Way to distant galaxies and even catch a glimpse of what the Universe might have looked like in the first few seconds after the "Big Bang."

Paul E. Garber Facility

The Paul E. Garber Preservation, Restoration and Storage Facility, located in Suitland, Maryland, houses NASM's reserve collection of historically significant air- and spacecraft and enables enthusiasts to view many more of the Smithsonian's renowned collection than exhibit space permits at the National Air and Space Museum on the Mall. The facility is named in honor of Paul E. Garber, Historian Emeritus and Ramsey Fellow of the National Air and Space Museum, who joined the Smithsonian Institution in June, 1920, and was responsible for acquiring a large portion of the Smithsonian's current aeronautical collection. Formerly called the Silver Hill Museum, its name was changed in recognition of Paul E. Garber's many years of devoted service.

Although the facility has been used as an artifact storage and restoration center by the Smithsonian Institution since the mid-1950s, it was not until 1977 that some of the buildings were opened to the public as a "no frills" museum. On display are approximately 90 aircraft as well as numerous spacecraft, engines, propellers, and other flight-related objects. Among the collection exhibited are a MiG-15, the Soviet swept-wing jet fighter that fought the North American F-86 Sabre in the skies over Korea; a Hawker Hurricane IIC, the World War II British fighter of the type used in the Battle of Britain; a one-third-scale model of the Vostok Spacecraft which, on April 12,

This SPAD XIII, seen here awaiting restoration at the Garber Facility, is now in the Museum's World War I gallery.

472

1961, carried Yuri Gagarin, the first man into space; and a huge J-2 engine, one of the powerplants for the Saturn launch vehicle. Both military and civil aircraft are simply displayed with brief identification labels. Trained guides conduct free tours of the warehouse/exhibition areas, and visitors can enter the restoration area where the technicians can be seen at work preserving aircraft.

Deterioration occurs even when storage conditions are good. Over the course of time some of the fabric and wood early aircraft in the Museum's collection, for example, had fallen into pitiable condition because of the havoc wrought by variations in temperature and humidity. And such problems of upkeep are confronted on a daily basis by the expert craftsmen and conservators at the Garber Facility.

However, the facility's aircraft conservators are such demanding craftsmen that, at times, they have to be restrained from restoring aircraft to a better shape than they were in when they were first built. Air frames are brought back to flight status, engines are not—although they are restored to a point where they *could* become operable. The Garber Facility's philosophy is that it is better to spend the time and money to arrest corrosion in a dozen aircraft than to spend the time bringing just one all the way up to flight status. Still, about twenty artifacts (including propellers, landing gears, and engines) are preserved each year and about four aircraft restored.

One of the major problems facing the Garber Facility craftsmen is that many of the airplanes needing restoration are military. Military aircraft, mass-produced as part of the war effort, were expected to pass inspection, but not to last for an eternity. Their average life expectancy was about ninety days. A good example confronting the Garber Facility's workers was the recently restored Messerschmitt Me 262, the world's first operational jet fighter, now on display in the Museum's Jet Aviation gallery. It was manufactured by the Germans during the latter stages of World War II, and although some 1,400 were made, only about 300 saw combat.

In order to stop the corrosion process, the Me 262 was completely taken apart, cleaned, then reassembled. The work was doubly difficult because German manufacturers used composite construction techniques combining aluminum with steel in the fighters. The corrosion that had occurred in the Me 262 was considerably more extensive than had been envisioned and 6,077 manhours were required in this aircraft's restoration. However, as former NASM director Walter J. Boyne pointed out in a Smithsonian publication that details the restoration of the Museum's jet:

> The number of manhours is a very poor way to record the ingenuity, effort and even love that went into the restoration. In the abstract, the concept of restoring a historic World War II fighter is one that excites almost every aviation buff. The reality of spending hundreds of hours sanding, drilling, riveting, washing, masking and painting the aircraft is quite another thing, and it is a tribute to the morale of the men [at the Garber Facility] . . . that no one avoided any job, no matter how tedious or unpleasant. . . .
>
> —*Messerschmitt Me 262: Arrow to the Future,*
> by Walter J. Boyne

One visitor particularly impressed by the restoration had been an engineer in Germany who had worked on the development of the original Me 262. An interesting fact that he pointed out was that the 18½° sweep of the wing had been designed not for higher speed, as one might have thought, but to shift the aircraft's center of gravity rearward. The sweep did, of course, delay the onset of compressibility and permitted the Me 262 the exceptionally high limiting Mach number of .83.

Visitors to the Garber Facility tend to be either aviation buffs or persons with some direct relationship to a particular aircraft they have come to see. One such was Don Berlin who designed the famous Curtiss P-40. He had been project engineer on the

Northrop Gamma and went out to the facility with Walter J. Boyne to look at the Northrop Gamma *Polar Star* that Lincoln Ellsworth had flown across the South Pole. (The plane is now on display in the Museum's Golden Age of Flight gallery.) Berlin had not seen the Gamma for about forty years and, as he looked at the silver plane with the slight crumple aft of the engine nacelle caused by a particularly hard landing on the Antarctic ice, he turned to Boyne and said, "I knew it was going to happen. I *knew* then what should have been done to strengthen that structure!" And,

according to Boyne, Berlin was fully prepared to redraw the Gamma's design on the spot.

Art Gallery

The first works of space art created in strict accord with existing scientific knowledge were the illustrations published in Jules Verne's 1865 novel *From the Earth to the Moon.* Not quite one hundred years later, in 1963, the National Aeronautics and Space Administration, in cooperation with the

Allarme General des Habitants Gonesse, 1783. *1827. Engraving with watercolor on paper, 11¾ × 16". Gift of the American Institute of Aeronautics and Astronautics, New York*

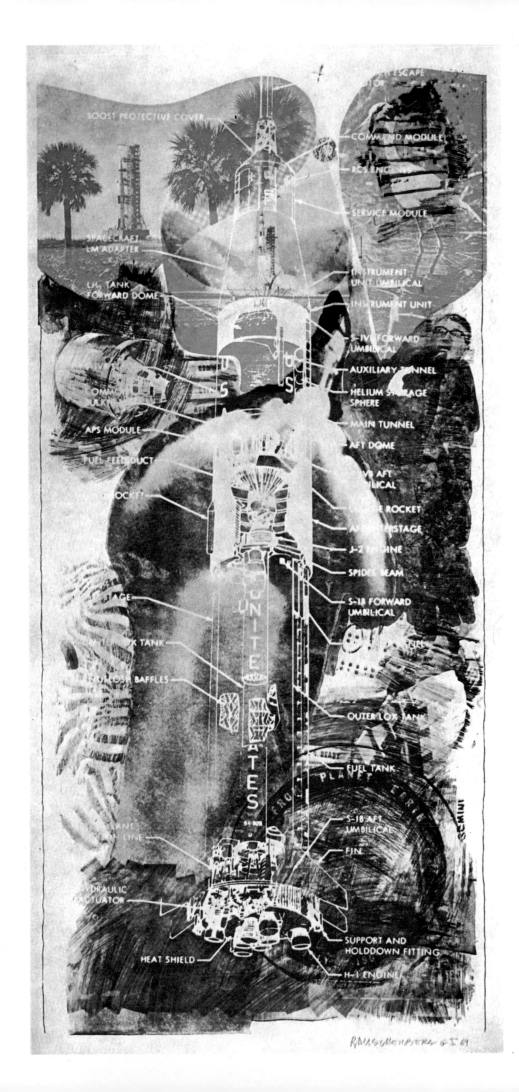

*Robert Rauschenberg. Sky Garden.
1969. Color lithograph on paper,
89 × 42". From the Stoned Moon series
reproduced courtesy of Gemini G.E.L.,
Los Angeles, California.*

Ralph Goings. Fly with Patco. *1973. Oil on canvas, 36 x 52". From the Stuart M. Speiser Collection*

Audrey Flack. Spitfire. 1973. Acrylic with oil glazes on canvas, 70×96". Gift of Stuart M. Speiser

Alma Thomas. Blast Off. 1970. Acrylic on canvas, 74×54" (framed). Gift of Vincent Melzac

National Gallery of Art, invited some of this country's foremost artists to contribute their imagination, perceptions, and talent to document the space program in the most ambitious art project since the WPA artists' projects of the Depression. Much of the resulting art appeared in the Museum's inaugural exhibit.

Today, the Museum's permanent art collection—containing paintings, drawings, prints, sculpture, textiles, and crafts relating to the theme of flight ranging historically from early 18th- and 19th-century balloon images through the development of aircraft and on to space exploration—is so extensive that its works of art overflow the Flight and the Arts Gallery and can be found in almost every exhibit area, office, hallway, and corner of the Museum.

Library and Research

The National Air and Space Museum Bureau Library is part of the system of Smithsonian Institution Libraries that supports the research and exhibit programs of the Institution and the specialized interests of the staff and the public it serves. The NASM collection contains over 40,000 books, 7,000 bound journals as well as microforms of books, journals, and technical documents covering the history of aviation and space exploration, flight and lighter-than-air technology, the aerospace industry and biography, rocketry, earth and planetary sciences and astronomy.

The rare and scarce aeronautica and astronautica of the Museum are housed in the Admiral DeWitt Clinton Ramsey Room. There one finds over 1,500 pieces of aeronautical sheet music in the Bella Landauer collection in addition to the William A.M. Burden collection of early ballooning works and some first editions autographed by notable pioneers of flight.

The Museum's Information Management Division holds a historical archives documenting the collection and containing more than two million photographs and drawings of aircraft and spacecraft, as well as extensive materials on aerospace personalities and events.

Computer Gallery

"Beyond the Limits: Flight Enters the Computer Age," a major new gallery exhibition depicting the computer revolution in aerospace, is scheduled to open at the Museum in 1989.

From initial design to navigation and daily maintenance, computer developments during the past ten years have so drastically altered every facet of air and space flight that today engineers no longer even consider drawing up blueprints for a new aircraft or spacecraft. Although computers were modified in the early 1950s to perform engineering tasks, the "computerization" of aerospace was not complete until the 1970s when microchip technologies permitted the building of computers small and lightweight enough to be carried on board air- and spacecraft.

Among the earliest and most dramatic uses of the on-board computer was the one on Mariner 10 launched in 1973. Controlled from Earth, Mariner 10's computer enabled the spacecraft to use the gravity of Venus to hurtle it through space toward Mercury.

One important element in this new gallery will be one of the first-generation production model supercomputers, a CRAY-1. This computer, capable of performing billions of computations in seconds, also generated graphics that enabled engineers to "flight test" an engine or wing design before building costly prototypes. One such product of supercomputer design is the X-29, a radical swept-forward wing layout that engineers have long known was highly efficient. However, until computer-aided design and manufacture was available, they were unable to build a wing strong enough and lightweight enough in this configuration to carry an aircraft.

Aerospace Chronology

November 21, 1783: The first flight in history is made by de Rozier and d'Arlandes in a Montgolfier hot-air balloon at Paris.

December 1, 1783: First flight in a hydrogen balloon (by Charles and Robert) at Paris.

January 7, 1785: Blanchard and Jeffries, in a balloon, make the first Channel crossing by air.

January 9, 1785: Jean Pierre Blanchard makes the first flight in America. Launching his balloon from a Philadelphia prison yard, he flies to Gloucester County, New Jersey.

October 22, 1797: Garnerin, from a balloon, makes the first parachute descent from the air.

1799: Cayley designs the first modern configuration airplane, incorporating fixed wings, tail-unit control surfaces, and an auxiliary method of propulsion.

1804: First modern configuration airplane, Cayley's model glider, flies.

1809—10: Cayley publishes his classic triple paper on aviation, which lays the foundations of modern aerodynamics.

1853: First man-carrying flight in a heavier-than-air craft (Cayley's "coachman-carrier" glider)—but not under control by the occupant.

1876: Otto invents the four-stroke petrol engine.

1881—96: Lilienthal's first successful piloted gliding flights.

1889: Lilienthal publishes his classic. *Der Vogelflug als Grundlage der Fliegekunst.*

1896: Langley obtains his first success with two of his steam-powered tandem-wing models.

1899: The Wright Brothers invent a system of wing-warping for control in roll, and fly a kite incorporating it.

1901: The Wrights fly their No. 2 glider near Kitty Hawk.

1902: The Wrights make nearly 1,000 glides on their No. 3 glider and invent coordinated warp and rudder control, i.e., combined control in roll and yaw.

December 17, 1903: The first powered, sustained, and controlled airplane flights in history, by Wilbur and Orville Wright, near Kitty Hawk; the first of four flights lasts for 12 seconds, the last for 59 seconds.

September 20, 1904: First circle flown, by Wilbur Wright; this is witnessed and described by A.I. Root.

June, 1905: First fully practical powered airplane, the Wright Flyer III, flies.

October—November, 1906: Santos-Dumont makes the first official powered hopflights in Europe: the best (November 12) covers 721 feet in 21⅕ seconds.

August 8, 1908: At Hunaudières, France, Wilbur Wright flies in public for the first time in the first practical two-seat airplane, and transforms aviation by his display of flight controls.

September 3, 1908: Orville Wright first flies in public at Fort Myer, Virginia.

October 30, 1908: Henri Farman completes the first cross-country flight, flying 16½ miles.

July 25, 1909: On his No. XI monoplane, Louis Blériot makes the first Channel crossing from near Calais to Dover.

August, 1909: First great aviation meeting is held at Reims, and has widespread influence, showing the airplane is now a practical vehicle.

February, 1910: Hugo Junkers receives a patent for the design of a cantilever flying wing aircraft.

November 14, 1910: Eugene Ely takes off from the USS *Birmingham* in a Curtiss Pusher—the birth of the aircraft carrier.

January 18, 1911: Eugene Ely lands his Curtiss Pusher aboard the armored cruiser USS *Pennsylvania*, at anchor in San Francisco Bay—the first landing of an airplane on a ship.

1912: Ruchonnet and Bechereau introduce monocoque construction on the record-setting *Monocoque Deperdussin* racer, which wins the Gordon Bennett Cup by setting a world's speed record of 108.18 mph.

The first enclosed cabin airplanes (British Avro) and all-metal airplanes (the *Tubavion* of Ponche and Primard) complete their initial flights.

1913: The first multiengine aircraft having four engines, the Sikorsky Bolshoi, completes its initial flights in Russia.

1915: First flight of the Junkers J 1, the world's first all-metal cantilever monoplane.

May 16—17, 1919: First transatlantic flight via the Azores by the NC-4 flying boat of the United States Navy.

December, 1919: Smithsonian Institution publishes Robert H. Goddard's classic paper, *A Method of Reaching Extreme Altitudes.*

1920: First flight of an airplane having a practical retracting landing gear, the Dayton-Wright R.B. racer. This aircraft also had the first variable-camber wing.

March, 1920: First flight-test verification of the advantages of equipping an airplane with wing slots.

September 4, 1922: First transcontinental flight across the United States in a single day, by Lt. James H. "Jimmy" Doolittle, in a D.H. 4B biplane, from Pablo Beach, Florida, to Rockwell Field, San Diego, a distance of 2,163 miles.

January 9, 1923: First successful flight of the Cierva C.3 Autogiro, at Madrid. The Autogiro subsequently has great influence on the development of other rotary-wing aircraft, especially helicopters.

May 2—3, 1923: First nonstop transcontinental flight across the United States, from New York to San Diego, by Lts. Oakley Kelly and John A. Macready, in the Fokker T-2 monoplane, a distance of 2,520 miles in 26 hours, 50 minutes.

1924: The trimotor airliner (Junkers G.23) and the Fowler wing flap make their first appearance.

April 6—September 28, 1924: Two Douglas World Cruiser aircraft of the United States Army Air Service complete the first around-the-world, first transpacific, and first westbound Atlantic crossing, flying 26,345 miles in a flying time of 363 hours.

March 16, 1926: Robert Goddard demonstrates successful operation of a liquid-fuel rocket at Auburn, Massachusetts. His rocket attains a distance of 184 feet in 2½ seconds, the "Kitty Hawk" of rocketry.

May 20—21, 1927: Charles Lindbergh crosses the Atlantic from New York to Paris, the first solo nonstop crossing and the first by a single-engine aircraft.

July 4, 1927: First flight of the Lockheed Vega, a trend-setting cantilever-wing aircraft having a monocoque fuselage with stressed-skin construction, which imparts lighter weight and larger volume to the aircraft, as well as reducing drag and, thus, boosting performance.

June 11, 1928: First flight of a manned rocket-propelled airplane is made by Friedrich Stamer in a modified canard glider.

September 24, 1929: James H. Doolittle makes the first blind flight in aviation history, flying a specially instrumented Consolidated-Guggenheim NY-2 research airplane.

1930: Frank Whittle takes out his first patents on turbojet engine design.

September 29, 1931: RAF Flight Lt. George Stainforth completes the world's first flight faster than 400 mph in the Supermarine S.6B racing floatplane.

1933: Introduction of the controllable-pitch propeller on regular airline aircraft, beginning with the Boeing 247. This development, the result of work by Frank Caldwell, greatly improves airplane performance at both low and cruising speeds.

February 8, 1933: First flight of an advanced all-metal monoplane transport, the Boeing 247.

July 1, 1933: First flight of the Douglas DC-1, an innovative all-metal monoplane transport that serves as the basis for the future DC-2 and DC-3 series, which dominate American and much foreign air transport services.

November 11, 1935: Balloonists O.A. Anderson and A.W. Stevens reach a world altitude record for manned balloons of 72,395 feet, in the balloon *Explorer II*.

June 26, 1936: First flight of the world's first practical helicopter, the double-roter Focke-Achgelis FW-61.

1938: The National Advisory Committee for Aeronautics develops a family of low-drag laminar-flow airfoil sections ideally suited for high-speed aircraft designs.

December 31, 1938: First flight of a passenger airplane having a pressurized cabin, the Boeing 307 Stratoliner.

June 30, 1939: First flight of an airplane equipped with a liquid-fuel rocket engine, the Heinkel He-176.

August 27, 1939: Erich Warsitz completes the first jet flight in aviation history, flying the experimental Heinkel He-178, a Von Ohain turbojet.

May 13, 1940: Igor Sikorsky completes the first flight of the Sikorsky VS-300 helicopter, the first successful single-rotor helicopter in the world.

May 15, 1941: First flight of the Gloster E. 28/39, the first British turbojet airplane, powered by a Whittle engine.

February, 1942: First flight of the Douglas DC-4 transport, which sets the future design configuration for postwar four-engine airliners.

October 1, 1942: First flight of the Bell XP-59A Airacomet, the United States' first turbojet aircraft, at Muroc Dry Lake, California.

October 3, 1942: First successful flight of the German A-4 (V-2) liquid-fuel rocket-propelled ballistic missile, from Peenemunde, Germany.

December, 1943: American military and civilian aeronautical research directors complete initial discussion on using manned research aircraft to fly faster than the speed of sound.

1944: The first turbojet-propelled fighters, the Gloster Meteor and Messerschmitt Me 262, enter service, as does the first operational rocket-propelled interceptor, the Messerschmitt Me 163 *Komet*.

September 20, 1945: The first flight of a turboprop-driven airplane is made in England by a modified Gloster Meteor powered by two Rolls-Royce Trent turboprops.

September 29–October 1, 1946: The Lockheed P2V Neptune *Truculent Turtle* establishes a world's record distance without refueling for 11,235 miles (18,088 kilometers) in 55 hours and 17 minutes.

October 14, 1947: Capt. Charles E. Yeager becomes the first pilot to exceed the speed of sound, flying the air-launched experimental Bell XS-1 rocket-propelled research airplane to Mach 1.06, 700 mph (1,127 kph) at 43,000 feet (13,106.40 meters), over Muroc Dry Lake, California.

April 21, 1949: First flight of a ramjet-powered airplane—the French-built experimental air-launched Leduc 010—which flies for 12 minutes and attains a speed of 450 mph (725 kph) using only half of its available power.

August 25, 1949: First emergency use of a partial-pressure pilot-protection suit, by Maj. Frank K. Everest, on board the Bell X-1, following loss of cabin pressurization at 69,000 feet (21,031 meters).

June 20, 1951: First flight of the Bell X-5 variable-wing-sweep testbed, by Jean Ziegler, at Edwards Air Force Base.

December, 1951: Richard Whitcomb verifies the Area Rule concept to reduce aircraft drag characteristics at transonic and supersonic speeds. This concept, popularized as the so-called "Coke bottle" or "wasp waist" shape, is first verified by flight testing on the Convair F-102.

April 21, 1952: The world's first production turbojet transport, the de Havilland Comet, enters airline service with BOAC.

November 20, 1953: Research test pilot A. Scott Crossfield becomes the first pilot to exceed Mach 2, twice the speed of sound, in an experimental air-launched rocket-propelled Douglas D-558-2 skyrocket. The plane attains Mach 2.005, approximately 1,328 mph (2,138 kph) over Edwards Air Force Base, California.

July 15, 1954: First flight of an American jet transport, the Boeing 707, prototype for the extremely successful Boeing 707 transport and KC-135 transport/tanker aircraft.

September 27, 1956: Capt. Milburn Apt, United States Air Force, becomes the first pilot to fly three times faster than the speed of sound, reaching Mach 3.196, 2,094 mph (3,371 kph) in the Bell X-2. Apt is killed, however, when the plane tumbles out of control into the Mojave Desert.

November 28, 1956: Peter Girand, flying the experimental Ryan X-13 Vertijet VTOL airplane, completes the world's first jet vertical takeoff and transition to level flight.

October 4, 1957: Sputnik I, the first man-made earth satellite, is placed in orbit by the Soviet Union—the dawn of the Space Age.

November 3, 1957: Launch of Sputnik 2, carrying dog Laika, first living creature to orbit the Earth.

January 31, 1958: Explorer 1, the first United States satellite, is successfully launched.

September 12, 1959: Soviet Union launches Luna 2, the first man-made object to impact the Moon.

September 17, 1959: First powered flight of the North American X-15 hypersonic research airplane, by test pilot A. Scott Crossfield, at Edwards Air Force Base, California.

October 4, 1959: Soviet Union launches Luna 3, the first spacecraft to photograph the lunar farside.

April 1, 1960: TIROS I, the first weather satellite, is launched by the United States.

August 16, 1960: Capt. Joseph W. Kittinger, Jr., makes a record parachute descent by jumping from the balloon *Excelsior III* at an altitude of 102,800 feet and free-falling 17 miles before opening his parachute at 17,500 feet.

April 12, 1961: Maj. Yuri Gagarin completes the first manned space flight by making a one-orbital mission aboard the Soviet spacecraft Vostok.

May 5, 1961: Alan B. Shepard, Jr., becomes the first American astronaut to enter space, making a 15-minute suborbital flight.

September 12, 1961: The Hawker P.1127 experimental vectored-thrust research airplane completes its first transition from vertical takeoff to horizontal flight, and back to a vertical landing. The P.1127 serves as the basis for the world's first operational VTOL fighter, the Hawker-Siddeley Harrier.

February 20, 1962: United States astronaut Lt. Col. John Glenn becomes the first American to orbit the Earth, aboard the Mercury spacecraft *Friendship 7*.

July 10, 1962: The United States launches Telstar I, providing the first transatlantic satellite television relay.

August 27, 1962: Mariner 2, the first spacecraft to conduct a fly-by of another planet (Venus), is launched by the United States.

June 16, 1963: Valentina Tereshkova becomes the first woman in space, aboard the Russian Vostok 6.

August 22, 1963: NASA research pilot Joseph Walker attains an altitude of 354,200 feet (67.08 miles) in the hypersonic North American X-15 research airplane, the highest flight to that date by a winged aircraft, at Edwards Air Force Base, California.

September 21, 1964: First flight of the North American XB-70A Mach 3 experimental research airplane. The two XB-70s built furnished much valuable information useful to the design of large supersonic aircraft.

March 7, 1965: A Quantas Airlines' Boeing 707 makes commercial aviation's first nonstop Pacific Ocean crossing, flying from San Francisco to Sydney in 14 hours and 33 minutes.

March 18, 1965: Alexei Leonov becomes the first person to perform an extra-vehicular activity (EVA, or spacewalk), during the Voskhod 2 mission.

June 3, 1965: Edward H. White II becomes the first American to perform an extravehicular activity, during the flight of Gemini 4.

November 16, 1965: Soviet Venus 3 spacecraft is launched, and on March 1, 1966, becomes the first man-made object to impact Venus.

January 31, 1966: The Soviet Union launches Luna 9, the first unmanned spacecraft to make a soft landing on the Moon.

June 1, 1966: The first American spacecraft to make a soft landing on the Moon, Surveyor 1, is launched.

August 10, 1966: The United States launches Lunar Orbiter 1, which provides high-resolution photographs for the selection of Apollo landing sites.

October 3, 1967: Maj. William J. Knight sets a new unofficial world airspeed record for winged aircraft of 4,534 mph (Mach 6.72) in the North American X-15A-2; this is the fastest winged flight and the fastest X-15 flight ever made.

December 21–27, 1968: Apollo 8, piloted by astronauts Frank Borman, James Lovell, Jr., and William Anders, becomes the first manned spacecraft to orbit the Moon.

December 31, 1968: First flight of a supersonic transport, the Soviet Tupolev TU-144.

May 18–26, 1969: Astronauts Thomas Stafford, Eugene Cernan, and John Young test the Lunar Module in lunar orbit in Apollo 10 mission.

July 20, 1969: Apollo 11 astronauts Neil Armstrong and Edwin Aldrin become the first humans to step on another celestial body when they land on the Moon.

November 24, 1970: First test flight of the NASA supercritical wing, developed by

Index

494

Credits

Grateful acknowledgment is made for permission to quote from the following works:

Airpower Magazine, September, 1978, Granada Hills, California. *They Flew The Bendix Race* by Don Dwiggins, © 1965 Allied Signal Aerospace Company. *The American Heritage History of Flight*, © 1962 American Heritage Publishing Co., Inc. *Air Spy* by Constance Babington-Smith, 1957, © Chatto and Windus Ltd. Oral histories from the Oral History Research Office, Columbia University, copyright as follows: Lt. Macready © 1974 Trustees of Columbia University and The City of New York; Lt. Leslie P. Arnold © 1979 Trustees of Columbia University and The City of New York; Frank Coffyn © 1979 Trustees of Columbia University and The City of New York. "Cross-Channel Flight" by Louis Blériot from *The Saga of Flight*, 1961 John Day Co. *Samurai* by Saburo Sakai with Martin Caidin, Bantam Books, © 1957 Martin Caidin, by permission of Martin Caidin. Excerpt from *Early Birds*, Arch Whitehouse. Copyright © 1965 by Arch Whitehouse. Reprinted by permission of Doubleday, a division of Bantam, Doubleday, Dell Publishing Group, Inc. The Economist Newspaper, Ltd., July 26, 1969, © The Economist, London. *Miracle at Kitty Hawk*, edited by Fred C. Kelly, © 1951 Fred C. Kelly; copyright renewed © 1979 by Brian Kelly and Jean Kelly by permission of Farrar, Straus & Giroux, Inc. *Carrying the Fire* by Michael Collins, © 1974 Michael Collins,

by permission of Farrar, Straus & Giroux, Inc. *A House In Space* by Henry S.F. Cooper, Jr., © 1976 Henry S.F. Cooper, Jr., by permission of Holt, Rinehart and Winston, Publishers. *The Wright Brothers* by Fred C. Kelly, © 1943 and 1950 Fred C. Kelly. *Falcons of France* by James Norman Hall and Charles Nordhoff, © 1929 Little, Brown & Co., by permission of Little, Brown & Co. in association with The Atlantic Monthly Press. *First On The Moon* by Neil Armstrong, Michael Collins, Edwin E. Aldrin, Jr., with Gene Farmer and Dora Jane Hamblin, © 1970 Little, Brown & Co., by permission of the publisher. *The Sword Over The Mantle* by J. Bryan III, 1960 McGraw-Hill Book Company, by permission of the author. *The First and The Last* by Adolf Galland, © 1955 Methuen & Co., Ltd., by permission of the author. "Double Eagle II Has Landed," by Ben L. Abruzzo, © National Geographic Magazine, Dec., 1978. "My Flight Across Antarctica," Ellsworth Lincoln, © National Geographic Magazine, July, 1936. *A History of the World's Airlines* by R.E.G. Davies, © 1964 Oxford University Press. *The Spirit of St. Louis* by Charles A. Lindbergh, © 1953 Charles Scribner's Sons, by permission of Charles Scribner's Sons. *We Seven, By The Astronauts Themselves*, © 1962 Simon & Schuster, Inc., by permission of Simon & Schuster, a Division of Gulf &

Western Corporation. By permission of the Smithsonian Institution Press: from *Contact: The Story of the Early Birds* by Henry Villard, © 1987 Smithsonian Institution, Washington, D.C.; from *Messerschmitt Me262: Arrow to the Future* by Walter J. Boyne, © 1980 Smithsonian Institution, Washington, D.C.; from *Kelly: More Than My Share of It All* by Charles E. "Kelly" Johnson and Maggie Smith, © 1985 Smithsonian Institution, Washington, D.C. *The Flying Key Brothers and Their Flight To Remember* by Stephen L. Owen, © 1985 Southeastern Printing Company. Text copyright © 1986 Walter Boyne from *The Leading Edge* published by Stewart, Tabori & Chang, New York. Reprinted with permission of the publisher. From Epic of Flight: *The Jet Age* by Robert J. Sterling and the Editors of Time-Life Books, © 1982 Time-Life Books, Inc. Life Science Library, *Man and Space* by Arthur C. Clarke and the Editors of Time-Life Books, © 1964 and 1969 Time, Inc. *The Washington Post*, June 27, 1976, © *The Washington Post*. "Voyagers," *The Washington Post*, Dec. 24, 1986, © *The Washington Post*. *Always Another Dawn: The Story of a Rocket Test Pilot* by A. Scott Crossfield and Clay Blair, Jr., © 1960 A. Scott Crossfield and Clay Blair, Jr., by permission of The World Publishing Company.

Addendum:

Air Spy by Constance Babington-Smith, 1957, Chatto and Windus, Ltd., reprinted by permission of A.D. Peters & Co. Ltd.

Falcons of France by James Norman Hall and Charles Nordhoff, Copyright 1929 by Little, Brown and Company. Copyright © renewed 1957 by Mrs. James Norman Hall and Mrs. Laura G. Nordhoff. By permission of Little, Brown and Company.